Designing Controls for the Process Industries

Designing Controls for the Process Industries

Wayne Seames

CRC Press
Taylor & Francis Group
Boca Raton London New York

CRC Press is an imprint of the
Taylor & Francis Group, an **informa** business

CRC Press
Taylor & Francis Group
6000 Broken Sound Parkway NW, Suite 300
Boca Raton, FL 33487-2742

© 2018 by Taylor & Francis Group, LLC
CRC Press is an imprint of Taylor & Francis Group, an Informa business

No claim to original U.S. Government works

Printed on acid-free paper

International Standard Book Number-13: 978-1-138-70518-0 (Hardback)

This book contains information obtained from authentic and highly regarded sources. Reasonable efforts have been made to publish reliable data and information, but the author and publisher cannot assume responsibility for the validity of all materials or the consequences of their use. The authors and publishers have attempted to trace the copyright holders of all material reproduced in this publication and apologize to copyright holders if permission to publish in this form has not been obtained. If any copyright material has not been acknowledged please write and let us know so we may rectify in any future reprint.

Except as permitted under U.S. Copyright Law, no part of this book may be reprinted, reproduced, transmitted, or utilized in any form by any electronic, mechanical, or other means, now known or hereafter invented, including photocopying, microfilming, and recording, or in any information storage or retrieval system, without written permission from the publishers.

For permission to photocopy or use material electronically from this work, please access www.copyright.com (http://www.copyright.com/) or contact the Copyright Clearance Center, Inc. (CCC), 222 Rosewood Drive, Danvers, MA 01923, 978-750-8400. CCC is a not-for-profit organization that provides licenses and registration for a variety of users. For organizations that have been granted a photocopy license by the CCC, a separate system of payment has been arranged.

Trademark Notice: Product or corporate names may be trademarks or registered trademarks, and are used only for identification and explanation without intent to infringe.

Visit the Taylor & Francis Web site at
http://www.taylorandfrancis.com

and the CRC Press Web site at
http://www.crcpress.com

This work is dedicated to the loving memory of

my father, Albert Edward Seames, PE.

Contents

List of Figures xiii
List of Tables xxvii
Preface xxix
Author xxxiii

1. **Processing System Fundamentals** 1
   1.1 Continuous Processing 3
   1.2 Batch Processing 4
   1.3 Semi-Batch Processing 4
   1.4 Unit Operations 7
   1.5 Dynamics: The Speed of Information Flow in Processes 8
   1.6 Recycle Loops 11
   1.7 Documenting Process Flow 12
   1.8 Batch and Semi-Batch Process Drawings 19
   1.9 Distributed Control Systems 22
       1.9.1 Elements of the Distributed Control System 26
       1.9.2 Programmable Logic Control Systems 28
   1.10 Symbology for Control Systems 29
       1.10.1 Other, Less Common, Symbols 35
   Problems 36
   Reference 38

2. **Control System Fundamentals** 39
   2.1 Overview of Basic Unit Operations 39
       2.1.1 Motive Force (Momentum Transfer) 39
       2.1.2 Heat Transfer 39
       2.1.3 Mass Transfer 40
       2.1.4 Reactions 40
   2.2 Independent and Dependent Process Variables 40
       2.2.1 Control (Independent) Variables 40
       2.2.2 Measurement (Dependent) Variables 42
   2.3 Feedback Control Loop 43
   2.4 Disturbance Variables 49
   2.5 Feed Forward Contributions to Control 53
   2.6 Related Variables 56
   2.7 Ratio Control Loops 58
   2.8 Cascade Loops 61
   2.9 Feed Forward/Feedback Cascade Control Loops 68
   2.10 Process and Safety Systems 69
   2.11 Sequential Logic Control 74
   Problems 78

vii

3. Motive Force Unit Operations Control 83
- 3.1 Incompressible Fluids: Pumps 83
- 3.2 Compressible Fluids: Compressors, Blowers, and Fans 92
  - 3.2.1 Other Compressible Fluid Devices: Expanders and Turbines 97
- 3.3 Solids Handling Devices 100
  - 3.3.1 Solids Mixed with Gases and Liquids 100
  - 3.3.2 Physical Motive Force Devices 102
  - 3.3.3 Using Physical Pressure or Gravity Directly on the Solid for Short Distance Transport 103
- 3.4 Startup and Shutdown for Large Motive Force Systems 107
- 3.5 Equipment Protection Systems 107
- 3.6 Switching Controls for Parallel Motive Force Units 108
- Problems 110

4. Heat Transfer Unit Operations Control 113
- 4.1 Fluid-Fluid Heat Transfer 113
  - 4.1.1 Heating or Cooling a Process Stream to a Specified Temperature Using a Utility Stream 114
  - 4.1.2 Vaporizing a Liquid Process Stream Using a Utility Stream 117
  - 4.1.3 Condensing a Gaseous Process Stream Using a Utility Stream 119
  - 4.1.4 Process-Process Heat Transfer 122
    - 4.1.4.1 Cross Exchanger with Phase Change 125
- 4.2 Direct Mixing Heat Transfer 130
- 4.3 Electrical Resistance Heat Transfer 133
- 4.4 Fired Heaters 134
- 4.5 Monitoring and Adapting for Heat Exchanger Fouling and Scaling 138
- 4.6 Switching Controls for Parallel Heat Transfer Units 139
- Problems 141

5. Separation Unit Operations Controls 145
- 5.1 Single-Stage Separation 145
  - 5.1.1 The Flash Drum 145
  - 5.1.2 The Phase (Gravity) Separator 149
- 5.2 Multistage Distillation Overall Concepts 150
  - 5.2.1 Overhead System Controls for Distillation 157
    - 5.2.1.1 Overhead System with a Partial Condenser 157
    - 5.2.1.2 Overhead Control Scheme with Total Condenser 158
  - 5.2.2 Bottoms System Controls for Distillation 160

Contents

    5.3   Liquid-Based Absorption/Adsorption/Extraction/Leaching System Controls 163
            5.3.1   Single-Stage, Once-Through Absorption.......................... 164
            5.3.2   Single-Stage, Once-Through Extraction 166
            5.3.3   Multistage Absorption with Solvent Recovery 168
    5.4   Solid-Based Absorption/Adsorption/Extraction/Leaching System Controls 175
            5.4.1   Once-Through Solid Sorbent Systems 175
            5.4.2   Fluidized Bed Solid Sorbent Systems 177
            5.4.3   Fixed Bed Regenerated Solid Sorbent Systems 181
    Problems 190

6. Reaction Unit Operations Control 195
    6.1   Continuous Flow Reactors 195
            6.1.1   A Heterogeneous Binary Reaction of Two Liquid Reactants 197
            6.1.2   Safety Control Systems 201
    6.2   CSTR-Type Reactors 203
    6.3   Batch Reaction Systems 207
    6.4   Batch Reactors in Continuous Processes 209
    Problems 215

7. Other Control Paradigms 219
    7.1   Using Intermediate Calculations in Control Loops 219
            7.1.1   Using CALC Blocks in Blending Applications 221
            7.1.2   Using CALC Blocks to Monitor Heat Exchanger Performance 223
    7.2   Inferential Control 227
    7.3   High and Low Select Controls 231
    7.4   Override Control 233
    7.5   Split Range Control 233
    7.6   Allocation Control 235
    7.7   Constraint Control 235
    Problems 238

8. Controller Theory 241
    8.1   On/Off Control 241
    8.2   PID Control 243
            8.2.1   Proportional Response Control 246
            8.2.2   Integral Control 248
            8.2.3   Derivative Control 250
            8.2.4   Combined PID Control Algorithm 251
    8.3   Modifications to Minimize Derivative Kick 252

8.4 Nonlinear Control Strategies for PID Controllers 253
    8.4.1 Modified PID Control Algorithms 254
    8.4.2 Tuning Parameter Scheduling 255
    8.4.3 Nonlinear Transformations of Input or Output Variables 258
8.5 Adaptive Controllers 259
Problems 261

9. Higher-Level Automation Techniques 263
9.1 Plant Automation Concepts 263
    9.1.1 Field Instrumentation 265
    9.1.2 Regulatory Controls: The Backbone of Real-Time Process Controls 267
    9.1.3 Supervisory (Advanced Process) Controls 267
    9.1.4 Online Models 268
    9.1.5 Area Data Reconciliation 268
    9.1.6 Area Optimizer 268
    9.1.7 Plant-Wide Operational Control Applications 268
9.2 Advanced Process Controls 269
    9.2.1 Multiple Input/Single Output Controls (MISO) 270
    9.2.2 Fuzzy Logic Controllers 272
    9.2.3 Multiple Input/Multiple Output APC 275
    9.2.4 Multivariable Controls 276
9.3 Plant-Wide and Process Area Automation 279
    9.3.1 Data Reconciliation 279
    9.3.2 Plant-Wide and Process Area Optimization 280
    9.3.3 Planning and Scheduling 281
    9.3.4 Enterprise and Supply Chain Management 282
9.4 Higher-Level Automation of Batch Processes and Semi-Batch Unit Operations 282
Problems 284

10. Instrumentation (Types and Capabilities) 287
10.1 Pressure 287
10.2 Flow/Mass 291
10.3 Temperature 299
10.4 Level 301
10.5 Composition 305
10.6 Vibration 307
10.7 Throttling Control Valves 307
10.8 Speed Control Systems 310
10.9 Remotely Operated Block Valves 311
10.10 Pressure Relief Valves 313
Problems 315

Contents

11. Automation and Control System Projects 317
- 11.1 Specification and Design Concepts 317
  - 11.1.1 Applications 319
    - 11.1.1.1 Objectives-Based Controls: Changing the Process Control Paradigm 321
  - 11.1.2 Physical Automation Systems 322
  - 11.1.3 Human User Interfaces 323
  - 11.1.4 Physical and Systems Security 324
  - 11.1.5 Automating Batch Processes 326
  - 11.1.6 Safety Automation Systems 327
    - 11.1.6.1 Basic Principles 328
    - 11.1.6.2 Hazards Analysis 330
    - 11.1.6.3 Process Safety Layers 330
    - 11.1.6.4 Emergency Automation Systems 331
    - 11.1.6.5 Redundancy 333
- 11.2 Guidelines for Automation Projects 334
  - 11.2.1 Project Life Cycles 335
  - 11.2.2 How to Proceed with Automation Design for Process Plants 335
- Glossary 336
- Problems 338

12. Process Dynamic Analysis 343
- 12.1 Dynamic Models for Common Unit Operations 344
  - 12.1.1 Modeling a Liquid Knockout Drum: A Dynamic Model for Changes in Flow and Level 344
  - 12.1.2 Dynamic Modeling of a Single Pass Shell and Tube Heat Exchanger: A Dynamic Model for Energy Balances and Temperature 348
  - 12.1.3 Dynamic Modeling of a Fixed Bed Reactor With an External Cooling Jacket: Dynamic Modeling of Composition Change and Elemental Balance Variations ... 353
- 12.2 Dynamic Models for Common Instrument and Control System Components 356
  - 12.2.1 Control Variables 356
  - 12.2.2 Measurement Variables 357
  - 12.2.3 Controllers 357
- 12.3 Incorporating Simplified Process Changes into Dynamic Models 358
- Problems 361

Appendix A: Transform Functions and the "s" Domain 363

Appendix B: PID Controller Tuning 373

Appendix C: Controller Script 381

Index 385

List of Figures

Figure 1.1	Generic input/output diagram of a chemical process	1
Figure 1.2	Simplified generic block flow diagram of a chemical process with recycle.	2
Figure 1.3	The plant automation system is to the process as the brain and central nervous system are to a person.	3
Figure 1.4	Two-step batch process	5
Figure 1.5	(a, b) Two generic block flow diagrams showing alternative configurations to generate SO_2 from solid sulfur. Simple BFDs consisting of only reaction and separation steps are helpful in evaluating different process configurations before designing all of the details into the process.	13
Figure 1.6	A complete block flow diagram of the selected process from Figure 1.5a showing the process's initial, simplified material balance.	14
Figure 1.7	Use of the block flow diagram format to depict developed processes.	15
Figure 1.8	Typical process flow sheet showing the production of acetic acid from methanol.	16
Figure 1.9	One sheet of a typical process flow diagram.	17
Figure 1.10	One sheet of a typical piping and instrumentation diagram showing a pressure vessel used to knock out tar and ash from a process gas stream using a "quench water" stream.	18
Figure 1.11	Typical symbols used in a logic flow diagram for a batch process.	19
Figure 1.12	Simple batch process to mix, heat, and react two components, A and B, in a stepwise manner.	20
Figure 1.13	Logic diagram for the batch process described in Figure 1.12.	21
Figure 1.14	PFD sheet depicting a two-step semi-batch filtration unit operation embedded in a continuous process.	23

Figure 1.15	P&ID depiction of a four-step semi-batch filtration process step embedded in a continuous process.	25
Figure 1.16	Generic distributed control system hierarchy demonstrating many of the features of DCS.	27
Figure 1.17	A typical programmable logic controller bank.	28
Figure 2.1	Control scheme depictions of independent variables used in process control.	41
Figure 2.2	Using a recycle to control a pump (shown) or compressor (not shown) using a fixed speed motor.	42
Figure 2.3	Pressure vessel used to remove entrained liquid from a gas stream.	44
Figure 2.4	The basic regulatory control scheme for a pressure vessel used to remove entrained liquid from a gas stream.	48
Figure 2.5	An alternative regulatory control scheme for a pressure vessel used to remove entrained liquid from a gas stream.	49
Figure 2.6	Pressure vessel used to blend two fluids into a single fluid.	50
Figure 2.7	The control of a blending drum when both inlet streams are controlled by upstream process operations.	51
Figure 2.8	Pressure vessel used to add a base (fluid B) to fluid A to control its pH.	52
Figure 2.9	Process-utility heat exchanger used to cool a process fluid to a specified temperature.	53
Figure 2.10	The control scheme for a process-utility heat exchanger used to cool a process fluid to a specified temperature.	54
Figure 2.11	Process-utility heat exchanger used to cool a process fluid to a specified temperature with a feed forward/feedback control scheme.	56
Figure 2.12	Liquid knockout drum with a feed forward/feedback control scheme.	57
Figure 2.13	Mixing drum with a ratio controller for fluid A	59
Figure 2.14	Neutralization drum with a ratio controller for fluid B.	60

List of Figures

Figure 2.15 Boiler or fired heater used for steam generation. 61

Figure 2.16 Controlling the air-to-fuel ratio for a boiler or furnace............62

Figure 2.17 Typical control scheme for the overhead system for a distillation column employing a partial condenser and a fixed speed reflux pump 63

Figure 2.18 Cascade level to flow control loop on the overhead system for a distillation column employing a partial condenser and fixed speed reflux pump....................................64

Figure 2.19 The reflux accumulator after a total condenser for a distillation column 65

Figure 2.20 Controlling the reflux accumulator outlet streams with a cascade level-to-flow control loop and a nested cascade composition-to-flow-to-speed control loop................................66

Figure 2.21 Controlling the air-to-fuel ratio for a boiler or furnace with a feedback trim cascade control loop. CO stands for carbon monoxide. 67

Figure 2.22 Feed forward/feedback cascade control scheme to control gas phase composition by manipulating the cooling water flow rate, which in turn changes the fraction of the inlet vapor that condenses and thus will change the product composition, in a partial condenser for a system where process flow rate changes are significant....69

Figure 2.23 Feed forward/feedback cascade control scheme when the disturbance variable is nonresponsive. 70

Figure 2.24 Instead of using the absolute value of the measurement, a Rate of Deviation alarm uses the Rate of Change of the Deviation, the first derivative of the measurement value.......71

Figure 2.25 The independent safety system instrumentation and controls to protect a liquid knockout drum from potentially hazardous or damaging scenarios..........................74

Figure 2.26 A simple batch process to produce component C by adding a series of small increments of component B to a reactor containing component A..75

Figure 2.27 Detail of the control scheme used to insure that the heater operation is only regulated by the continuous temperature control loop during the correct step in the batch process. 77

Figure 2.28	Simplified flow sheet for a Claus sulfur plant.	81
Figure 3.1	A typical pump curve for a fixed speed pump.	84
Figure 3.2	A more advanced pump curve for a variable speed pump.	85
Figure 3.3	A basic fixed speed pump control using: (a) flow or (b) pressure as the dependent (measurement) variable.	86
Figure 3.4	A fixed speed pump control with a simple minimum flow recycle.	86
Figure 3.5	A pump and piping manifold to send liquid from the upstream process steps to two different downstream process steps at different flow rates and/or pressures.	87
Figure 3.6	A control scheme for a pump and piping manifold to send liquid from the upstream process steps to two different downstream process steps at different flow rates and/or pressures with fixed speed pumps.	88
Figure 3.7	An on/off control scheme for a fixed speed pump used to remove liquid from a drum based on the activation of low- and high-level switches.	90
Figure 3.8	A control scheme for a pump and piping manifold integrated with an upstream pressure vessel that sends the vessel outlet liquid to two different downstream process steps at different flow rates with fixed speed pumps while maintaining the liquid level in the vessel.	91
Figure 3.9	A control scheme for a variable speed cooling water supply pump integrated with a heat exchanger used to cool a process fluid.	91
Figure 3.10	A control scheme using outlet flow and outlet pressure for a compressor with a steam turbine driver.	93
Figure 3.11	A control scheme for a two-stage compressor with an electric motor.	95
Figure 3.12	A compressor map	96
Figure 3.13	Antisurge control with dedicated recycle cooler for a compressor with an electric motor.	97
Figure 3.14	Antisurge control for a nontoxic gas with insufficient value to warrant a recycle; a gas that can be vented to atmosphere.	98

List of Figures xvii

Figure 3.15	A compression/expansion system to generate very cold air.	99
Figure 3.16	A typical control scheme to blend a solid into a liquid to obtain a slurry in a mixing vessel, where the solids flow rate is set by an upstream unit operation	101
Figure 3.17	A typical control scheme to add a solid to a liquid process stream in a mixing vessel, where the flow rate is set by a downstream process and viscosity (μ) is used as an indirect measure of the quality of the blended product.	103
Figure 3.18	Using pressurized lock hoppers to load and unload a solid into a reaction vessel.	104
Figure 3.19	A typical control scheme for a batch process system that uses pressurized lock hoppers to load and unload a solid into a reaction vessel.	105
Figure 3.20	Typical piping and control configurations for a pump set where one pump is in operation and one is an installed spare	109
Figure 4.1	Using an incompressible utility liquid to cool a process stream to a specific temperature.	115
Figure 4.2	Control scheme for a heat exchanger that is generating steam to cool a very hot process stream.	116
Figure 4.3	Control scheme for a heat exchanger that is condensing utility steam to heat a process stream to a specific temperature and employing a steam trap.	118
Figure 4.4	Process scheme for a heat exchanger that is condensing a utility steam to vaporize a process stream.	118
Figure 4.5	Control scheme for a heat exchanger that is condensing a utility steam and employing a steam trap to vaporize a process stream.	120
Figure 4.6	Outlet configurations when condensing a process fluid in a heat exchanger.	120
Figure 4.7	Control scheme for a heat exchanger that is condensing a process stream using cooling water and employing a condensate drum on the process outlet stream.	122
Figure 4.8	Exchanging energy between the inlet and outlet streams of a reactor in a cross exchanger.	123

Figure 4.9   A more generalized process scheme for exchanging energy between two process streams in a cross exchanger. ... 123

Figure 4.10 A cross exchanger where the energy available in process fluid B exceeds the required duty for process fluid A; a bypass is placed on the process fluid B side of the exchanger to avoid overcooling stream A. 124

Figure 4.11 The complete heat exchanger system for the example of heating a hexane stream and cooling a weak acid stream when the weak acid stream has a higher duty requirement than the hexane stream. 126

Figure 4.12 An alternate configuration for the heat exchanger system for the example of heating a hexane stream and cooling a weak acid stream when refrigerated water is unavailable or too expensive. 127

Figure 4.13 Process and control configuration for the heat exchanger system for the example of vaporizing a hexane stream and cooling a weak acid stream where the hexane stream has the smaller duty requirement. 128

Figure 4.14 Exchanging energy between two process streams in a cross exchanger with a supplemental trim exchanger configured in parallel for the stream with the higher duty requirement to insure total condensation of the stream and employing a common condensate drum 129

Figure 4.15 Exchanging energy between two process streams in a cross exchanger with two supplemental trim exchangers. ... 129

Figure 4.16 Using a hot inert gas to directly heat a solid. 131

Figure 4.17 Using a hot inert gas to directly heat a solid with bulk dryer temperature control. 131

Figure 4.18 Scheme showing one way to control a hot inert gas used to directly heat a solid with weigh-in-motion solids flow rate used as a feed forward input to the hot gas flow rate control loop (WIM, weigh-in-motion solids flow measurement). 132

Figure 4.19 Using a hot inert gas to directly heat a solid via fluidization. 132

Figure 4.20 Using an external heat exchanger to vaporize liquid separated out of a gas stream. 134

List of Figures xix

Figure 4.21 Using an electrical resistance heating coil to vaporize small quantities of liquid accumulating in a surge drum. 135

Figure 4.22 Control scheme when heating a process fluid in a direct fired heater...136

Figure 4.23 Direct fired heater with both primary and secondary process fluids; adding additional tube banks increases the amount of energy that can be recovered from the fired heater...136

Figure 4.24 Heating a process fluid in a direct fired heater with excess heat routed through two secondary sections to heat two additional fluids..137

Figure 4.25 Heating a process fluid in a direct fired heater with excess heat used to generate utility grade steam....................138

Figure 4.26 A simple heat exchanger fouling/scaling monitoring system using a XA, a calculated alarm block. 139

Figure 4.27 A bank of parallel heat exchangers with four in service and one serving as an installed spare..140

Figure 4.28 Scheme to monitor heat exchanger fouling or scaling and automatically swap in the spare for the fouled exchanger....141

Figure 5.1 Control scheme for an adiabatic flash drum with a liquid level pot using the analysis of a key component and flow rate as the dependent variables. 148

Figure 5.2 Control scheme for a two-phase separator where the organic phase has a lower density than the aqueous phase..150

Figure 5.3 Vapor-liquid traffic in a trayed distillation column. 151

Figure 5.4 The bottoms system with adequate liquid level to provide pressure for the vapor to return to the distillation column 152

Figure 5.5 The overhead system for a distillation column.......................154

Figure 5.6 Control scheme for the overhead portion of a distillation system employing a total condenser, having a reflux ratio ≥1.0, and two independent pumps with variable speed drivers. HK is the heavy key. 159

Figure 5.7 The bottoms system of a distillation column with a partial forced reboiler system (employing a pump to provide the pressure to get the vapor back into the column) and bottoms product transfer pump. 161

Figure 5.8	Control scheme for the distillation bottoms system with a partial thermosyphon reboiler and a mass ratio of reboil vapor to bottoms liquid product greater than 1. LK denotes the light key.	162
Figure 5.9	Single-stage gas absorption system.	165
Figure 5.10	Control scheme for a single-stage gas absorption system. C_s denotes the concentration of the solute.	166
Figure 5.11	Single-stage once-through liquid-liquid extraction system.	167
Figure 5.12	Control scheme for a single-stage once-through liquid-liquid extraction system.	169
Figure 5.13	Gas extraction system with solute recovery and solvent recycle.	169
Figure 5.14	Controls for a gas absorption system with solute recovery and solvent recycle	173
Figure 5.15	A once-through mixer/settler type adsorption system.	176
Figure 5.16	Control scheme for a once-through mixer/settler type adsorption system.	178
Figure 5.17	A continuous fluidized bed adsorption/regeneration system.	179
Figure 5.18	Control scheme for a typical continuous fluidized bed adsorption/regeneration system.	182
Figure 5.19	A typical semi-batch fixed bed adsorption/regeneration system.	183
Figure 5.20	The logic flow diagram for one of the absorber beds in the semi-batch fixed bed adsorption unit operation shown in Figure 5.19.	184
Figure 5.21	Control scheme for a semi-batch fixed bed adsorption system.	189
Figure 6.1	Plug-flow-type reactors.	196
Figure 6.2	A multistage reactor system.	196
Figure 6.3	Maintaining isothermal reaction conditions using (a) a jacket or (b) heating/cooling tubes (in this example the shell is filled with catalyst).	197

List of Figures xxi

Figure 6.4	A fixed bed reactor for an exothermic reaction with two reactants, both of which are preheated.	198
Figure 6.5	Control scheme for an isothermal (using cooling tubes) fixed bed reactor with an exothermic reaction and two reactants, both of which are preheated.	200
Figure 6.6	Typical safety system for a fixed bed reactor with an exothermic reaction and two reactants, both of which are preheated.	202
Figure 6.7	A two-stage CSTR reactor system with an intermediate distillation product purification step.	205
Figure 6.8	Typical controls for a two-stage CSTR reactor system with an intermediate distillation product purification step.	206
Figure 6.9	Typical batch cycle for biological reaction systems.	208
Figure 6.10	Typical seed reactor configuration for the production of lactic acid.	210
Figure 6.11	Control scheme for the first stage of a semi-batch seed reactor configuration for the production of lactic acid.	212
Figure 7.1	Using CALC blocks to correct a flow rate reading for temperature.	221
Figure 7.2	Using CALC blocks to determine the correct rate of blending of two variable streams.	222
Figure 7.3	An alternate CALC block configuration.	223
Figure 7.4	Another alternate CALC block configuration.	224
Figure 7.5	Another alternate CALC block configuration.	225
Figure 7.6	A blend system where B is premixed with an inert fluid prior to mixing with fluid A to increase the stability of the overall control system to changes in composition or flow of either fluid.	226
Figure 7.7	System to monitor the heat transfer efficiency in a heat exchanger prone to fouling or scaling where the temperature of one or more of the inlet streams varies widely.	227
Figure 7.8	An alternative system to monitor the heat transfer efficiency in a heat exchanger prone to fouling or scaling where the temperature of one or more of the inlet streams varies widely.	228

Figure 7.9	Inferential control scheme to protect a compressor from changes in inlet gas density.	229
Figure 7.10	Example of how to depict a *soft sensor* that uses multiple laboratory-generated input data.	230
Figure 7.11	Using the quantity of cooling water consumed per unit quantity of reactant A to infer the optimum quantity of reactant B to feed to a reactor.	231
Figure 7.12	Symbols for (a) low select and (b) high select CALC blocks.	231
Figure 7.13	Reactor temperature control scheme employing a high-temperature select or CALC block.	232
Figure 7.14	Override control scheme to insure that the flow of slurry through the pump meets or exceeds the settling velocity of the solids in the slurry. OR denotes an override controller.	234
Figure 7.15	Use of a split range controller to improve control under two drastically different flow conditions.	234
Figure 7.16	Allocation controller used to control the flow rate through four parallel heat exchangers with one installed spare.	236
Figure 7.17	Typical control scheme for a distillation column with both overhead and bottoms composition control specifications. LK denotes the light key. HK denotes the heavy key.	237
Figure 7.18	Revised control scheme for the Figure 7.17 distillation column when the column overhead condenser is constrained at maximum cooling water flow and the overhead product purity is more important than the bottoms product purity.	238
Figure 8.1	A typical on/off electrical resistance heater controller.	242
Figure 8.2	A typical on/off level controller.	243
Figure 8.3	A typical on/off level controller with (a) a single level switch and (b) a two switch arrangement. O/H/C: denotes an open, hold, close type on/off controller	244
Figure 8.4	Impact of loop tuning on the response of a PI controller.	249
Figure 8.5	Demonstrating the derivative of the error function.	250

List of Figures

Figure 8.6	Truncating to the linear region of a nonlinear function.	254
Figure 8.7	The temperature profile through a reactor subject to uniform catalyst deactivation at (a) initial operation and (b) near end of life operation.	256
Figure 8.8	Adaptive controllers can be used to isolate a measurement signal in a noisy environment (a) to provide useful information (b).	261
Figure 9.1	A hierarchy of automation elements in a process facility.	264
Figure 9.2	A typical data/information transfer model for the various layers of a plant automation system.	265
Figure 9.3	Process automation applications/layers for (a) the entire plant and (b) an individual process area.	266
Figure 9.4	System to monitor the overall heat transfer in a heat exchanger prone to fouling or scaling where the temperature of one or more of the inlet streams varies widely using a MISO APC instead of calculation blocks.	270
Figure 9.5	The bottom portion of a separation column using a cross exchanger type partial reboiler.	271
Figure 9.6	Depicting a MISO APC to use heat flux as the measurement variable for the intermediate control loop in a nested cascade scheme for a cross exchanger type partial reboiler.	273
Figure 9.7	A 3D correlation plot to select the best temperature and residence time combination to maximize conversion for a single reactant ratio.	276
Figure 10.1	A Bourdon-tube-type pressure gauge	288
Figure 10.2	An electromechanical pressure sensor.	289
Figure 10.3	A strain gauge transducer for pressure measurement.	289
Figure 10.4	Manometers can be configured for (a) pressure measurement or (b) flow measurement.	290
Figure 10.5	An orifice-type flowmeter. (a) Schematic and (b) image.	292
Figure 10.6	A venturi-type flowmeter. (a) Image and (b) schematic.	292
Figure 10.7	A Coriolis-type mass flowmeter. (a) Image and (b) schematic.	293
Figure 10.8	Pitot tube flowmeters are often used for very low flow, low pressure gas flow measurements.	294

List of Figures

Figure 10.9   Rotameter-type flowmeter. 295
Figure 10.10 An electromagnetic-type flowmeter. 296
Figure 10.11 An ultrasonic-type flowmeter. 297
Figure 10.12 Direct mechanical flowmeters: (a) paddle meter image, (b) turbine meter image, (c) turbine meter schematic. 297
Figure 10.13 A vortex-type flowmeter. 298
Figure 10.14 A weigh-in-motion mass flowmeter for solids measurement. 299
Figure 10.15 (a) Image and (b) schematic showing how a thermocouple measures temperature. 300
Figure 10.16 A resistance temperature device (RTD). 301
Figure 10.17 An infrared sensor for temperature measurement. 301
Figure 10.18 Level measurement using (a) float position measurement and (b) a conductive tape and float system. 302
Figure 10.19 A bubble-type level measuring device measures the pressure required to push bubbles through a fluid. 303
Figure 10.20 A displacer-type level measurement device. 304
Figure 10.21 A wave-based level sensor. 304
Figure 10.22 Examples of sensors for a direct measurement type online analyzer 305
Figure 10.23 For sampled type online analyzers, a slip stream of the process fluid is routed through a sample conditioning system such as is shown here. 306
Figure 10.24 A pneumatically actuated throttling control valve. 308
Figure 10.25 Globe-valve-type control valve in the (a) closed and (b) open positions. 309
Figure 10.26 (a) Reverse acting (AFC), (b) direct acting (AFO) pneumatically driven actuators, and (c) a typical actuator body. 310
Figure 10.27 A remotely operated ball valve. (a) Schematic and (b) image. The hole is full width in a block valve, but smaller when used as a throttling control valve. 311
Figure 10.28 A remotely operated butterfly valve. 312
Figure 10.29 A remotely operated gate valve. 312

List of Figures   xxv

Figure 10.30	A spring-loaded pressure relief valve in both (a) normal (closed) and (b) event (open) positions.	313
Figure 10.31	A rupture disk.	314
Figure 11.1	A functional view of the operational section of the plant automation system; the section of the system that has been the focus of this textbook.	318
Figure 11.2	A physical view of the computer software and hardware necessary to provide and support the functions of the plant automation system.	318
Figure 11.3	The applications and supporting layers in an integrated plant automation system can be classified according to their general functions, such as the overview classification shown here.	320
Figure 11.4	The fire triangle.	328
Figure 11.5	Qualitative classification of event probability and consequence during hazards analyses	331
Figure 11.6	Relief valve protection of a pressure vessel.	341
Figure 12.1	The energy balance around a segment of a tube in a heat exchanger.	351
Figure 12.2	Fixed bed reactor with cooling jacket.	353
Figure 12.3	The plug flow condition through a fixed bed reactor.	354
Figure 12.4	A simple model of the process and control system.	358
Figure 12.5	Simple models of disturbances or changes in a process and control system	359
Figure B.1	The measurement response for an ideally tuned PID controller.	374
Figure B.2	The controller output and measurement behavior for the relay auto-tuning method.	375

List of Tables

Table 1.1	Sequential Logic Table for the Process Scheme Shown and Described in Figures 1.12 and 1.13	21
Table 1.2	A More Complete Sequential Logic Table for the Process Scheme Shown and Described in Figures 1.12 and 1.13	24
Table 1.3	Common Control and Instrument Symbols for Process Drawings	30
Table 1.4	Measurement Parameter Designations	31
Table 1.5	Device Type Designations	32
Table 1.6	Mathematical Symbols Used in Control System Information Boxes	33
Table 1.7	Common Higher-Level Control System Symbols for Process Drawings	35
Table 3.1	Sequential Event Table Associated with Figure 3.19	106
Table 5.1	The Sequential Events Table for the Semi-Batch Fixed Bed Adsorption Control Scheme Shown in Figure 5.20	186
Table 6.1	Sequence of Events Table for SLC-100 for the First Stage of a Semi-Batch Seed Reactor Configuration for the Production of Lactic Acid	213
Table 8.1	An Example of Regional Gain Scheduling	257
Table 9.1	Classification of Treatment Regions within a Metal Curing Furnace	274
Table 11.1	Important Differences between Continuous and Batch Processes	327
Table A.1	Laplace Transforms for Commonly Used Process Dynamic Functions	365
Table B.1	PID Controller Settings Based on the Ziegler-Nichols Ultimate Gain and Period	376

Preface

A few years ago, I finally had the opportunity to take over a one semester senior (fourth year) course entitled "Process Dynamics and Controls." I was looking forward to this as I had developed substantial applied controls experience during my 16-year industrial career. During that time, process controls underwent a complete transformation from electronic instruments to digital distributed control-based systems. I was excited to see new textbooks that reflected this "revolution" in process controls.

Imagine then my disappointment when I found that all the major textbooks in this field were still following the same format and with essentially the same content as textbook published in the 1960s and 1970s! These books emphasize simplified mathematical descriptions of process dynamics using time-dependent linear ordinary differential equations (ODEs) and their analytical solutions using Laplace transform solution methodologies.

The primary goal of these textbooks appears to be to help the reader understand the dynamics of the proportional-integral-derivative (PID) controller mathematically, so that the stability of control loops could be properly evaluated. This was an appropriate approach to the subject matter when process control was performed using a suite of stand-alone electronic controllers but is much less important now that control is most commonly performed using sophisticated, integrated distributed control and automation systems. While stability analysis is still of some interest to process control engineers, modern control system algorithms have reduced the importance from primary to secondary for control engineers.

Another big gap in the current literature is a lack of coverage of batch processes and, more importantly, the integration of batch unit operations within continuous processes (such a step being known as a semi-batch unit operation). Of the 17 most common textbooks that I reviewed, only one gave this topic any coverage at all. Yet almost every continuous commercial process pathway includes at least one semi-batch step, and batch processing represents a sizeable minority of process pathways employed (particularly in certain industries such as pharmaceutical).

"Designing Controls for the Process Industries" was conceived to address these deficiencies in the currently available literature. The goal is to completely transform chemical engineering process control and process dynamics education to focus on those aspects that are most important for process engineering in the twenty-first century.

Instead of starting with the controller, the book starts with the process and then moves on to how basic regulatory control schemes can be designed to achieve the process' objectives while maintaining stable operations. Without a deep understanding of the process itself, the power of the modern plant automation system cannot be fully enabled.

As much as possible, I have tried to follow the International Society of Automation's (ISA) guidelines for process control and instrumentation documentation. Some adjustments to the ISA guidelines were made where these improved the clarity of the concepts presented in the text. Most importantly, all of the process control schemes assume that field signals will be converted into digital form at the field device and that control will be accomplished in a distributed control system or programmable logic controller module(s).

In addition to continuous control concepts, I have embedded process and control system dynamics into the text with each new concept presented. I have also included sections on batch and semi-batch processes within new concept areas where appropriate. Finally, sections on safety automation are also included within concept areas.

The four most common process control loops—feedback, feed forward, ratio, and cascade—are introduced in Chapter 2, and the application of these techniques for process control schemes for the most common types of unit operations is provided in Chapters 3 through 6. For the practicing engineer, these chapters may prove to be the most useful for designing new control schemes or to help troubleshoot existing process instabilities. By comparing the schemes in these chapters to an existing situation, the engineer may be able to identify poorly designed control schemes. Modification of poorly designed control schemes may be an easy and cost-effective way to solve process instability problems. This is often a better approach than to try to "tune" your way out of a problem.

More advanced and less commonly used regulatory control options are presented in Chapter 7 such as override, allocation, and split range controllers. These techniques provide additional ways to increase the overall safety, stability, and efficiency for many process applications.

Chapter 8 introduces the theory behind the most common types of controllers used in the process industries. For those instructors that prefer to start with a "what's inside the box" approach, you might want to go through Chapters 1 and 2 and then jump to Chapter 8 prior to Chapters 3 through 7. For those instructors who are uncomfortable making a complete transition from the older course formats to that presented in this text, Appendix A provides content on how to solve simple linear ODEs using Laplace transforms, while Appendix B provides information on PID controller tuning.

Chapters 9 through 12 provide various additional plant automation–related subjects. An instructor in a one semester course is unlikely to be able to use all of this material but has the opportunity to emphasize those aspects that they feel are most important. Personally, I use Chapters 9 and 10. I then use Chapter 11 in a capstone design course. Chapter 12 is probably more appropriate for a graduate-level class in process dynamic modeling or as part of an advanced transport phenomena course. However, instructors who want to emphasize process modeling in their course may wish to use this material.

Preface

Supplemental material available to instructors includes complete PowerPoint™ slide files for each chapter. Over 700 multiple-choice questions are also available for flipped class mode of instruction.

My thanks to the University of North Dakota and the Fulbright Foundation for supporting this work. I thank Peter Martin of Schneider Electric (formerly Invensys/Foxboro) for his advice and encouragement when I was data gathering for the project. I also thank my former colleagues at Saudi Aramco who helped me to gain most of my knowledge of process controls and process control projects. Special thanks to M. Jim Dunbar, Mark Barbee, and Hamdi Noureldin.

Thanks also go out to my teaching assistant Ian Foerster for all of his suggestions, especially with the homework problems and solutions, and to Will Nielsen for his careful editing of the text. My thanks to all of the students, both local and via distance, who provided feedback on the material during the two trial years of instruction at the University of North Dakota and to my colleagues for their enthusiasm and support for the project. No one catches typos like university students! I also thank Jessica Mann for drawing some of the more complicated figures. My thanks to CRC Press's Taylor & Francis Group led by Senior Editor Allison Shatkin. Last but not least, my thanks to my wife, Janet, for her patience and support.

The following companies provided images that were used in the text: Dwyer Instruments Inc., Michigan City, IN, USA; Emerson Automation Solutions, Houston, TX, USA; Badger Meter, Inc., Milwaukee, WI; Compressor Controls Corp., Des Moines, IA, USA; Vande Berg Scales, Sioux Center, IA, USA; and LT Industries/process NIR analyzers, Gaithersburg, MD, USA.

Author

Wayne Seames is a Chester Fritz Distinguished Professor of chemical engineering at the University of North Dakota (UND), Grand Fork, North Dakota. An Arizona native, Seames received his BS in chemical engineering at the University of Arizona, Tucson, Arizona, in 1979. His 16-year industrial career included assignments as a process engineer, controls project engineer, and process control group leader. In 1992, he was assigned as project manager for plant automation systems, for the Ras Tanura Upgrade and Expansion project, one of the largest process control–related projects in the world. In 1995, Seames returned to Arizona where he earned his doctorate in chemical engineering in July 2000. Amongst his academic awards are the 2014/2015 Fulbright Distinguished Chair and a Visiting Professorship at the University of Leeds, UK; the 2013 UND Faculty Scholar Award for Excellence in Scholarship, Teaching, and Service; the 2012 UND Award for Interdisciplinary Collaboration in Research or Creative Work; the 2007 UND Foundation/Thomas Clifford Faculty Achievement Award for Individual Excellence in Research; and the 2006 "Professor of the Year" award from UND School of Engineering and Mines. He was elected a fellow of the National Academy of Inventors in 2017 and is a named inventor on eight U.S. patents.

1
Processing System Fundamentals

A process is a series of functions or steps (unit operations) that transform one or more substances entering the process (raw materials or inputs) into one or more other substances (outputs). The goal of the process is to provide substances with advantageous qualities compared to the raw materials from which they derive. Those substances that are the goal of the process are known as products. Other output substances that have value or can be used elsewhere are known as by-products, and those that cannot be used elsewhere are known as wastes.

Processes may also require chemicals that do not end up as part of the products or by-products. These are known as recyclable and/or consumable materials or chemicals. If reactions are involved, they are often facilitated by catalysts, which are a type of consumable material, since they must eventually be replaced.

The most common use of a process is to convert a raw material into a saleable product such as those described by Figures 1.1 and 1.2. Examples include the conversion of crude oil into transportation fuels, the conversion of monomers into polymers, and the conversion of nutrients into antibiotics. Another common use of a process is to convert a harmful waste into a less-harmful waste (these are known as environmental processes). Examples include the

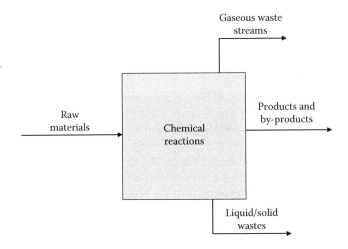

FIGURE 1.1
Generic input/output diagram of a chemical process.

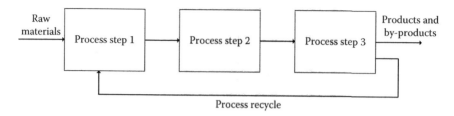

FIGURE 1.2
Simplified generic block flow diagram of a chemical process with recycle.

removal of sulfur dioxide (SO_2) from coal-fired power plant flue gas, organics from wastewater, and toxic metals from mine tailings.

Regardless of the specific process chosen, every process facility has three fundamental objectives:

1. To maximize profit
2. To minimize environmental impact
3. To meet acceptable health and safety standards

These objectives are accomplished through the people working in the installation, aided by the facility's plant automation system (PAS). When well designed and implemented, the PAS acts like the central nervous system and subconscious parts of the body, while the people represent the conscious parts of the body's control system (Figure 1.3). The foundation of the PAS consists of the field instruments and the regulatory controls, the nerves and motor control centers of the PAS. More advanced controls and higher-level automation functions use the features and data of this regulatory control foundation to perform more complicated tasks that assist in meeting the facility's objectives; the subconscious portion of the PAS. Finally, the PAS has human user interface (HUI) capability so that control operators can assess the operation of the process and make manual changes where required; the conscious portion of the PAS.

The goal of this textbook is to help the reader learn the concepts and basic techniques associated with the design of the PAS, with specific emphasis on the foundational level of the PAS: field instruments and regulatory controls. In this introductory chapter, we review how processes work—continuous, batch, and combinations of both. We define the unit operation concept and illustrate how unit operations are used to build processes. Then we look at information flow within processes and the challenging case of recycle loops. We introduce the most common tools used for documenting process flow—flow sheets, block flow diagrams, and process flow diagrams. Then we provide an overview of physical process control systems for both continuous and batch processes. Finally, we include an introduction to more complete versions of the PAS.

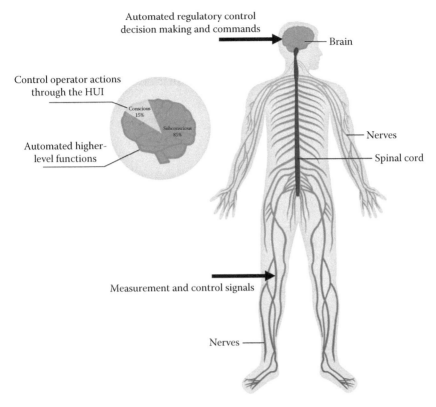

FIGURE 1.3
The plant automation system is to the process as the brain and central nervous system are to a person.

1.1 Continuous Processing

Process facilities are expensive. World-scale plants can cost billions of dollars and take many years to build. To get the most out of these facilities, most companies endeavor to maximize the time process facilities operate. Under best practices, most facilities in the process industries can achieve a 95% operating factor. That is, they are in operation an average of 24 h per day, 347 days per year. Certain industries such as agricultural processing plants may have to accept lower operating factors due to a lack of raw materials. But the general concept, maximize time in operation, is always desirable to maximize profits.

To maintain high levels of operation, most process facilities are designed to operate continuously. The goal is to operate at the steady-state conditions that most efficiently transform the raw materials into primary products for a given target production rate. The production rate is not usually based on the process facility's capacity, but on external demands. For example, the production rate may be based on projected or committed sales orders for the primary products. For an environmental process, the production rate may be based on the quantity of waste raw materials received. Historically, process plant production rates average 60%–75% of their design operating capacity.

1.2 Batch Processing

While continuous processing is usually used, there are certain circumstances where batch processing is preferred. For example, specialty chemical manufacturers may be able to make a number of different but similar products, using the same basic processing equipment. So they may configure the plant for production of product A for a month, then shut down and reconfigure the facility for the production of product B.

Another common use of batch processing is for those processes that employ biological transformations, reactions, or production steps. This is very common in the pharmaceutical industry, where drugs, such as many antibiotics, are synthesized by an organism. Biological transformations often involve a seven-step cycle: (1) inoculation (seeding the reactor with the organisms), (2) accommodation (allowing time for the organisms to adjust to their new environment), (3) growth, (4) production (when the organisms synthesize/expel the target product material), (5) death, (6) removal of biomass from the reactor, and (7) sterilization of the reactor (to prevent mutated or diseased strains from developing). When the biological step dominates the entire process, it often makes sense to operate the entire process in batches.

1.3 Semi-Batch Processing

There are many unit operations that must be used in batch mode, even though we may want to use them in a continuous process. Consider, for example, a process stream that must be filtered to remove solid material from a liquid (Figure 1.4). Most filtration methods operate as batch operations. The solids build up on the filtration media to a certain thickness or pressure drop, and then the material is physically removed from the filter and collected. Another example is the use of a physical adsorption agent to remove one or more chemical constituents from a gaseous or liquid mixture. The feed material flows through a fixed bed of the adsorbing material until the entire bed reaches saturation. The flow is then stopped and the adsorbent is regenerated, removing the adsorbed constituents that are recovered.

Processing System Fundamentals

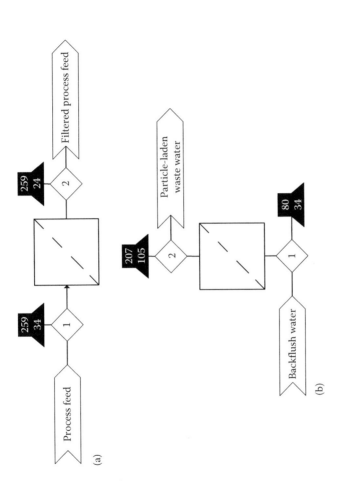

FIGURE 1.4
Two-step batch process: (a) Filtration of solids out of a liquid stream and (b) removal of the solid particles collected on the filter. *(Continued)*

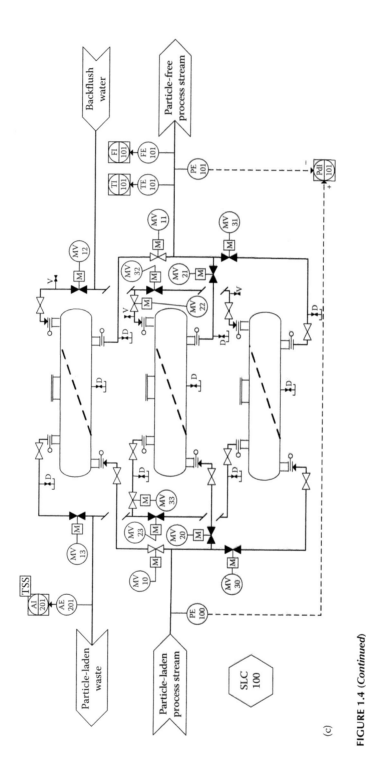

FIGURE 1.4 (Continued)
Two-step batch process: (c) in a semi-batch process, two or more filters (3 in this example) are placed in parallel so that liquid can be continuously filtered even when one or more filters are taken out of filtration service for particle removal.

When batch steps are used in a continuous process, multiple copies of the batch equipment are provided in parallel with connecting manifolds to allow the process stream to feed into the desired unit(s) while the other unit(s) is in a different portion of the batch process. For example, we can have two filters in parallel. While one filter is removing the solids from the liquid, the parallel filter can be undergoing the steps to remove the solids from the media and prepare the filter for its next filtration cycle. This type of operation is known as a semi-batch operation. The configurations can be as complicated as needed. For example, in an ethanol plant, the bioreactors use yeast to produce ethanol via fermentation of corn meal. The facility uses four or more reactors in parallel to insure that there is a continuous input of corn meal into this step and a continuous production of ethanol out of this step. However, each bioreactor operates in batch mode, going through the biological lifecycle steps described earlier.

1.4 Unit Operations

In the early days of the process industries, it was observed that certain steps were used frequently and that the use of these steps was similar for a large number of applications. These steps became known as unit operations. Organizing the process into unit operations simplifies initial process design. It also simplifies the design for the process controls necessary to the facility.

The most common unit operations are incompressible fluid momentum transfer (pumps), compressible fluid momentum transfer (compressors, blowers, fans), heat transfer (heat exchangers, furnaces), gas/liquid separation (flash drums, knockout drums), liquid/liquid separation (two phase separators), liquid/solid separation (filters, settling drums), constituent separators based on vapor-liquid equilibrium (distillation, strippers, flash drums, evaporators), constituent separators based on chemical driving force (absorbers, extractors, leaching, membranes, adsorption), constituent separators based on bubble point or freeze point (partial condensers, driers, ovens, crystallizers), chemical reactors, and solid momentum transfer (conveyors, extruders).

In Chapters 3 through 7 we will learn how to design simple, reliable control strategies for the most common unit operations. This makes regulatory control system design much simpler. As we design the process, we can find the appropriate control strategy for the unit operation application and use that strategy, instead of having to invent new strategies for every different process step we encounter. One of the goals of this book is to provide you with a menu of control schemes that can be selected and applied in process systems.

1.5 Dynamics: The Speed of Information Flow in Processes

The term *process dynamics* typically refers to one or both of the following types of information: (1) time dependences through the process and (2) changes in the state of process compared to steady state. Let us start with the first. Raw materials enter the process and are transformed through a series of unit operations into products, by-products, and wastes. As the material moves through the process, it contains information about its contents and properties, such as the temperature of a fluid.

Ideally, we would like our process to operate at the perfect steady-state conditions that achieve the objectives of our facility. However, there are always minor deviations from this condition within the process. We can measure these deviations by comparing the value of certain properties within the process to the ideal condition. We can then use this information to control the process in such a way that we obtain the products we want while maximizing profit, minimizing environmental impact, and staying within acceptable safety limits.

For both of these types of process dynamics, we access this process information through measurements. The most common are temperature, pressure, flow rate, level, and composition. We supplement these measurements with selected equipment-related measurements such as rotational speed, amperage, voltage, and vibration. These additional measurements provide indirect information about the process as well as help us to stay within acceptable safety limits.

Once a key property of the process material is measured, we can assess how well the process is performing compared to certain benchmarks or standards. In control terminology, these standards are known as *setpoints*. We can then adjust the process conditions to move the measurement closer to the standard. We will discuss how this is accomplished in Chapter 2. However, we do not stop the process and wait to make the adjustment. The process continues to operate and the material we just sampled for a measurement has moved onward in the process by the time we are ready to make a correction. The gap between when we sample and when we adjust the process is known as the *time lag*.

It is not enough just to design the process control system to operate at the desired steady-state conditions defined by the measurement standards. For continuous processes and for the processing step within a single step of a multistep batch process, dynamics are continuously introduced into the process. Some of these occur because we make a change to the operating conditions of the facility. Others are introduced by the control system itself. As we will see in Chapter 2, we actually control the process using the difference between the desired conditions and the actual conditions. This difference is known as the *error*. Other dynamics come about because of small perturbations that can be caused by any of a number of factors such as a temporary equipment malfunction, a change in the conditions of a catalyst or heat exchanger surface, etc. So we must be able to accommodate these dynamic changes in our control system.

In order to properly design our system, we must consider the following factors:

1. The speed at which dynamic information moves through a real-time process, which varies based on the physical properties of the process material
2. The distance the process material travels from the point at which the information in the process is changed to the point at which the information is measured
3. The speed at which the process information is measured, translated into a useable data point, and processed in the control system to generate an adjustment to the process

Process pressure dynamics move at the speed of sound. Pressure measurement also occurs very quickly, on the order of milliseconds for the most common measurement types. Modern control systems can process real-time measurements within tens of milliseconds. Therefore, reaction to process pressure dynamics is fast. As a consequence, pressure is the most *responsive* variable that can be used in the control system.

The process dynamics associated with the rate of flow of incompressible fluids also move at the speed of sound. The dynamics of flow measurement devices varies more than that of pressure measurement devices (see Chapter 10 for a description of the most common measurement methods) and is typically slightly slower. However, incompressible fluid flow dynamics are very responsive and this parameter is considered to be a responsive variable when used in the control system.

The process dynamics associated with temperature are a combination of the speed of radiative heat transfer, convective heat transfer, and conductive heat transfer. In general, changes in temperature are dominated by convective heat transfer with minor contributions from the other two mechanisms. Therefore, process temperature dynamic information typically moves at slightly greater speed than that of the fluid. Temperature measurement is fast, approaching the speed of pressure measurement. So, the speed of the fluid typically sets the responsiveness of this variable. Gaseous streams typically move at 15–30 m/s (50–100 fps) in process piping, while liquid streams move at 1.5–3 m/s (5–10 fps) (McCabe et al. 2005). In general, temperature is considered to be a responsive variable.

Compressible fluids have one additional degree of freedom compared to incompressible fluids. A dynamic change in momentum can change not only the flow of the fluid but also its pressure/density. The portion that impacts pressure moves at the speed of sound, while the portion that impacts flow moves at the speed of the fluid. Compressible fluid flow dynamics can be classified as responsive.

At selected points in the process, liquid may be allowed to accumulate before proceeding to the next unit operation. Sometimes, these liquid hold-up spots are incidental to the unit operation's function. An example is a shell and tube heat exchanger. If a liquid phase is present on the shell side of the exchanger, the area in the shell must be partially (if the liquid is vaporizing) or fully filled with liquid in order to insure efficient heat transfer across the tubes (see Figure 4.4). This liquid level can be measured and the measurement used to adjust the flow rate in or out of the shell in order to change the level. By uncovering some of the tubes, the heat transfer can be made less efficient and the amount of heat transferred between the shell and tube phases can thus be controlled. This is also true on the tube side of the exchanger. To insure that the tubes are full or partially full of liquid, the liquid level in one of the heads of the exchanger (the area where the fluid is divided between the tubes or collected from all of the tubes) can be measured and used to control the flow into or out of the tube side of the heat exchanger.

Another example is the bottoms (the liquid outlet stream taken below the lowest tray) of a trayed distillation column. Often a thermosyphon reboiler is used for liquid boil-up of some of the bottoms liquid leaving the column. This type of reboiler avoids the need for a bottoms pump. Instead, the liquid head (the pressure force generated by the height and density of a fluid) in the bottom of the distillation column provides sufficient pressure to insure that the vaporized material leaving the reboiler can reenter the column just below the bottom tray. While the purpose of the liquid is primarily to provide a pressure head, the liquid level can also be measured and used to manipulate any of a number of column parameters (e.g., the bottoms product draw) to insure that the distillation system separation meets its objective.

Sometimes a wide spot in the piping, a surge drum, is added to provide liquid hold-up. This is most commonly used in recycle loops. We will discuss this in more detail in Section 1.6.

Because of the residence time of the liquid at hold-up spots, the transfer of dynamic process information is delayed. Residence times (the average time a given molecule of the fluid stays within the vessel) vary from seconds to hours. There are many methods to measure level in a vessel, but measurement and signal processing is much more rapid than the residence time delay built into the process. Since the residence time in the vessel is long by control standards, level dynamics are *not responsive*. Sometimes, having a slow response time is good (see Section 1.6), but sometimes it gets in the way of good control. We will learn how to overcome this problem through the use of cascade control loops in Section 2.9.

It is often useful to know the concentration of selected constituents in a process stream or unit operation. Some of these can be measured directly, while others can be inferred through measurement of an indirect parameter that can be correlated to the target constituent. Here are some examples:

1. The density of a mixture might provide indirect measurement of the purity of a liquid product stream.

2. The heat of combustion generated from a mixture might provide an indirect measure of the purity of a combustible gaseous stream.
3. The conductivity of an aqueous liquid stream may provide an indirect measurement of the ion content in a wastewater stream.
4. The pH may infer the concentration of acids in an aqueous liquid.

All of these measurements are referred to as analytical measurements.

Some analytical measurements can be made directly in the process stream, such as pH or conductivity. For these types of measurements, the analysis time may be comparable to the analysis time of flowmeters. For these systems the process dynamics of analysis can be considered to be responsive.

Other analytical measurements require that a sample be extracted from the process stream. The sample may be processed directly or sent through a sample conditioning system prior to analysis. If the analyzing instrument is rapid, such as an infrared (IR) detector type instrument, then the measurement speed is on the order of seconds to minutes. However, if the actual analysis takes more time, such as for a gas chromatograph, the measurement speed will be on the order of tens of minutes to an hour. For these systems, the process dynamics of analysis must be considered to be *not* responsive. Thus, indirect or sampled analytical measurement systems may require special techniques (the same may hold for level measurement systems) to insure that the dynamics of their systems are handled by the control system.

Each analytical measurement method has a different analysis time and these can vary widely, so be sure to investigate this before classifying the relative responsiveness of this parameter.

1.6 Recycle Loops

Partial or total recycle loops are very common in continuous processes. For example, a distillation system includes two recycle loops. One of these loops is associated with the overhead stream. The stream leaving the top of the distillation column, know as the overhead, is partially or totally condensed. All or part of the condensed liquid is then recycled back to the top of the column as the column reflux liquid. Similarly, the bottoms liquid from the column is partially reboiled back to vapor, which is then recycled back to the bottom of the column as the column boil-up gas.

A common unit operation is solvent absorption/extraction. In this unit operation, one or more components are selectively removed from the process fluid using a solvent. The component(s) to be removed, known as the solute, has an affinity for the solvent which provides the chemical driving force for mass transfer from the original process fluid into the solvent. In most processes, the solvent is then routed to a regenerator where the solute is removed from the solvent by manipulating the properties of the solute-rich solvent stream (e.g., the stream might be cooled if solubility decreases with decreasing temperature or heated if the solubility decreases with increasing temperature). The solute-lean solvent is then recycled back to the absorber/extractor.

Most chemical reactions do not proceed to 100% conversion of all reactants. It is common practice to recover the unreacted material in downstream separation processes. This recovered material is then recycled back to the reactor. There are many other examples.

Process dynamics are particularly important when some or all of the process fluid is recycled. Unless properly managed, even minor fluctuations in the process fluid are magnified when the dynamic information is recycled in the process. The recycled dynamic may cancel out the dynamic in the original process step. But more commonly, the recycled dynamic adds to the original dynamic, creating an even larger fluctuation. This larger fluctuation then contributes an even larger dynamic and so forth until the process fluctuates completely out of control.

The key to managing recycle loop dynamics is to have at least one liquid hold-up spot in the recycle loop that is controlled via level control. As we saw in Section 1.5, level control is *not responsive*. This low-response feature can be enhanced by selecting controller settings that increase the sluggishness of the corrective response to level changes. In effect, we provide the process with a location where dynamic fluctuations can be smoothed out or integrated over a larger volume of process fluid. This allows the outlet liquid from the hold-up spot to proceed with very minimal dynamic information which in turn, provides smoother controls. We will learn in Section 2.9 how to use a level-to-flow cascade control scheme to decrease the dynamics in the liquid outlet stream to an even greater extent.

1.7 Documenting Process Flow

Processes are efficiently documented using drawings. The simplest is the input/output (I/O) diagram (Figure 1.1). This is a very simple diagram where the entire process is depicted as a single "black box" or block. Any key reactions that occur in the process are usually provided inside of the block. Inlet streams should be limited to only those inputs that contribute to the final primary product(s). For example, if the process is a Claus sulfur plant, the inlet stream would be the flue gas containing the hydrogen sulfide (H_2S) that is to be converted into elemental sulfur. Oxygen used in the initial thermal oxidizer would not be included because none of the oxygen ends up in the primary product. Outlet streams are all products plus any waste streams generated from a portion of the inlet streams. By convention, the inlet streams are shown on the left-hand side, the products and by-products on the right-hand side, and the wastes on the top (gases) and bottom (liquids and solids) sides of the block.

Excluded from the diagram are consumed chemicals (inputs that do not end up in any of the saleable products), catalysts, utilities, and wastes generated from consumed chemicals. For example, if natural gas and air are burned to generate steam and flue gas from the combustion is emitted, then none of these streams would be shown on the I/O diagram for the process (but might be shown on an I/O diagram of the utilities area). The exception to this would be if the energy from the combustion ultimately ended up in the facilities' primary product(s) such as in a power plant where the steam would be used to generate the electricity that is the primary product of the facility.

I/O diagrams may be qualitative or may include the overall mass balance (quantitative). These are used to define the overall capacity of the facility as well as the raw material transformation efficiency (the fraction of the inlet raw material(s) that contributes to usable products).

Another simple type of drawing is the block flow diagram (BFD). The BFD (Figure 1.2) is originally used to assist in the design of process schemes. For this usage, all reaction and separation steps are depicted as blocks on the diagram. In addition to the raw material, product, and waste streams, the BFD includes all intermediate process streams. It also includes any consumable chemicals. Any process recycles are included on the diagram.

Qualitative BFDs (Figure 1.5) allow various combinations and variations in the flow scheme to be evaluated quickly. Once one or more flow schemes have been chosen for further study, a quantitative BFD (Figure 1.6) can be used to document the process material balance. This material balance should match the balance from the process' I/O diagram.

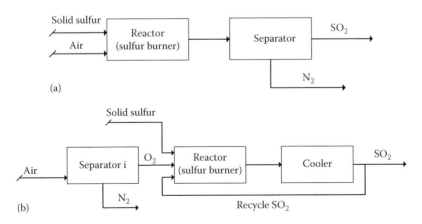

FIGURE 1.5
(a, b) Two generic block flow diagrams showing alternative configurations to generate SO_2 from solid sulfur. Simple BFDs consisting of only reaction and separation steps are helpful in evaluating different process configurations before designing all of the details into the process.

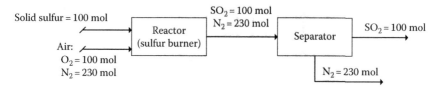

FIGURE 1.6
A complete block flow diagram of the selected process from Figure 1.5a showing the process's initial, simplified material balance.

BFD-type drawings are also used as simplified process flow diagrams (explained below) to show just the essence of a process that has already been developed in greater detail. In this role, the BFD (Figure 1.7) may include heat exchange and momentum transfer steps in addition to reaction and separation steps or may depict the process at a higher level, where each "block" represents many process steps. On a BFD, each unit operation is depicted as a block or "black box" within the overall process.

Most process design textbooks employ the flow sheet (Figure 1.8) as the primary development tool for the design, optimization, and documentation of processes. Flow sheets are also often employed to depict the process graphically on control system graphic displays. However, for design and documentation of a process, the more common drawing is the process flow diagram (PFD), shown in Figure 1.9. These two drawings are similar. The PFD denotes the actual unit operations that will be utilized in the process. This drawing includes all streams utilized in the process including inlet, intermediate, and product process streams, consumable chemicals, wastes, and utilities. The energy balance is defined by including the power rating of momentum unit operations (e.g., the motor power requirements of a pump) plus the temperature and pressure in each process stream. Preliminary key unit operations specifications are also included.

Some designers include simple control valves on PFDs and flow sheets. However, this is not recommended unless the control valve is being used as a unit operation as opposed to a process control system component. The PFD does not contain adequate information to correctly specify the regulatory controls for the process. So adding control valves does not really contribute to the information contained on the diagram and can easily be specified incorrectly.

An example of using a control valve as a unit operation is when the valve is being used in an adiabatic flash operation. One way to use mechanical work to remove internal energy from a fluid is to: (1) place the fluid in the liquid state at low pressure, (2) pressurize the liquid to a very high pressure, and (3) adiabatically (using insulation around the control valve and associated piping) depressurize the liquid across a control valve. Some or all of the liquid will vaporize as the pressure is reduced. The energy required for the latent heat of vaporization will come from the fluid's internal energy, reducing the temperature of the fluid. This is the principle behind refrigeration loop systems.

Processing System Fundamentals 15

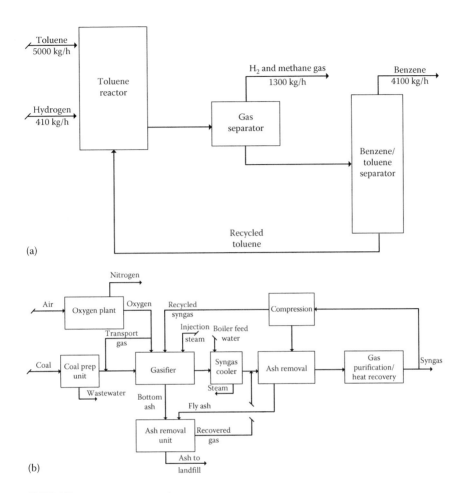

FIGURE 1.7
Use of the block flow diagram format to depict developed processes: (a) The basic process steps to produce benzene from toluene and (b) the overall process to gasify coal into syngas.

The next level of functional design detail is typically documented on piping and instrument diagrams (P&ID or PID). These diagrams are a functional representation of the equipment, piping, block valves, isolation valves, vent valves, drain valves, measurement instruments, controllers, control valves, and safety automation system elements needed for the process (Figure 1.10).

During the early phases of design, default equipment types are used. These are either the most probable type or the most common type employed in the process industries. Common examples are centrifugal pumps and compressors, shell and tube heat exchangers, trayed distillation columns, and fixed bed plug flow reactors with cooling/heating jackets. In later design phases when specific types of equipment (e.g., a shell and tube heat exchanger vs. a plate and frame exchanger) have been determined, the drawing is updated

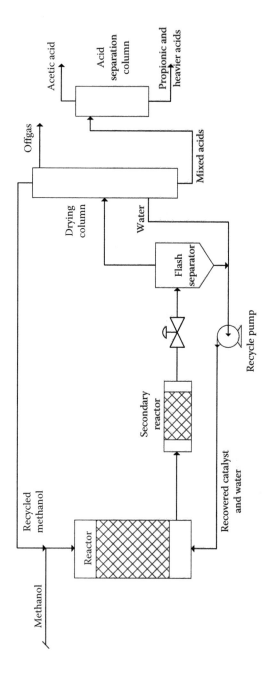

FIGURE 1.8
Typical process flow sheet showing the production of acetic acid from methanol.

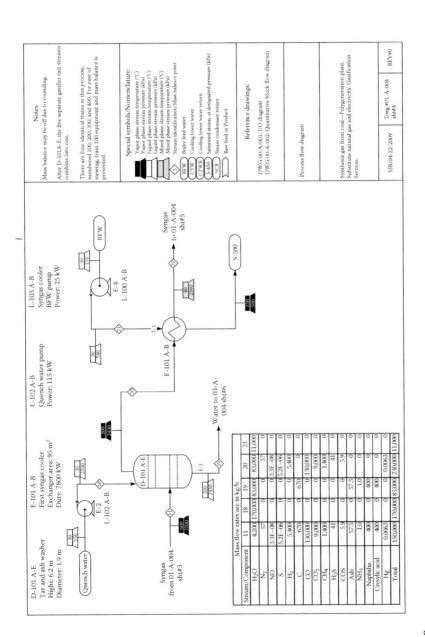

FIGURE 1.9
One sheet of a typical process flow diagram. (Drawing courtesy of Stacy Bjorgaard, Mitch Olson, and Travis Waind, University of North Dakota Chemical Engineering Program Alumni, Grand Forks, ND.)

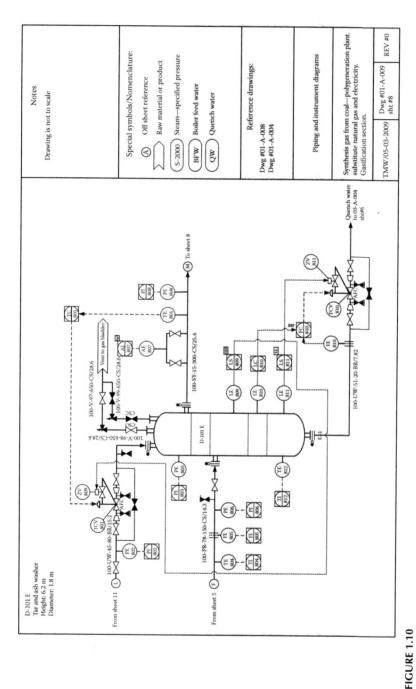

FIGURE 1.10
One sheet of a typical piping and instrumentation diagram showing a pressure vessel used to knock out tar and ash from a process gas stream using a "quench water" stream. (Drawing courtesy of Stacy Bjorgaard, Mitch Olson, and Travis Waind, University of North Dakota Chemical Engineering Program Alumni, Grand Forks, ND.)

Processing System Fundamentals

to depict the actual specific equipment type utilized. The motors for pumps, compressors, mixers, conveyors, etc. should also be shown on the P&ID and given distinct equipment numbers.

The P&ID includes every pipe used within the process area associated with the unit operations depicted on the diagram. Also included are all isolation (block) valves, drain valves, vent valves, and isolation blinds (a blind is a piece of solid metal inserted between two flanges in a piping system to temporarily block the flow of fluid through the pipe). More importantly for our purposes, the P&ID also includes depictions of all field instruments, regulatory control schemes, and control valves. Safety system instruments and controls are also included. Simple supervisory control schemes, those that can be easily depicted on the drawing, such as cascade controls, are also included. We will review control symbology and learn how to read the instrumentation and control system depiction shown on the P&ID in Section 1.10.

1.8 Batch and Semi-Batch Process Drawings

The drawings described above work well for documenting continuous process flow. They are also adequate for documenting the flow within a single step in a batch or semi-batch process. They are not as well adapted for assisting with the design and documentation of the sequence of steps that are encountered in batch processes and batch unit operations embedded in continuous processes.

The initial design of a batch process can be facilitated through the use of a logic flow diagram. This diagram, borrowed from the computer programming world, shows the sequence of steps and decisions that occur between steps in a batch process. Typical logic flow diagram symbols and their uses are provided in Figure 1.11. Consider the simple batch process described in Figure 1.12. A logic flow diagram for this process is shown in Figure 1.13.

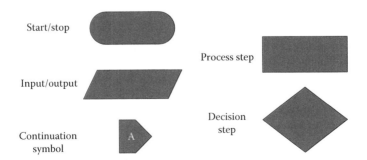

FIGURE 1.11
Typical symbols used in a logic flow diagram for a batch process.

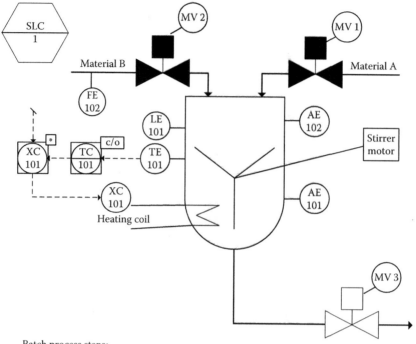

Batch process steps:

1. Open MV1 and fill the tank with "A" until LE 1 reaches the desired height, LH.
2. Close MV1 and start the stirrer motor.
3. Open MV2 and add a fixed amount of "B" as measured by FE 2.
4. Close MV2.
5. Energize XC1 until TE 1 reaches the desired temperature.
6. Control the temperature using XC1 at this temperature.
7. When all "B" is consumed as measured by AE1, deenergize XC1.
8. Is all "A" consumed as measured by AE2? If yes go to 9, if no go to 3.
9. Stop the stirrer motor, open MV3 until LE1 reaches its low level limit.

FIGURE 1.12
Simple batch process to mix, heat, and react two components, A and B, in a stepwise manner. For some reversible reactions, adding only a small quantity of one reactant to the other will facilitate the forward reaction.

A sequential logic table is another design tool for batch and semi-batch processes. This table reflects the logic diagram in stepwise tabular format (Table 1.1). The preliminary sequential logic table will include a description of the step and the functional termination criterion for each step. For more complicated termination decision making, "if," "then," and "go to" statements may be employed. As an example, consider a batch biological process with two bioreactors in parallel. A termination criterion might be that when the activity of the bioreactor declines to a certain point, the bioreactor is taken out of service and drained. However, we might want to keep the bioreactor in service, even at declining production, if the parallel bioreactor's biomass

Processing System Fundamentals

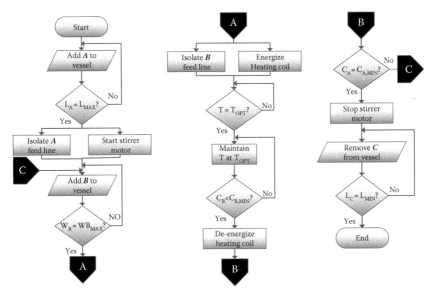

FIGURE 1.13
Logic diagram for the batch process described in Figure 1.12.

TABLE 1.1

Sequential Logic Table for the Process Scheme Shown and Described in Figures 1.12 and 1.13

Step	Description	Termination Criteria	Next Step #
1	Fill vessel with A	$L = L_{H,A}$	Go to 2
2	Start stirrer motor	Motor on	Go to 3
3	Fill vessel with B	$L = L_{H,B}$	Go to 4
4	Energize heating coil and heat up vessel	$T = T_H$	Go to 5
5	Control temperature until all B is consumed	$A_B = A_{B,Min}$	Go to 6
6	De-energize heating coil	Heater off	If $A_A > A_{A,min}$ Go to 3.else.7
7	Stop stirrer motor	Motor off	Go to 8
8	Drain vessel	$L = L_L$	If another batch go to 1.else.9
9	Stop batch process		

This version would be included on the process flow diagram.

is not yet ready to produce product at an acceptable level. In this case, the termination criteria might be if "activity in A < X and activity in B > Y, then go to step Z."

Using the logic flow diagram and the preliminary sequential logic table, a series of PFDs can be developed for each step in the sequence. For semi-batch processes, these diagrams are very simple and multiple steps are often depicted on a single PFD sheet. For each batch step, the PFD shows the relevant process flow scheme, the material balance that occurs during the batch step, and the temperature and pressure of each stream. Inputs and outputs not utilized by that batch step are typically omitted to maximize drawing clarity (Figure 1.14).

A full sequential logic table also includes the open/close status of appropriate isolation valves and the on/off status of key motors such as those on pumps or mixers (Table 1.2). This more complete table is typically developed in conjunction with the P&IDs. For batch processes, isolation valves are typically operated remotely. These valves are distinct from the throttling control valves used in continuous processes or during the continuous operation within a batch step.

Control valves are not designed to completely isolate a system; they leak and as the valves age, their seals erode and their rate of leakage increases. So block-type, open/close valves should be used. These are most commonly operated with a motor operator, but may also be air operated. These motor operators should be depicted on the P&ID (see Figure 1.15) and given a distinct valve number. To show a valve in the open position, only show the outline of the shape (an example is MV-510 on Figure 1.15). To show the closed position, make the symbol a solid black color (an example is MV-514 on Figure 1.15). Each of the devices (e.g., valves and motors) shown on the full sequential logic table should be connected to a sequential logic controller (SLC) symbol on the P&ID. Sometimes the connecting lines to the SLC are left off the drawing because there are so many connections (as on Figure 1.15, where the lines are left off).

1.9 Distributed Control Systems

In the 1980s, instrument vendors began adapting computers for use in process control systems. One distinct difference compared to almost all other applications was the need for very high levels of device reliability and data integrity. While it might be annoying to have your personal computer freeze up or inexplicitly shutdown, this same situation in a complex process plant could result in a major disaster, with significant financial loss and even loss of life. As a consequence, manufacturers had to develop methods that would allow them to obtain the reliability of the then current electronic systems with digital replacements.

The key breakthrough was the development of operating systems that would accommodate backup or slave computers that would have the same configuration and contain the same data as the operating or master computers. This function is known as tracking or *check-pointing*. At a frequent, specific time interval, all of the configuration, settings, and data resident in the master computer are copied onto the slave computer, replacing the slave's

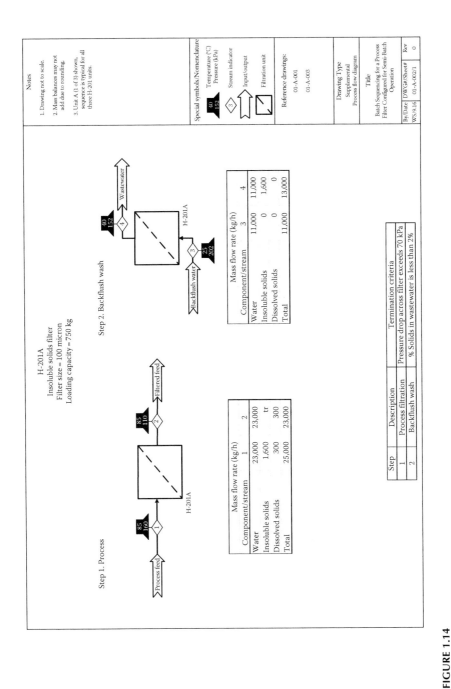

FIGURE 1.14
PFD sheet depicting a two-step semi-batch filtration unit operation embedded in a continuous process.

TABLE 1.2

A More Complete Sequential Logic Table for the Process Scheme Shown and Described in Figures 1.12 and 1.13

Step	Description	MV 1	MV 2	MV 3	MC 1	XC 101	Termination Criteria	Notes
1	Fill vessel with A	O	C	C	OFF	OFF	$N = N + 1$. IF $LE-101 = LE-101_{H,A}$ Go to 2	N counts the number of batches. High alarm on LE-101 can be used to determine when correct A is in vessel.
2	Start stirrer motor	C	C	C	ON	OFF	$FTE-102 = 0$. If MC-1 = ON Go to 3	FTE-102 totals the flow measured by FE-102.
3	Fill vessel with B	C	O	C	ON	OFF	If $FTE-102 = FTE-102_{H,B}$ Go to 4	High alarm on FTE-102 can be used to determine when correct B is in vessel.
4	Energize heating coil and heat up vessel	C	C	C	ON	ON	If $TE-101 = T_H$ Go to 5	TC-101 can be used to automatically keep the coil energized until the correct temperature is read by TE-101.
5	Control temperature until all B is consumed	C	C	C	ON	ON	If $AE-101 = AE101_{B,Min}$ Go to 6	TC-101 can be used to automatically keep the correct temperature. AE-101 measures the concentration of B in the vessel.
6	De-energize heating coil	C	C	C	ON	OFF	If $AE-102 > AE-102_{A,Max}$ Go to 3.else.7	AE-102 measures the concentration of A in the vessel. If A remains, Steps 3–5 are repeated until all A is consumed.
7	Stop stirrer motor	C	C	C	OFF	OFF	If MC-1 = OFF Go to 8	When all A is consumed, the stirrer is halted.
8	Drain vessel	C	C	O	OFF	OFF	If $LE-101 = LE-101_L$ and $N < N_{max}$ Go to 1.else if $LE-101 = LE-101_L$ Go to 9	When the vessel is drained, a check is made on whether there are more batches scheduled. If yes, Steps 1–7 are repeated. If no, the process is stopped by Step 9.
9	Stop batch process	C	C	C	OFF	OFF	$N = 0$	The batch counter, N, is reset for the next time this sequence is used.

This version would be included on the piping and instrument diagram. The instrument tag numbers are from Figure 1.12.

Processing System Fundamentals 25

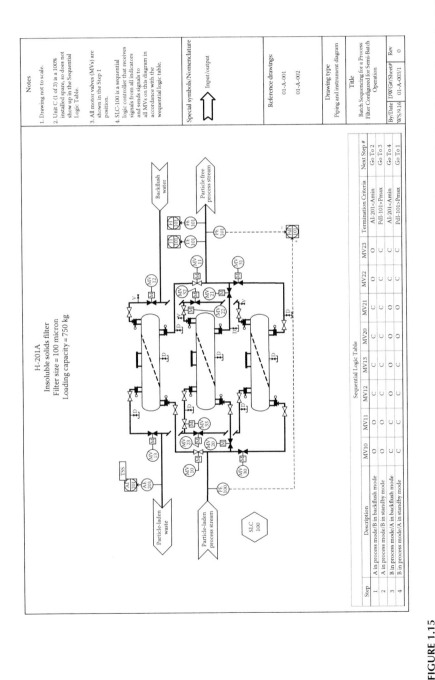

FIGURE 1.15
P&ID depiction of a four-step semi-batch filtration process step embedded in a continuous process. In this example, the valves are denoted only by their number without the "MV" or "MOV" Prefix. Note, one of the filters is an installed spare and is not needed to complete the sequence.

current version. If the master computer fails, the computer operating system automatically and instantaneously reverses the roles of the computers: slave to master and master to slave. This operation must have sufficient transparency to avoid initiating upsets into the process. By that, I mean that as far as the rest of the distributed control system is concerned, there has been no impact from the change in the operating computer.

These new control computers also needed to communicate with a wide variety of devices. In fact, they needed to be able to accommodate thousands of inputs per second from hundreds of separate measurement devices and to output hundreds of commands to control devices. Because of the high level of data traffic, the communication bus and operating software had to be very robust. Most vendors developed their own proprietary communications firmware and software to accommodate these rigorous requirements.

Points of commonality were needed so that devices developed by different vendors could communicate with each other. The result was a series of fairly standard languages and the development of translator programs to convert from one system to another. The legacies from these early developments still exist in today's distributed control systems, although more standardization has occurred.

1.9.1 Elements of the Distributed Control System

A generic hierarchy demonstrating many of the features that can be employed in a distributed control system (DCS) is shown in Figure 1.16. Field instruments are used to measure parameters associated with the process, such as temperature, pressure, flow rate, level, and composition, as well as those associated with the processing equipment, such as motor speed and vibration (see Section 1.5). Another category of field instruments are the devices used to control the process such as control valves, relays, and power regulators.

Field measurement instruments typically contain an electronic package that converts the actual measurement into a digital output signal of sufficient power to insure that the signal can be easily read by the DCS communications system. Similarly, control devices typically have an electronic package that converts the incoming digital input signal into the proper format for use in the control device. In some older systems, the signals are analog electronic signals following a specific standard, with the most common being one in which an amperage reading of 4 milliAmps represents a value of zero and 20 milliAmps represents a value of 100. In even older systems, the conversion from analog to electronic (or the reverse from electronic to analog) or digital is handled by a separate device attached to the field instrument. These devices are known as field transducers. Another device, known as the field transmitter, is then used to boost the signal for use in the DCS.

When the field instruments generate digital signals that are all in the same format (such as the Foundation Fieldbus format), a data highway is attached

Processing System Fundamentals 27

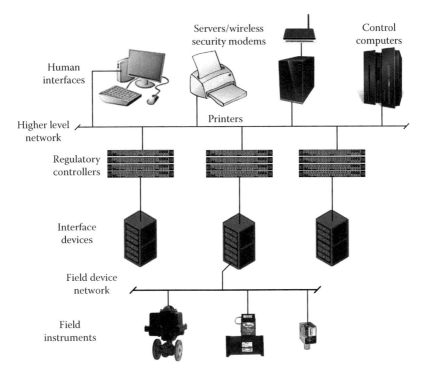

FIGURE 1.16
Generic distributed control system hierarchy demonstrating many of the features of DCS. (Field instrument images: Courtesy of Dwyer Instrument Company, Michigan City, IN.)

to each field instrument either by cable (wire or fiber optic) or microwave. Wireless systems are now available, but are not widely applied since a cable is still required to provide power to the instrument, making the incremental costs of wired instruments over wireless instruments relatively small. Wired systems have less potential problems due to signal interference than wireless systems. If the field instruments do not generate digital signals in the same format or generate analog signals, then separate wires are run from each instrument to an interface device. The interface device converts the input signal from the field instrument format into a digital signal in the DCS' format (and the reverse for control devices).

At the heart of the DCS are the regulatory controllers. These devices hold the configuration of all the regulatory controller functions used in the process. How regulatory controls are accomplished will be introduced in Chapter 2 with additional details provided in Chapter 8. Each controller operates like a separate computer, hence the name *distributed control*. A special type of controller, known as the sequential logic controller (SLC) or sequential events controller, is used for batch steps within processes. SLC is typically accomplished in a special device known as a programmable logic controller (PLC), which is described in more detail in Section 1.9.

A data highway connects all of the regulatory controllers to each other and to other devices included in the DCS. For example, this highway allows the rest of the DCS to interface with human user interface devices to allow human access to controller and indicator data. A control computer is used for higher-level functions—those that cannot be performed in an individual regulatory controller. These types of functions are introduced in Chapter 9.

For inputs that are not used in controllers, such as those that are simply provided as information to the control operator, separate input interface modules are used to convert the signal from the field data highway to the control system data highway (if required).

Another higher-level data highway may also be provided, which allows the DCS to communicate with external devices such as corporate information systems. This highway is typically kept separate from the critical elements of the DCS for security reasons.

1.9.2 Programmable Logic Control Systems

PLCs are designed to execute sequential logic. There are many different types of PLCs with a wide range of capabilities (Figure 1.17). The simplest PLCs just perform sequential steps for a signal device. The most complex ones act as the primary device in a DCS and even have the ability to perform simple continuous regulatory control functions. PLCs were first developed to replace physical relays used in sequential logic systems for industrial manufacturing. By moving the logic to software, PLCs eliminated the need to rewire the system and/or add additional hardware when the sequential logic was modified.

Early PLCs used a programming language based on ladder logic to determine the actions for outputs based on inputs. Ladder logic contains functions

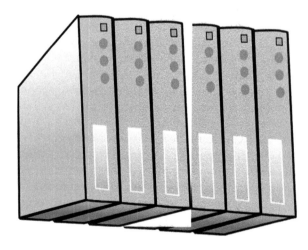

FIGURE 1.17
A typical programmable logic controller bank. Each box represents a separate controller unit.

Processing System Fundamentals   29

that will simulate an open/close contact, a counter, a timer, math operations, etc. As more computational and memory capabilities became available, PLC functionality was expanded to include reusable function blocks where control algorithms, including continuous control algorithms, such as those described in Chapter 8, could be developed and stored for use during operation.

PLCs are most commonly used to control the operation of a single piece of equipment, particularly equipment where startup and shutdown logic is extensive, such as a major gas compressor or a fired heater. For these types of unit operations, a PLC (or in some cases multiple PLCs) is usually provided by the vendor that is designed solely to control that specific piece of equipment.

1.10 Symbology for Control Systems

The International Society of Automation (ISA, formerly the Instrument Society of America) has established a guideline to provide uniformity in documenting field instruments, control system components, and control valves on drawings [ANSI/ISA-5.1-2009 Instrumentation Symbols and Identification]. The complete guideline provides a great deal of latitude and flexibility in how controls are depicted. No doubt, as control systems continue to evolve, this symbology will also need to evolve. Based on today's version of the ISA guideline, here is a basic, simplified system that will provide clear documentation of the vast majority of control schemes. If you have a situation that does not fit within these simplified guidelines, please refer to the full ISA guideline for assistance.

Drawing symbols to depict basic control system instruments and controller functions are summarized in Table 1.3. When drawing a control scheme, each individual instrument and control function is given a unique identifier of the form XY-aaa, where "X" indicates the type of dependent (measurement) variable associated with the instrument and "Y" indicates the action associated with the device. "aaa" is the unique instrument identifier. Measurement parameter designations are shown in Table 1.4. Note that X is used for any type of measurement that does not fit into any of the other categories. It is also used as a placeholder when any of a number (or all) of the symbols could be relevant. Typical device designations are shown in Table 1.5. As with measurement parameters, X is used for any type of device that does not fit into any of the other categories or to represent a circumstance where multiple symbols could be relevant.

Each symbol is described briefly below (symbol numbers refer to the numbers in Table 1.3):

1. A circle represents a function that is located within the process area near to the physical device.
2. If there is a horizontal line bisecting the center of the circle, this indicates that a readout is accessible to human users.

TABLE 1.3

Common Control and Instrument Symbols for Process Drawings

Ref. Number	Description	Image
1.	A function that is located within the process area near to the physical device. "X" indicates the type of dependent variable associated with the readout (F = flow, T = temperature, etc.) and "Y" indicates the action associated with the device (I = Indicator, C = Controller). "aaa" is the unique instrument identifier.	○ XY aaa
2.	A horizontal line bisecting the center of the circle indicates that a readout is accessible to human users. This is an indicator or controller located in the process area near the physical measurement device.	⊖ XY aaa
3.	A local remote indicator. When the indicator is located in or at the edge of the process area, remote from the physical measurement device, but not within a digital process control system (PCS), the single horizontal line is replaced by a double line.	XY aaa (circle with double horizontal line)
4.	A circle encased inside of a square indicates that the function is located within the PCS.	□○ XY aaa
5.	A circle encased inside of a square with a horizontal line bisecting the center of the circle indicates that the function is located within the PCS and has a Readout that is accessible to human users.	□⊖ XY aaa
6.	An information box placed on the upper right-hand corner of the symbol is used to provide additional information about the function of the controller.	Circle in square with small box at upper right, XY aaa

(Continued)

Processing System Fundamentals 31

TABLE 1.3 (*Continued*)

Common Control and Instrument Symbols for Process Drawings

Ref. Number	Description	Image
7.	A PCS control function with info box and operator access.	
8.	A controller block used for an intermediate operation with no operator access. The specific operation would be defined using the info box.	

TABLE 1.4

Measurement Parameter Designations

Symbol	Measurement Type
A	Analysis
F	Flow
I	Current
L	Level
P	Pressure
R	Resistance
S	Speed
T	Temperature
V	Vibration
X	Miscellaneous

3. If the indicator is located in or at the edge of the process area, remote from the physical measurement device, but not within a digital process control system (PCS), the single horizontal line is replaced by a double line. Note that local remotes will typically only be present in facilities with older control systems. In modern systems, the measurement device sends a digital signal into the PCS right at the field device.

TABLE 1.5

Device Type Designations

Symbol	Measurement Type
E	Element[a]
C	Controller
CV	Control Valve
d	Differential
dT	Pneumatic to Current Transducer
I	Indicator
M	Motor
MV	Motor Actuated Block Valve
S	Switch
T	Transmitter or Transducer
Td	Current to Pneumatic Transducer
V	Valve
X	Miscellaneous
Z	Safety System Device

[a] Or element with integrated transducer and/or transmitter.

4. A circle encased inside of a square indicates that the function is located within the PCS. If there is no line within the circle, the controller readings are not accessible by the human users of the systems—only system administrators would have access to these functions.

5. If the controller readings are accessible by the human users of the system, a horizontal line is included in symbol 4.

6. An information box can be placed on controller blocks to provide additional information.

7. If the operator has access to the information in the block, a horizontal line is included. One use of this box is to define the type of controller specified. The default is a standard PID feedback controller (we will learn about the most common controller types in Chapter 8). If any other controller function is specified, it should include an information box. For example, if an on/off (open/close) controller is being specified, it would be designated by:

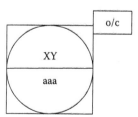

Processing System Fundamentals

The information box is also used to provide additional information. For example, the symbol "A" is used for all analysis/composition measurements. However, an analysis loop can mean any of a number of chemical measurements, so an information box should be placed on every analyzer controller (AC) and analyzer indicator (AI) block to indicate what chemical or property is being measured by the AE. For example, if the pH of a stream is being measured and used in a control loop, it would be designated by:

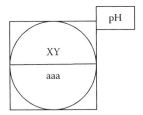

8. Controller blocks can also be used for intermediate operations. Only one operation should be included in each block. Designators for the most common mathematical calculation block operations are shown in Table 1.6. For example, two flow rates, a and b, might be added together prior to use in a controller block. This would be designated by:

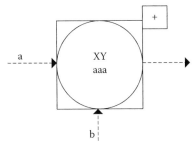

Some of these intermediate functions are so common that a shorthand designation has been developed for them. In these cases, the intermediate operation is integrated into the controller block. For example, to monitor the pressure drop across a filter, the pressure

TABLE 1.6

Mathematical Symbols Used in Control System Information Boxes

Symbol	Definition
+ or Σ	Addition/summation
× or *	Multiplication
/	Division or ratio

measurement downstream of the filter would be subtracted from the measurement upstream of the filter. This would be designated by:

where PdC stands for pressure differential control.

In cascade control schemes, the output from one controller will be used to adjust the setpoint of another controller (see Section 2.9 for cascade controls). In this case, a line is drawn from the master control block (usually level or analysis) to the slave control block (commonly flow, temperature, or pressure) and the term "ESP" or "RSP" is placed on the line near the slave control block. ESP stands for *external setpoint* and RSP stands for *remote setpoint*. Both are commonly used. In this textbook, we will use the term "ESP" exclusively. For example, a level controller cascade to a flow controller would be designated by:

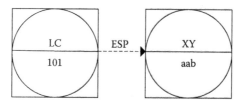

If you have a control block that performs multiple functions in a single step, a multisided block such as a hexagon or octagon is used instead of a square. In the ISA standard, there is no discrimination amongst the various uses of this symbol. You have to figure it out from context and, if you are lucky, notes added to the drawing. To add clarity, we will deviate slightly from the ISA standard by using symbols with different numbers of sides to denote different classes of use of this symbol, as shown in Table 1.7.

9. One use of this symbol is for a sequential logic controller (SLC, see Section 2.11). These controllers are used to manage the steps in batch and semi-batch processes. The SLC may receive multiple inputs in order to ascertain when it is time to terminate a step in the sequence. It then usually must send multiple outputs to block valves and motors to reconfigure the process for the next batch step. For clarity, we will use a hexagon block to uniquely denote an SLC in this textbook.

10. Multifunction control blocks are also used on P&IDs to depict advanced control strategies that include multiple equation steps. An example might be a control strategy that balances the flow of

Processing System Fundamentals 35

TABLE 1.7

Common Higher-Level Control System Symbols for Process Drawings

Ref. Number	Description	Image
9.	Sequential logic controller. These controllers are used to manage the steps in batch and semi-batch processes.	SLC aaa (hexagon)
10.	Advanced control strategies performed in a control computer instead of in controller blocks.	XC aaa (octagon)
11.	Any safety system controller that includes multiple outputs or multiple calculations to perform its function.	ZC aaa (pentagon)

a process fluid through parallel heat exchangers that are prone to fouling. Multivariable controllers are also depicted with this symbol when they are documented on P&IDs. For clarity, we will use an octagon block to uniquely denote any advanced control strategies that require multiple calculations.

11. Another use is for emergency management systems, such as an emergency shutdown system for a compressor/turbine train. For clarity, any safety system controller that includes multiple outputs or multiple calculations to perform its function will be uniquely denoted by a pentagon in this textbook.

1.10.1 Other, Less Common, Symbols

In the mid-twentieth century, electronic control systems began to supplant pneumatic control systems. A remotely indicating measurement in these electronic control systems included the measurement element (XE, where X denotes the type of measurement), a device known as a transducer to convert the measurement from the original format (often pneumatic) into electronic (XdT), a device known as a transmitter to boost the electronic signal to sufficient strength for transmission to the remote location where the signal would be viewed (XT), and the analog indicator where the board operator could read the measurement in engineering units (XI). When the signal was used in a control loop, an analog controller replaced the indicator (XC). If the operator could read the measurement value in addition to its use in the control loop, it was known as an indicating controller (XIC). The output from the XC was transmitted to another transducer (to convert from electronic into pneumatic), and this transducer (XTd) then provided a pneumatic pressure that applied pressure to a diaphragm within the control valve (XCV). This diaphragm pushed against a spring of constant force and the balance between the diaphragm and the spring adjusted an insert in the fluid flow path, setting a restriction for the flow through the XCV.

In the late twentieth century, digital control systems began to replace the electronic systems and vendors began integrating the transducer and transmission functions with the measurement device. The transducer required for the control valve was also integrated with the CV. As a consequence, today's XE, XdT, and XT are all one device and the XTd and XCV are integrated into a single device. There are two common conventions used for the first case: to call the integrated measurement system an XE or an XT. Usage is about even between these two conventions. For the purposes of consistency, we will use the first convention, XE, throughout this textbook. Virtually everyone uses XCV as the nomenclature for the second integrated system.

Problems

1.1 Define the following terms: setpoint, time lag, a responsive variable, and a nonresponsive variable.

1.2 Explain the difference between a continuous process, a batch process, and a semi-batch process.

1.3 Provide five examples of a semi-batch unit operation (*Hint*: not an entire process, just a single unit operation).

1.4 For each of the following measurement variables, classify the variable as "responsive" or "nonresponsive": temperature, level, pressure, direct in-process analysis, extracted process analysis, incompressible flow, and compressible flow.

1.5 How do you keep a recycle loop from going out of control?

1.6 Gaseous butylene and isobutane react to form gaseous alkylate per the following reaction:

$$C_4H_8 + C_4H_{10} \rightarrow C_8H_{18} \qquad (1.1)$$

Feed streams are:
  a. Butylene feed containing 74 vol% butylenes, 17 vol% sobutene, 8 vol% n-butane, and 1 vol% propane.
  b. Isobutane feed containing 90 vol% isobutane, 9 vol% n-butane, and 1 vol% propane.

1.6.1 Draw an input/output diagram for a process that will produce 1700 m^3/day of alkylate (*Hint*: at mild conditions, mole% and volume% are essentially the same for gases).

1.6.2 If the reaction conversion is 25%, there is a substantial recycle stream in the process. Assuming perfect separations allowing you to produce pure products: propane, n-butane, and alkylate, draw a quantitative block flow diagram for the new unit.

1.7 Using external resources such as the Internet or process textbooks, find the information you need and draw an input/output diagram (without the material balance) for the following processes:

1.7.1 The production of polyethylene from ethylene

1.7.2 The separation of crude oil into light gases, LPG, pentanes, naphtha, kerosene, diesel, fuel oil, and tars

1.7.3 The conversion of methane into methanol

1.7.4 The removal of CO_2 from a power plant flue gas to generate a nearly pure CO_2 waste stream

1.7.5 The production of cooking oil from soybeans

1.8 Draw a generic block flow diagram for the following processes:

1.8.1 The production of polyethylene from ethylene

1.8.2 The conversion of methane into methanol

1.8.3 The removal of CO_2 from a power plant flue gas to generate a nearly pure CO_2 waste stream

1.8.4 The production of cooking oil from soybeans

1.9 Processes requiring SO_2 may generate the feedstock by combustion of sulfur:

$$S + O_2 \rightarrow SO_2 \qquad (1.2)$$

However, to maintain a reasonably low temperature in the burner, sufficient amounts of cool and inert gas must dilute the oxygen before it is fed to the burner. The oxygen mixture should be about 70 mol% inert and 30 mol% oxygen.

1.9.1 Generate an I/O diagram to produce 100 mol of SO_2 from sulfur and air.

1.9.2 Develop two separate and distinct generic BFDs for this process assuming perfect separations.

1.10 A fixed amount of material A is added to a tank; then a fixed amount of material B is added to the tank containing material A. Upon heating and stirring, the two react to form C. The reaction works best if just a small amount of B is added to A at a time.

    1.10.1 Draw a logic diagram of this batch process.

    1.10.2 Add the appropriate instrumentation to the diagram and develop a sequential events table of the process.

1.11 A simple batch process is described as follows:

    a. Use pump #1 to fill a tank up to a given level with liquid; then turn off the pump.

    b. Energize a heating coil in the tank and heat the liquid until a given temperature is reached; then turn off the heater.

    c. Use pump #2 to drain the tank down to a given level; then turn off the pump.

    1.11.1 Draw a process sketch and logic diagram of this batch process.

    1.11.2 Add the appropriate instrumentation to the diagram and develop a sequential events table of the process.

Reference

McCabe, W.L., Smith, J.C., and Harriott, P., *Unit Operations of Chemical Engineering*, 7th Edition, McGraw-Hill, Dubuque, IA, 2005.

2
Control System Fundamentals

In this chapter, we introduce the most common unit operations used to build processes. We will also introduce the basic building blocks of regulatory control for both continuous and batch unit operations. This information will be used in Chapters 3 through 6 as a starting point to present and discuss regulatory control strategies for the most common unit operations.

2.1 Overview of Basic Unit Operations

2.1.1 Motive Force (Momentum Transfer)

In a real-time process, material moves through the various unit operations where it is transformed from its original state into usable products, by-products, and wastes. For fluids—gases and liquids—movement through the process usually occurs due to a pressure difference between the upstream and downstream unit operations as fluids will always flow from an area of high pressure to an area of low pressure. The pressure of the fluid is increased using pumps (liquids), compressors (gas, high pressure), blowers (gas, moderate pressures), and fans (gas, low pressure) at strategic points in the process to facilitate this movement.

Solids are more challenging. Usually they must be physically moved, such as by using a conveyor or an extruder. Sometimes gravity is used to move a solid while at other times the solid is entrained in a gas (fluidization) or liquid (slurry) so that it can be moved using the entraining fluid.

2.1.2 Heat Transfer

In commercial-scale processes, heat transfer is most often accomplished using indirect heating with heat transfer fluids. Steam is most commonly used for heating and water is most commonly used for cooling. For very high temperatures, a fired heater may be used. In this case, a fuel (natural gas, fuel oil, etc.) and air are combusted together using a burner and the resulting hot flue gas is used to heat the process fluid or an intermediate heat transfer fluid. For very low temperatures, refrigerated water or liquid nitrogen may be used. Direct electric resistance heating is not usually feasible in large systems but is commonly used in small-scale facilities.

2.1.3 Mass Transfer

There is a diverse suite of techniques that can be applied to change the composition of process streams and/or to separate one process stream, whether gas, liquid, or solid (or a combination of these), into two or more process streams with different compositions. Some of the most common techniques are:

- Separation by phase (gas, polar liquid, nonpolar liquid, solid)
- Separation by differences in the equilibrium concentration of multiple phases (e.g., distillation)
- Separation by chemical solubility differences (absorption of one or more gas components from the gas phase into a liquid, extraction of one or more liquid components from a polar liquid phase into a nonpolar liquid phase or the reverse, leaching of one or more solid components into a liquid or a gas)
- Separation by physical attraction (adsorption of one or more gas or liquid components within or onto a sorbent)

2.1.4 Reactions

Chemical reactions transform the molecular makeup of one or more substances into one or more different substances. Chemical reactions occur when two or more substances are mixed together in the correct ratio and in the presence of the correct combination of temperature, residence time, and pressure. Most commercial chemical reactions are facilitated by the presence of a catalyst.

2.2 Independent and Dependent Process Variables

2.2.1 Control (Independent) Variables

There are only a limited number of ways to control the movement of materials between unit operations and to control the unit operations themselves. For gases and liquids, the flow through a pipe can be manipulated using an adjustable throttle. By changing the size of the opening of a choke point in the pipe, the amount of fluid that can flow through the pipe for a given pressure drop can be changed. Conversely, the pressure required to force a given quantity of fluid through an opening can be adjusted. The device used to accomplish these is known as a throttling control valve or simply a control valve, CV (Figure 2.1a). The control valve is the most common independent variable used in commercial process plants. A physical description of control valves is provided in Section 10.7.

Control System Fundamentals

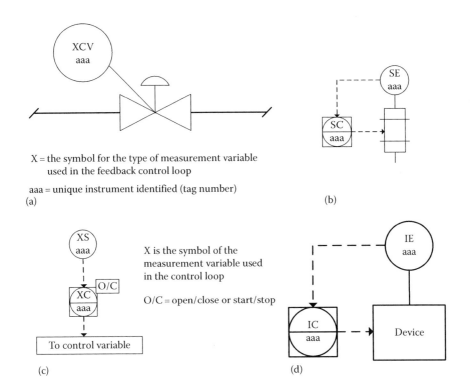

FIGURE 2.1
Control scheme depictions of independent variables used in process control: (a) the throttling control valve; (b) regulating momentum control to a pump, compressor, mixer, or solids handling device with a speed controller (S, speed; note speed is not the true independent variable, which is usually the power applied to the motor or driver); (c) using a switch to regulate the open/close contact time for an electrical circuit or the open/close state of a block valve to allow/prevent fluid flow to/from a unit operation and (d) adjusting the resistance in an electrical circuit for an electric heater (I, current).

Another way to control the flow of fluids is to vary the momentum introduced into the fluid by a pump or compressor. This is usually accomplished by varying the electric current supplied to the motor that drives the device, which, in turn, varies the speed (rotation, piston travel speed, etc.) of the pump or compressor (Figure 2.1b). If the device has a fixed speed motor, a recycle stream can be used (Figure 2.2). The recycle allows the pump/compressor unit to process the same quantity of fluid while also allowing the quantity of fluid that actually travels to the next unit operation to be adjusted by varying the fraction of the total flow that is recycled compared to the amount that flows to the next unit operation. A control scheme for this system is provided in Figure 3.4.

For solids movement, the speed of the conveyor or extruder can be manipulated. This is most commonly accomplished by varying the electric current supplied to the motor that drives the device (Figure 2.1b).

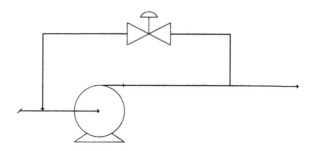

FIGURE 2.2
Using a recycle to control a pump (shown) or compressor (not shown) using a fixed speed motor.

If an electric resistance heater is used, there are two common ways to control the energy provided by the heater. The most common is to open and close a contact on the electrical supply so that the heater receives electric current for only a fraction of the time. By varying the fraction of time the heater is energized, the energy input to the heater can be adjusted, which adjusts the surface temperature of the heating element (Figure 2.1c). The other method uses an adjustable rheostat, a device that varies the resistance in the electrical circuit. By varying the resistance, the quantity of current that flows through the circuit, and thus through the heater, can be adjusted (Figure 2.1d).

In similar fashion to the open/close relay for an electrical circuit, a block valve can be opened or closed on a fluid contained in a pipe. In this case, the quantity of fluid moving from one unit operation to another has only two states: full flow or no flow (Figure 2.1c). This strategy is often employed for batch unit operations.

That's basically it. These represent the vast majority of the independent variables available in the process that can be used for control. To summarize, control valves, electrical supply interruption time, electric current resistance, and block valve open/close status are the independent variables that can be manipulated to control the process. In regulatory control design, we must identify all of the independent variables present for each unit operation. Each of these independent variables must be addressed in our control strategy design to insure that the process will meet its objectives.

2.2.2 Measurement (Dependent) Variables

To know what manipulations to make to the independent variables and whether those manipulations are successful, we need to measure the effect that each independent variable has on the process. We do this by measuring parameters that are affected by changes to the independent variables. These are known as dependent variables. The most common dependent variables are flow rate, pressure, temperature, level, and material properties such as chemical composition. In a well-designed control system, we try to couple each independent variable with a dependent variable that is sensitive to changes in that particular independent variable.

For example, we might couple the flow rate in a pipe to a control valve located on that same pipe. As another example, we might couple the temperature of the process fluid leaving a heat exchanger with a control valve controlling the flow rate of steam entering the utility side of that same heat exchanger. Strategies for pairing up independent and dependent variables for the most common types of unit operations are the principal subjects of Chapters 3 through 6.

2.3 Feedback Control Loop

The fundamental building block for process control is the single input/single output feedback control loop. This is the most common control strategy in the process facility's regulatory control system. In Section 2.2 we introduced the concept of the independent variable/dependent variable pair. Under this concept, we make a manipulation to an independent variable and then test the impact of that adjustment by measuring the dependent variable. The next step is to figure out how to use the dependent variable measurement to determine how much to adjust the independent variable. This is the topic of this section.

Consider the simple case of a pressure vessel embedded in the middle of a process. This unit operation is one of the simplest and most flexible available and is used for a wide variety of purposes. A very simple use is to remove entrained liquid droplets from a gas phase stream. The system is shown in Figure 2.3a. As the gas enters the pressure vessel, the velocity of the stream is reduced by at least one order of magnitude. A tangential inlet nozzle is often used to impart a swirl to the stream (Figure 2.3b) to force the liquid drops to the outer wall of the vessel by centripetal force. A coalescing pad, typically known as a mist eliminator or demister, is placed across the top of the vessel (Figure 2.3a). Any droplets remaining in the gas stream will tend to coalesce on this pad as the gas stream passes through on its way out of the pressure vessel. The liquid drops to the bottom of the vessel and leaves through a bottom opening.

A material balance of the system is:

$$M_I = M_{O1} + M_{O2} + M_A \tag{2.1}$$

where
M_I is the mass of material entering through the feed nozzle
M_{O1} is the mass of material leaving through the top outlet nozzle
M_{O2} is the mass of material leaving through the bottom outlet nozzle
M_A is the accumulation (or reduction) in material in the vessel

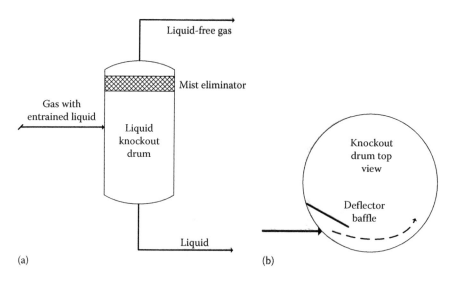

FIGURE 2.3
Pressure vessel used to remove entrained liquid from a gas stream: (a) process depiction used on P&IDs and process flow sheets, (b) top view of the gas entrance: on many knockout drums the entrance is designed to swirl the fluid around the outside of the vessel so that centripetal force can be used to assist in disengaging the droplets out of the gas.

Under normal operation, our goal is steady-state operation, and $M_a = 0$. But in real systems, we will never be entirely at steady state: $M_a \neq 0$. Under this scenario, Equation 2.1 shows that we can choose any three of these parameters as the independent variables of the system. The fourth will then be governed by Equation 2.1. Since our goal is to control the unit operation at near steady-state conditions, it is most common and easiest to use the following as the three independent variables associated with the liquid knockout drum, using the relationship between them (Equation 2.1) to specify the accumulation term:

1. The mass of material entering through the feed nozzle, M_I
2. The mass of material leaving through the top outlet nozzle, M_{O1}
3. The mass of material leaving through the bottom outlet nozzle, M_{O2}

In most cases, the inlet flow rate is specified by an upstream unit operation and thus is not a true independent variable. If we were to use M_I as a control variable for the liquid knockout drum, we will likely interfere with the operation of one or more upstream systems. This leaves the two outlet streams as the only independent variables. The simplest thing we can do is put a throttling control valve on each stream. By changing the open area in the valve, we can change the rate of flow that can pass through the valve for a given pressure driving force. In control terminology, an independent variable is known as a *control variable*.

Our objective is to set the valve opening area, known as the *valve position*, to achieve the process or *operational objectives* of the liquid knockout drum. The first objective is to minimize or eliminate all liquid from the gas stream that leaves the pressure vessel through the top outlet nozzle. At first glance, this might seem to be the only objective for this unit operation. After all, this is the reason we put the knockout drum in the process in the first place. But if we examine the process more carefully, a couple of other objectives are revealed. For example, we would like to minimize or eliminate the amount of gas that leaves the pressure vessel through the bottom outlet nozzle. We also want to avoid allowing the pressure in the vessel to get too high (the vessel might rupture at high pressure). Conversely, we'd like the pressure in the vessel to be high enough to provide sufficient driving force for flow to the next unit operation for both the gas outlet and liquid outlet streams.

Let's deal with the first objective. How do we insure that we minimize or eliminate liquid flow out of the top of the vessel? The knockout drum is designed to inherently meet this objective. The only situations where this would not occur would be: (1) the liquid was allowed to collect in the vessel to a high enough level that liquid was reentrained into the gas stream or (2) the mist eliminator became plugged, reducing or eliminating gas flow out the top of the vessel and forcing gas to flow out the bottom of the vessel. This second scenario is a physical impediment to the operation of the knockout drum. We cannot accommodate it in the control system directly. Later, we will learn how to use parameter measurements to monitor the health and safety of the system. But for now, we will eliminate this situation from our analysis.

Based on the above, as long as the liquid level in the bottom of the vessel is controlled, this objective will be met. Therefore, our control system should include the measurement of the liquid level in the vessel as a dependent variable.

Now let's deal with the second objective. How do we minimize the flow of gas out of the bottom of the vessel? This is straightforward; we insure that there is always at least some liquid in the bottom. As long as there is a sufficient quantity of liquid, no gas can leave the vessel through the bottom nozzle. Therefore, our control system should include the measurement of the liquid level in the vessel as a dependent variable. This is the same conclusion as for the last objective, which means we will be able to meet both of these operational objectives with the same *control objective*.

The third operational objective requires that sufficient pressure be maintained in the vessel to insure that both fluids will flow to their next destination. Since the overhead stream is compressible, this objective requires that we use the measurement of the pressure in the gas outlet stream (or in the vessel above the mist eliminator) as a dependent variable to insure that the gas outlet meets the desired pressure.

For the bottoms stream, the fluid is incompressible. The pressure in the liquid outlet stream will be based on: (1) the pressure exerted by the gas above the liquid in the vessel, (2) the pressure exerted by the height and density of the liquid in the vessel and outlet line, and (3) the friction in the various components of the system that resist fluid flow, including the pressure drop in the control valve. Turning to the first of these, we know from the last paragraph that the gas pressure will be controlled to make sure the outlet gas stream has the correct pressure. This leaves only the second contribution, the pressure exerted by the liquid in the vessel available for controlling the bottoms pressure. Thus, we need to include a liquid level measurement in the control scheme as a dependent variable to achieve this objective. Once again, another operational objective can be met with the same control objective. The third (the friction loss in the bottoms piping) is a physical constraint of the system that we cannot change using the control system.

Note: if the liquid outlet pressure available is insufficient to drive the liquid to the next unit operation, a pump can be installed in the bottoms outlet line. If a pump is installed, then the pressure necessary in the knockout drum must be the higher of the following: (1) the pressure needed to enable the overhead gas to travel to the next unit operation (unless a compressor is installed in the overhead gas line) and (2) the minimum pressure necessary for proper pump operation (e.g., for centrifugal pumps, the minimum net positive suction head pressure must be achieved to avoid cavitation in the pump). When a bottoms pump is used, the control variable is either a control valve (usually installed after the pump) or the speed of the pump motor (if a variable speed pump is used).

To summarize, even though there are three operational objectives, there are only two control objectives and only two measurements are needed: the overhead gas pressure and the vessel liquid level. We now have the proper basis for control. We started with three independent variables:

1. The mass of material entering through the feed nozzle, M_I
2. The mass of material leaving through the top outlet nozzle, M_{O1}
3. The mass of material leaving through the bottom outlet nozzle, M_{O2}

The first will usually be specified by the upstream unit operation. We could then design the installation of two control valves, one on the top outlet line and the other on the bottom outlet line, to serve as the other two independent variables. To adjust the valve position of the overhead CV, we specify the use of the overhead gas pressure (upstream of the overhead CV, but after the mist eliminator). To adjust the valve position of the bottoms CV, we specify the use of the liquid level in the vessel.

The overhead gas pressure and the liquid level are *dependent variables*. We can measure the overhead gas pressure. If the pressure is too high, we can increase the opening in the overhead CV. Then, we can measure the gas pressure again to see if our correction changed the gas pressure to the correct value. Similarly, we can measure the liquid level. If the level is too low, we can decrease the opening in the bottoms CV. In controls terminology, the dependent variables are known as the *measurement variables*.

The sequence just described is known as a *feedback loop*. Measure, adjust, and remeasure to test the effect of the adjustment. If the process were static, we could do exactly what we described in the last paragraph. Take a measurement, make an adjustment, take another measurement, make another adjustment, and so forth, until the two dependent variables were exactly at their desired values for effective operation of the liquid knockout drum. But the process isn't static. During the time that it took for us to measure and adjust, changes to the process can occur, either from changes in upstream or downstream unit operations or even from changes induced by previous adjustments of one or both control valves. To accommodate this feature, real-time processes must use some type of control algorithm to continuously make adjustments to the independent variables, based on the values of the measurement variables. The portion of the control system that performs the control algorithm operations is known as the *controller*.

Here is the basic concept for the controller. For each measurement variable, the human operator (or in more advanced cases, the control system itself) specifies the desired value of the measurement variable. The desired value is the value that allows the unit operation to best meet its process objectives. This desired value is known as the standard or *setpoint*. The controller compares the measured value to the setpoint and calculates the difference between these two values. This difference is known as the *measurement error*, or simply the *error e*. The controller then uses a mathematical algorithm to calculate a change to the control variable based on e. This is known as the *controller output*. We will learn about the most common types of controller algorithms in Chapter 8.

Figure 2.4 depicts our simple regulatory control system design using the symbology we described in Section 1.10. In our analysis above, we paired one measurement variable with one control variable. Now we add a controller to relate the measurement to the control valve position. This gives us the three basic elements of the feedback control loop: the measurement element, the controller, and control element. Further, we specified the number of feedback control loops to match the degrees of freedom for the unit operation. This is an important point. There must be exactly the same number of control loops as there are degrees of freedom in the process. We will learn later how to create additional degrees of freedom when the operational objectives of the unit operation require the use of additional control loops.

The question arises, could we have swapped the way we paired the measurement and control variables? That is, could we have specified the design as shown in Figure 2.5? Theoretically, the answer is yes. But practically, this would decrease the controllability of the system. When possible, we want to use the most direct measurement available to provide feedback for the control variable. In this case, adjusting the overhead valve position directly and immediately changes the pressure in the gas retained by the vessel. If I change

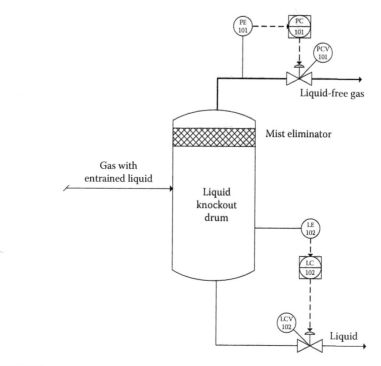

FIGURE 2.4
The basic regulatory control scheme for a pressure vessel used to remove entrained liquid from a gas stream.

the bottom valve position, the level of the liquid will change, which will then change the volume of gas in the vessel. The change in volume will change the gas pressure. This effect is not as direct. It would take much longer for the control loop to reach the desired control variable operating state.

Similarly, adjusting the bottoms valve position directly and immediately changes the rate of flow of liquid out of the vessel which in turn changes the liquid level. If the overhead valve position changes, it changes the pressure of the gas, which changes the pressure exerted on the liquid, which then changes the flow rate out of the bottoms (for a constant valve position). This change in flow then changes the liquid level. Again, the effect is not as direct, so the control will not be as responsive.

A control scheme like that shown in Figure 2.5 is an example of a *coupled* or *cross coupled control system*. In coupled systems, the actions of one independent variable have an immediate impact on the dependent variable of a separate independent variable's control loop. By contrast, Figure 2.4 is an example of a *decoupled control system*. Decoupled systems are easier to tune, are more stable, and are better able to handle disturbances in the process system. Therefore, decoupled control schemes are almost always preferred over coupled control schemes.

Control System Fundamentals

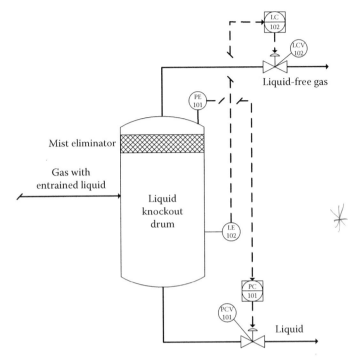

FIGURE 2.5
An alternative regulatory control scheme for a pressure vessel used to remove entrained liquid from a gas stream. This scheme is less stable and decreases the controllability of the process compared to the scheme shown in Figure 2.4.

not as good as a decoupled control system

2.4 Disturbance Variables

In the liquid knockout drum example in Section 2.3, we matched up two measurement variables, pressure and liquid level, to two independent variables. We assumed that the third independent variable would be controlled by the upstream process. Under this scenario, the inlet flow rate is an uncontrollable variable. Any changes, planned or unplanned, to this flow rate will affect or disturb the operation of the liquid knockout drum. This situation is common in tightly integrated processes. These uncontrollable variables are known as *disturbance variables*.

Note that disturbance variables must be independent variables that are not controlled by the unit operation system under consideration; they cannot be dependent variables. For example, our control strategy for the liquid knockout drum did not use the measurements of the flow rates of either of the outlet streams in the two feedback control loops. So, you might think these are disturbance variables. However, these measured parameters are not independent. Any adjustment in the overhead control valve will adjust the overhead flow rate even though a measurement of that flow rate is not used by the controller to make the adjustment. Similarly, any adjustment in the bottoms control valve will affect the flow rate in the bottoms pipe even though a measurement of that flow rate is not used by the controller.

Let's look at an example of a disturbance variable. Another use of a pressure vessel is to blend two liquid streams together. In this case, there are two inlet lines and one outlet line for the vessel (see Figure 2.6). There are three independent variables, which we can specify as:

1. The mass of material entering through the first feed nozzle, M_{I1}
2. The mass of material entering through the second feed nozzle, M_{I2}
3. The mass of material leaving through the outlet nozzle, M_O

If both of the inlet streams are controlled by upstream processes, then there is only one independent variable left for control in this unit operation, M_O. A common control strategy would be to link the measured level in the mixing vessel to the outlet control valve to form the only control loop needed (Figure 2.7). In this case, there are two disturbance variables: M_{I1} and M_{I2}. A change in the amount of either of these two flow rates will impact the operation of the mixing vessel.

Now let's look at a slightly more complicated example. A common application for a mixing vessel is to adjust the pH of an aqueous stream. This might

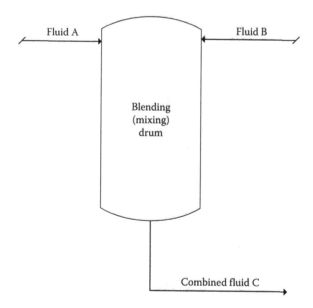

FIGURE 2.6
Pressure vessel used to blend two fluids into a single fluid.

Control System Fundamentals

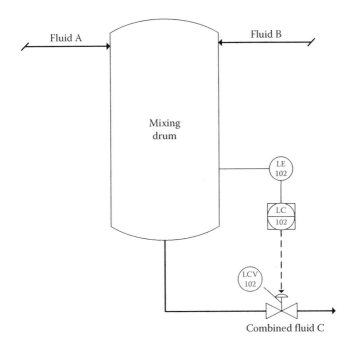

FIGURE 2.7
The control of a blending drum when both inlet streams are controlled by upstream process operations.

be necessary for the process or to allow the stream to be discharged as a wastewater stream. Let's assume that we have a mildly acidic wastewater stream that must be neutralized prior to discharge. So, the first inlet stream would be the process-generated acidic stream and the second inlet stream would be a basic aqueous stream. We still have the same three independent variables:

1. The mass of material entering through the first feed nozzle, M_{I1}
2. The mass of material entering through the second feed nozzle, M_{I2}
3. The mass of material leaving through the outlet nozzle, M_O

However, M_{I2} is no longer a disturbance variable. Instead, it becomes a control variable. We want to control M_{I2} such that the pH of the outlet stream is neutral (has a value of 7). Our control design for this neutralization mixing vessel will have two control loops. One will link the vessel liquid level to the bottoms control valve. The other will link a measurement of the pH in the bottoms liquid to the second feed line control valve (Figure 2.8).

While M_{I2} is no longer a disturbance variable, the function of the mixing vessel is more complicated than the simple blending case described earlier. As a result, there are actually two additional disturbance variables

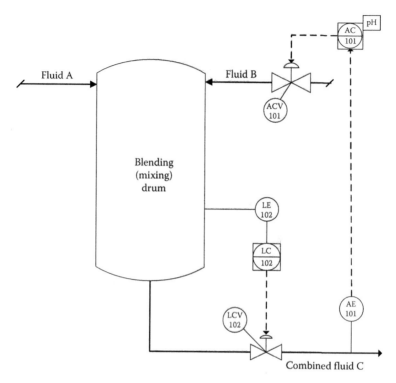

FIGURE 2.8
Pressure vessel used to add a base (fluid B) to fluid A to control its pH.

that should be of concern. The first is the acid concentration of the first inlet stream. If there is more acid in this stream, it will affect the performance of the neutralization mixing vessel. More caustic will be required to meet the desired pH in the outlet stream.

To find the other, we need to know more about the neutralization inlet fluid. The most common base used for this purpose is sodium hydroxide. This chemical is usually supplied as solid pellets that are mixed with water to give a specific base concentration. This aqueous basic fluid is then sent to the neutralization mixing vessel. Thus, the other additional disturbance variable is the concentration of base in the neutralization feed stream. There will be slight variations in the concentration from each batch of base solution. If there is an error in the mixing system, there may be a substantial change in concentration. Slight or substantial, these changes will affect the operation of the neutralization mixing vessel.

In the next section, we will learn the most common technique used to mitigate the effect of disturbance variables on the performance of a unit operation, the feed forward control loop.

2.5 Feed Forward Contributions to Control

Before we address disturbance variables, let's develop the basic control scheme for a simple heat exchanger. Consider a tightly integrated process that includes a heat exchanger as shown in Figure 2.9. Let's assume that the objective of the heat exchanger is to use a constant temperature cooling water stream to reduce the temperature of a process stream. For a shell and tube type heat exchanger, the unit operation can have up to four independent variables:

1. The process flow inlet to the heat exchanger
2. The process flow outlet from the heat exchanger
3. The utility stream inlet to the heat exchanger
4. The utility stream outlet from the heat exchanger

If we assume the process fluid completely fills the heat exchanger and that the fluid is incompressible, then the inlet and outlet process fluid streams are interrelated. Control of one flow rate sets the other flow rate, reducing the number of independent variables to three.

If we assume the process flow is controlled by the upstream unit operation, then we cannot put any controls on the process fluid into or out of the heat exchanger, and the number of independent variables available for control is reduced, again, to two. This leaves the two utility stream flow rates. If we were to completely fill the heat exchanger with cooling water, an incompressible fluid, then there is only one independent variable remaining: the flow rate of cooling water in or out of the exchanger. The most common design is to install the control valve on the cooling water outlet of the heat exchanger so that the valve can assist in keeping the heat exchanger full of cooling water. The objective of the heat exchanger is to cool the process fluid. To do this, we

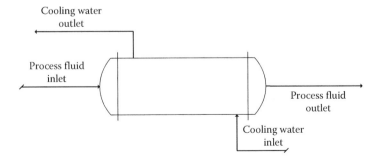

FIGURE 2.9
Process-utility heat exchanger used to cool a process fluid to a specified temperature.

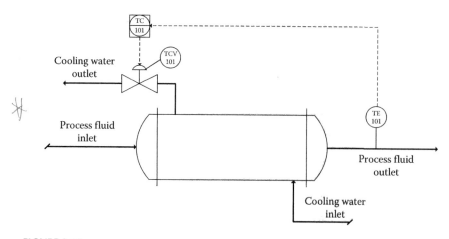

FIGURE 2.10
The control scheme for a process-utility heat exchanger used to cool a process fluid to a specified temperature.

specify a control loop that pairs the temperature measurement of the outlet process fluid to the control valve on the cooling water outlet line. The complete control scheme is shown in Figure 2.10. We'll learn how to control other heat transfer scenarios in Chapter 4.

Now that we have the basic feedback control scheme, let's turn our attention to disturbance variables associated with this unit operation. In this system, if there is a sudden change in the process fluid flow rate, it will impact the performance of the heat exchanger. But if the temperature in the process fluid is important, then this change will result in a process fluid at the wrong temperature. The heat exchanger control system will not sense the temperature change until it measures the temperature of the outlet process stream. By then it may be too late; the process temperature is now incorrect. This sudden change in temperature may then affect a unit operation downstream of the heat exchanger, which could propagate the disturbance and make the entire process harder to control.

We do not need to leave our process blind to changes in disturbance variables. We can use a technique known as feed forward control to mitigate the effect of these changes. Before we see how to apply the technique, let's step back and look at the concept.

In Section 2.2 we learned about the feedback control loop. In feedback control, we measure a dependent variable, compare its value to a standard (the setpoint), use an algorithm to calculate an output value, then use this output value to adjust the independent variable. The future measurements of the dependent variable provide feedback as to how close the measurement is to the standard based on the current state of the control variable (such as the valve position of a control valve). In other words, the control loop has a built-in feedback mechanism.

Control System Fundamentals

In a feed forward control loop, a disturbance variable is measured and compared to a standard (the setpoint). The standard is used in an algorithm to calculate an output value that is then used to adjust the independent variable. While a disturbance variable impacts the performance of the unit operation, the adjustment of the independent variable has no effect on the value of the disturbance variable. The *disturbance variable is independent*. So there is no direct feedback that the adjustment of the control loop's independent variable is properly controlling the unit operation. Another term for this type of control is *open loop control*. To determine if the adjustment made to the controller is useful, the control system must measure a separate parameter, one that is dependent upon the control variable.

So a feed forward control loop can be used to anticipate changes to the unit operation caused by a disturbance variable, but it cannot insure that the independent variable comes to the setting for optimum unit operation performance. Sometimes, but rarely, we do not care if the optimum setting is reached. Consider our liquid knockout drum example. Usually we do not really care what the actual height of the level in the vessel is, as long as there is some liquid in the vessel to insure that none of the process gas leaves through the bottom outlet line. So, as long as the level is within a certain range of levels, operation is correct. In this circumstance, it is possible to use a feed forward control loop pairing the inlet flow rate to the outlet bottoms valve position.

Most of the time, our application will require a fairly narrow range of operation for optimum performance. So usually the feed forward control loop is combined with a feedback control loop. There are two common methods for this:

1. The output from the feed forward controller and the output from the feedback controller are added together and the sum is used to adjust the independent variable.
2. The two controllers are nested in a cascade control scheme.

We will learn about cascade control loops and how to set up a feedback/feed forward cascade loop in Section 2.8. For now, let's focus on the first method.

Consider our heat exchanger example in Figures 2.9 and 2.10. We can measure the inlet process flow rate, compare that rate to a setpoint, and calculate an output (i.e., a control variable adjustment) in the feed forward controller. We can also measure the process outlet temperature, compare that temperature to a setpoint, and calculate an output in the feedback controller. Now we can add the two outputs together using a special regulatory control function known as a calculation (or calc) block (denoted by the label "XC") and send the sum to the control valve located on the cooling water outlet line (Figure 2.11). Note that the CV is tagged to correspond to the feedback dependent variable rather than the feed forward disturbance variable.

Similarly, for the liquid knockout drum, we can measure the inlet process flow rate, compare that rate to a setpoint, and calculate an output in the feed

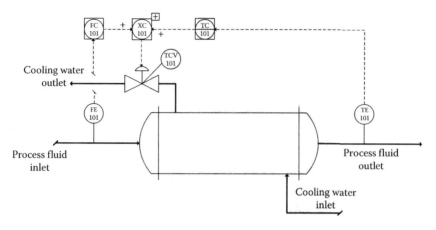

FIGURE 2.11
Process-utility heat exchanger used to cool a process fluid to a specified temperature with a feed forward/feedback control scheme.

forward controller. We can use the output from this feed forward controller for either of the two control valves associated with the pressure vessel: the gas outlet or the liquid outlet. Usually it is best to choose the feedback loop that will be impacted the most by a change in the disturbance variable. In the case described in Section 2.3, that would be the gas stream. So, we can sum the feed forward controller output with the pressure feedback controller output and send the sum to the gas outlet pressure control valve (Figure 2.12).

2.6 Related Variables

In Section 2.5, we saw how we could relate a disturbance variable to a dependent variable and use the error in both measurements to adjust an independent process variable. Those examples used one of a number of disturbance variables for feed forward control.

Just as there may be disturbance variables that are not used in regulatory controls, there are also usually more dependent (measurement) variables that can be evaluated for a given process than are used in the controls. In regulatory controls, we can only use the number of measurement variables that exactly matches the number of control variables in our control scheme. We also saw how some measurement variables were interrelated. For example, we saw how the flow rate in the liquid knockout drum bottoms stream was interrelated to the liquid level in the drum. All else being at steady state, an increase in the bottoms flow rate will result in a decrease in the drum's liquid level.

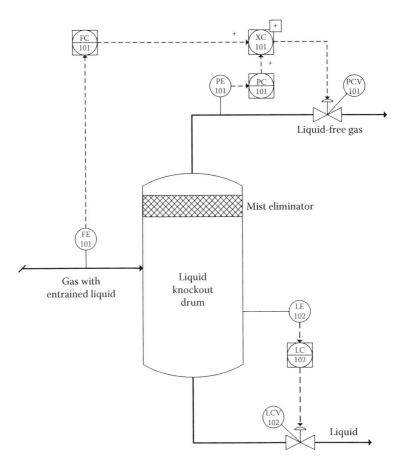

FIGURE 2.12
Liquid knockout drum with a feed forward/feedback control scheme.

A more complex set of interdependencies occurs in distillation columns. Distillation works because the compositions of a vapor and a liquid of heterogeneous mixtures at equilibrium will be different (unless they are at their azeotrope, when they are the same). This relationship is both temperature and pressure dependent. Thus, if we know the temperature and pressure of a vapor-liquid equilibrium mixture, the composition of both phases is fixed. Conversely, if we know the composition of the liquid and the pressure of system, the temperature is fixed. Because temperature and pressure are more responsive parameters than composition (and usually easier and cheaper to measure), it is very common to specify the overhead pressure and the bottoms temperature as the measured parameters in feedback control loops associated with distillation columns in lieu of composition.

A different type of interdependency occurs for the pumping of incompressible fluids. When energy is added to the pump through the motor, this change in fluid momentum can affect the fluid flow rate, fluid pressure, or both. So the increase in a fluid's pressure and an increase in the fluid's flow rate are interrelated. This relationship is typically defined by a function known as the pump curve. We will look at these in more detail in Chapter 3. Regardless of the actual pump curve, it is important to be aware of this interrelationship so that the correct control design can be specified. Since the variables are interrelated, only one can be used in a feedback control loop associated with the operation of the pump.

We can only use a few key dependent variables in the control scheme. However, sometimes we want to use information contained in more than one dependent variable. There are several techniques to do this. The two most common are the ratio controller and the cascade controller. These will be discussed in the next two sections.

2.7 Ratio Control Loops

Consider the simple mixing vessel introduced in Figure 2.6 and described in Section 2.4. For the simple two-stream mix, we constructed a level control loop for the only independent control variable associated with that unit operation (recall that the two inlet flow rates were disturbance variables) as shown in Figure 2.7.

We then made the mixing a little more complicated by assuming that we could control the flow rate of the second inlet stream, M_{I2}, which in that example was controlled by monitoring the pH of the outlet fluid stream, as shown in Figure 2.9. But this control scheme is susceptible to upsets if the composition of either inlet fluid changes frequently or if the flow rate of fluid A changes. This is because the liquid hold-up in the drum makes the control loops nonresponsive. Many times, we can overcome this limitation by using a ratio controller if we know how to relate fluid B to fluid A.

For example, if M_{I1} is 100% component A and M_{I2} is 100% component B, we might want to control the flow rate of M_{I1} to give 50% A and 50% B, or perhaps 90% A and 10% B. A simple way to do this is to pair the flow rate of M_{I2} to a control valve on M_{I1}. The setpoint for the controller would correspond to the correct ratio between the two components. This requires an external evaluation or calculation to provide the proper value of the setpoint.

A more powerful and accurate way to perform this control is to take the ratio of the flow of stream A to stream B and then to compare this ratio to a setpoint. This type of control is known as a *ratio control loop*. In the ratio controller, two separate measurements are input to the controller. The controller then divides the value of input A by the value of input B to yield the *measurement ratio*. This value is then compared to a setpoint, but now the setpoint is the desired value of the ratio rather than the desired value of either absolute measurement.

Control System Fundamentals

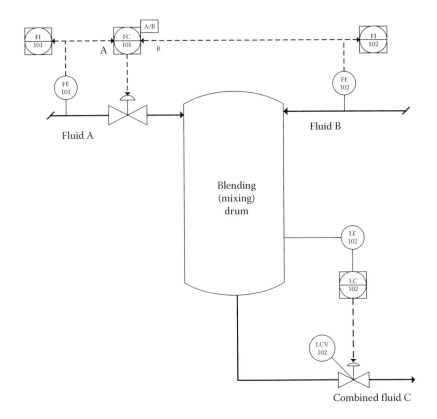

FIGURE 2.13
Mixing drum with a ratio controller for fluid A. *Note*: The inlet flow measurements are routed to indicator blocks so that the values can be independently read by the control operator. Also note the labels on the streams entering the ratio controller to make it clear how the ratio calculation is being performed.

The error between the measurement ratio and the setpoint is then used by the controller to generate the output value that will be sent to adjust the independent variable. Either measurement can be used as the numerator in the ratio calculation. However, if there is a big difference in the two values, the control is usually more stable if the small measurement is used as the numerator. If the values are of the same magnitude and one of the measurements is taken from the stream that has the control valve (if a control valve is used as the independent variable rather than adjusting motor speed), then control is usually more stable if the measurement from the control valve line is the numerator (Figure 2.13).

Returning to our example of an acid neutralization tank (Figure 2.8), the objective is to add a concentrated base solution, M_2, to a dilute acid solution, M_1, to yield a pH of 7. The flow rate of the neutralization stream will most likely be much smaller than the flow rate of the acidic stream, so the

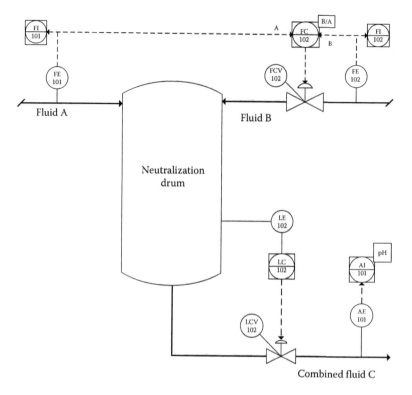

FIGURE 2.14
Neutralization drum with a ratio controller for fluid B. Note that the pH measurement is no longer used in the control scheme but is available to the operator who can adjust the ratio controller setpoint as required. We will see how to use the analyzer reading to do this automatically in Section 2.8.

ratio controller would divide M_2 by M_1 and use this measurement ratio as its measurement value. Figure 2.14 shows how this would be specified for this example.

Probably the most common use of the ratio controller is to insure the correct oxidant-to-fuel ratio for a combustion reaction. Take, for example, a steam boiler that uses natural gas (methane) as its fuel source as shown in Figure 2.15. Natural gas and air are fed to a burner where they are mixed into an explosive mixture and fed into a flame where the irreversible oxidation reaction occurs. The energy generated by this highly exothermic reaction substantially increases the temperature of the resulting reaction products, known as the flue gas. This hot gas then passes over a series of boiler tubes where heat transfer preheats boiler feed water, vaporizes it to steam, and then superheats the steam.

The objective of the steam boiler unit operation is to maximize the generation of steam per unit of fuel consumed. To accomplish this, the ratio of

Control System Fundamentals

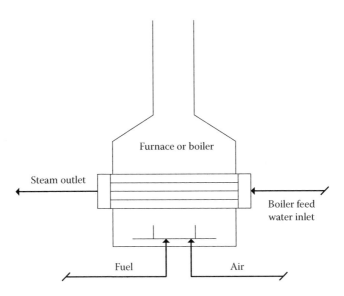

FIGURE 2.15
Boiler or fired heater used for steam generation.

air to fuel must be optimized. A high air-to-fuel ratio will insure maximum consumption of the fuel, but the air includes an inert, nitrogen, which will absorb much of the heat of reaction. This limits the temperature that will be reached by a given quantity of fuel reacted. The higher the flue gas temperature, the higher the heat transfer driving force across the boiler tubes and thus the higher the steam generation.

So there are two balancing parameters to consider in this control: consume as much of the fuel as possible but use as little air as possible. Therefore, the air-to-fuel ratio is a very important control parameter for this type of system. Typically, the fuel flow rate is set by the steam demand and a ratio controller is used to set the air rate at the desired ratio (see Figure 2.16). But how do we know what the best value is for the setpoint of the air-to-fuel ratio? It turns out the control system can do this automatically but we need another tool in our toolkit for this: the cascade control loop.

2.8 Cascade Loops

Recall in Section 1.6 that there must be at least one place in a recycle loop where process dynamics can be dampened to avoid having process changes lead to uncontrollable operation. The most common technique is to insure that there is at least one level control loop within the process recycle loop.

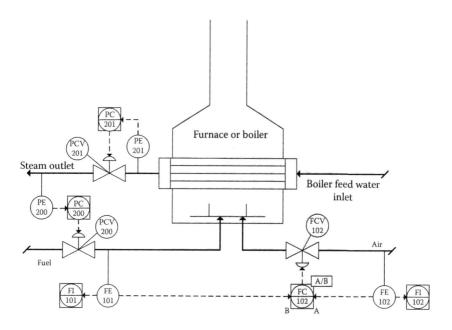

FIGURE 2.16
Controlling the air-to-fuel ratio for a boiler or furnace. In this example, the hot furnace gases vaporize boiler feed water to produce steam at a specified pressure, as controlled by PC-201. The quantity of steam required is determined by monitoring the pressure in the steam header downstream of PC-201 and using this measurement variable (PE-200) in PC-200 to control the quantity of fuel added to the boiler. A ratio controller, FC-102, is used to control the quantity of air added to the boiler in order to maintain optimum combustion of the fuel, as measured by FE-101.

Recall that level is not a very responsive parameter. This is because of the time lag inherent in the process due to the residence time in the vessel. It is this lack of responsiveness that we want to use to dampen the dynamics. However, to still retain good control, the level controller tends to make fairly large output changes. These large changes work against the dampening effect of the level control loop by upsetting the process stream. One way to handle this is to detune the controller. To do this, we set the control algorithm such that the output to the control variable only partially and slowly reacts to errors between the level measurement and the setpoint. In many cases this is adequate, but not in others.

Consider the case of a distillation column where the overhead product has a tight purity specification. Vapor leaves the top of the distillation column, is fully or partially condensed in the overhead condenser (heat exchanger), and fed to the reflux accumulator drum. For a partial condenser, the distillate product is taken off of this drum as a vapor product. For a full condenser, the liquid bottoms from the drum are split, with one portion recycled to the column as liquid reflux and the other portion comprising the distillate product stream.

Control System Fundamentals

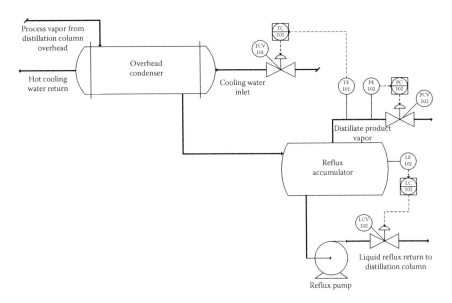

FIGURE 2.17
Typical control scheme for the overhead system for a distillation column employing a partial condenser and a fixed speed reflux pump. *Note*: the temperature is measured after the reflux accumulator rather than in the two phase flow stream out of the overhead condenser to give a more reliable reading.

One typical control scheme for the partial condenser case is shown in Figure 2.17. In this case, the reflux accumulator drum level is paired with a control valve on the reflux return line to the column. One of the purposes of the reflux accumulator drum is to dampen the dynamics in the overhead recycle loop. The drum is designed with sufficient residence time to perform this function. However, if the level controller makes too great a change to the control valve, it can lead to upsets in the column, negating the advantage of having the drum. So, it would be better if the level control loop were very responsive. Unfortunately, this would counteract the dampening function of the drum. So, we need a control loop that would give the smooth operation of a responsive control loop with the dampening effects of an unresponsive control loop. The solution is a cascade loop.

Recall from Section 2.6 that the flow rate out the bottom of a pressure vessel is interrelated to the liquid level in the drum. If everything else is at steady state, a decrease in flow rate will result in an increase in level. Only one of these two parameters can be used in the control loop. Flow control would provide the responsive characteristics we want, but level provides the dampening effects we want.

The ultimate objective of the reflux accumulator is to maintain an adequate liquid level in the drum to dampen oscillations. We can accommodate this by measuring the liquid level and comparing this measurement to a setpoint in a controller. But instead of sending the output to adjust the valve position of the control valve, we can send the output to adjust the setpoint of another controller!

So let's construct a flow control loop that measures the flow rate in the outlet line and uses that flow rate to set the reflux rate. But instead of using a setpoint that is manually specified by an operator in the controller, we will use an external setpoint (ESP). The value of the ESP can be adjusted by the level controller to insure that the liquid level in the accumulator gradually adjusts to the desired level setting. Since we don't need the liquid level value to be at an optimum point, we can make this level controller relatively slow acting. It won't make drastic adjustments to the ESP of the flow controller. Then, since the flow control loop is responsive, it will smoothly move the reflux flow rate from the current value to the desired value. The flow scheme is shown in Figure 2.18.

This type of control loop is known as a *cascade control loop*. To work correctly, the response dynamics of the master loop (in this example the level control loop) must be around an order of magnitude slower than the response dynamics of the slave loop (in this example the flow control loop). The difference in response dynamics can be due to the differences in the responsiveness of the measurement (see Section 1.5), due to a time lag in the process, or due to differences in how the master and slave controllers are tuned (which will be covered in Chapter 8). This difference in

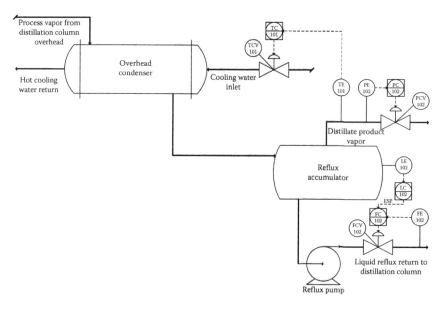

FIGURE 2.18
Cascade level to flow control loop on the overhead system for a distillation column employing a partial condenser and fixed speed reflux pump. *Note*: the signal from the master (level) controller to the slave (flow) controller has the notation ESP to denote that it is providing an external setpoint to the FC.

responsiveness allows the slave controller to nearly completely react to the adjustment to its ESP from the master controller prior to the next controller action by the master controller. This helps improve the stability of the control loop. For the example in Figure 2.18, level measurement is usually at least an order of magnitude slower than flow measurement, so the cascade loop will be stable.

Because of the lack of responsiveness of analyzer and level measurements, cascade control loops are commonly used to improve the controllability of a process that uses these measurements in control loops. For example, let's consider the other type of overhead configuration for a distillation system, one with a full condenser. In this case, all of the fluid entering the reflux accumulator leaves in the bottoms liquid pipe. The pipe then splits, with a fraction recycled to the column and a fraction leaving as the distillate product. The two independent variables may be control valves on each line or may be a control valve on one line and the motor speed of the reflux pump. Let's consider this second case, which is shown in Figure 2.19. Commonly, the important dependent variables are the level in the reflux accumulator drum and the purity of the distillate product.

One of the control loops can be identical to the level-to-flow cascade from Figure 2.18. The other control loop could consist of an online measurement of one component in the distillate product (usually the heavy key) with

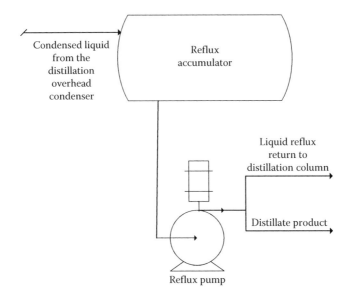

FIGURE 2.19
The reflux accumulator after a total condenser for a distillation column. The accumulator outlet is all liquid, which is divided between a reflux (recycle) stream and the distillate product stream. In the configuration shown here, both outlet streams are routed through a common variable speed pump.

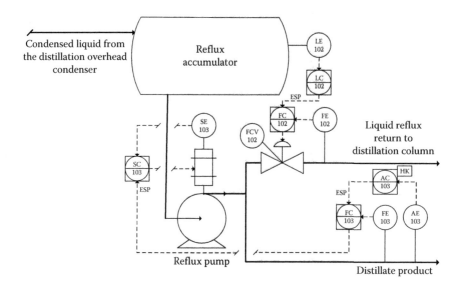

FIGURE 2.20
Controlling the reflux accumulator outlet streams with a cascade level-to-flow control loop and a nested cascade composition-to-flow-to-speed control loop. The distillation column has a total condenser and a variable speed reflux pump. HK denoted the heavy key which is the component selected for the control scheme.

the output from this analyzer control loop in a cascade arrangement with a flow control loop that varies the motor speed of the reflux pump (which then changes the fraction of the total liquid recycled to the column). This scheme is shown in Figure 2.20. Note, this second cascade is known as a nested cascade. The output of the analysis controller adjusted the setpoint of the flow controller. The output of the flow controller adjusts the setpoint of the speed controller on the variable speed pump. By changing the speed of the pump, the momentum imparted on the fluid as it passes through the pump is changed, which changes both the pressure and flow rate of the fluid. This relationship between changes in pressure and flow from a pump or compressor is discussed in more detail in Chapter 3.

We'll discuss other variations to the distillation schemes described here in Chapter 5. For the purposes of this introduction to cascade control, what is important to remember is that cascade control loops must involve *interdependent* variables. In the case just described, a change in the flow rate of the distillate stream will change the reflux-to-distillate ratio, which in turn will change the purity of the overhead separation in the column. Thus, distillate flow is interrelated to distillate composition.

As stated above, the difference in response dynamics can result from a time lag in the process rather than the responsiveness of the measurement variables. We can take advantage of this property to complete the steam boiler burner control system we left at the end of Section 2.7 on ratio controls. As shown in Figure 2.16, in this control scheme the flow rate of the fuel to the burner was controlled by the steam demand, as measured by the pressure in the steam header. The air flow rate was controlled using an air-to-fuel ratio controller. But how can we be assured that the air-to-fuel ratio setpoint is the optimum? One way is to measure one of the components in the flue gas right before discharge to the atmosphere. For example, if we measured the methane content, we could then use this measurement to adjust the air so that a very low level of methane remains in the flue gas. Or we could measure the oxygen content to meet a certain excess oxygen level that is known from experience to yield efficient combustion. A third option is to measure the level of a reaction product such as carbon monoxide. If combustion is efficient, there will be a low, but finite, level of CO present in the flue gas.

Once we have a measurement of a key component in the flue gas, we can compare this in an analyzer control loop to a desired setpoint. The controller output would then adjust the ESP of the ratio controller. This would have the result of adjusting the air-to-fuel mixture to reach optimum combustion conditions. The scheme, shown in Figure 2.21, is known as *ratio with feedback trim control*. Notice that in this case the feedback control loop is the analyzer or master loop. This loop provides an adjustment or *trim* to the ratio control loop operating on the air flow rate.

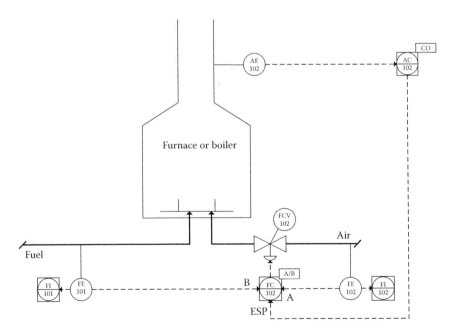

FIGURE 2.21
Controlling the air-to-fuel ratio for a boiler or furnace with a feedback trim cascade control loop. CO stands for carbon monoxide.

2.9 Feed Forward/Feedback Cascade Control Loops

In Section 2.5, we learned how to add the error from a feed forward controller associated with a disturbance variable and the error from a feedback controller associated with a dependent variable to improve the control when adverse effects from the disturbance variable are anticipated. Additive feed forward/feedback controllers work well when the responsiveness of both control loops is similar. However, when one of the control loops is much more responsive than the other, additive controllers become unstable unless the responsive loop is dampened to make it less responsive.

If the dependent variable loop is the less responsive loop, a cascade control scheme can be used. The dependent variable loop is the master and the disturbance variable loop is the slave loop. For example, consider the case where we want to partially condense a multicomponent stream in a water-cooled heat exchanger such that the gas phase concentration of a given component meets a specified value. After the heat exchanger, the vapor and liquid are separated in a pressure vessel. An online analyzer is used to measure the concentration of the key component in the gas stream leaving the pressure vessel. This concentration measurement is used as the dependent variable to control the cooling water outlet flow rate in the heat exchanger. If the inlet process flow rate changes frequently or by large amounts, significant disturbance will occur in the process since the control loop is not responsive. Since flow rate is much more responsive than analysis, a feed forward control loop using the disturbance variable (the inlet flow rate) is configured as the slave loop in a cascade with the analyzer control loop as the master. Changes in the inlet flow rate will cause rapid adjustment to the cooling water rate. How much this change in flow rate will affect the cooling water rate will, in the long term, be set by the gas phase analysis. This scheme is shown in Figure 2.22.

When the less-responsive loop is the disturbance variable loop, a direct cascade control loop cannot be employed because the disturbance variable does not provide feedback to insure that the control variable gets to the optimum setting. If this loop is the master loop, it would result in open loop control and the dependent variable would eventually drift away from the optimum unless the operator manually intervened in the controls. There are two options for this case. The most straightforward is when an indirect but responsive measure can be used to replace the less responsive disturbance variable. Consider the acid waste neutralization example described in Section 2.7. If the disturbance variable is the composition of the inlet waste stream, it may be possible to measure the pH of the inlet stream. A pH analyzer has sufficient responsiveness to be used as a disturbance variable in a feed forward/feedback control scheme.

The second way to handle this situation is to use an additive feed forward/feedback controller as the master and a responsive dependent variable as

Control System Fundamentals

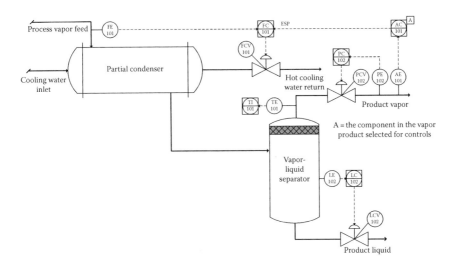

FIGURE 2.22
Feed forward/feedback cascade control scheme to control gas phase composition by manipulating the cooling water flow rate, which in turn changes the fraction of the inlet vapor that condenses and thus will change the product composition, in a partial condenser for a system where process flow rate changes are significant.

the slave. The dependent variable used in the master loop must be interrelated to the dependent variable used as the slave. Returning to the last example shown in Figure 2.22, let's assume that the inlet concentration of compound B varies considerably and affects the vapor-liquid equilibrium of the system (hence changes the target temperature at which a specific concentration of A in the outlet stream will be reached) but the flow rate is relatively stable. In this case, both the dependent variable, the concentration of A in the outlet stream, and the disturbance variable, the concentration of B in the feed stream, are nonresponsive variables. Therefore, it would be good to use a more responsive dependent variable as the slave feedback control loop. A good choice would be the temperature of the outlet stream. The control scheme for this configuration is shown in Figure 2.23.

2.10 Process and Safety Systems

The process control system's primary objective is to efficiently control the process. When it is correctly doing its job, the process will also be operating in a safe condition. Although rare, there are times when the process control system is not operating correctly. This could be due to an instrument failure or a data error. There also may be an occasional incident where the process is highly upset. This can be due to an equipment failure or an operator error.

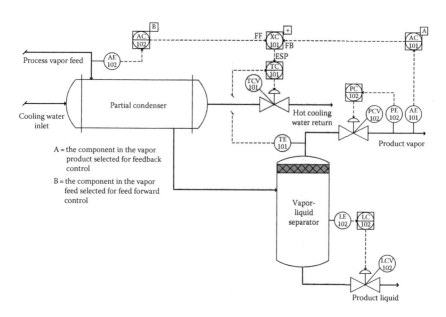

FIGURE 2.23
Feed forward/feedback cascade control scheme when the disturbance variable is nonresponsive.

The consequences when these types of incidences occur may be quite severe. These include injury to personnel; the release of toxic, hazardous, or flammable materials; and/or damage to expensive equipment. When these more severe upsets occur, the control operator may not have time to act to avoid having the process get out of control.

The process control system software includes the capability to provide alarms for every indicator and controller configured into the system. Usually these alarms give a visual and audible signal when a low or high reading is received from the measurement. While used infrequently, most process control system software versions also include the capability to use a *rate of deviation* alarm. Instead of being triggered by the absolute value of the measurement, the alarm is triggered by the first derivative of the measurement versus time curve (Figure 2.24). If the rate of change exceeds a certain value, the alarm is activated.

In all cases, the process control system alarms notify the control operator (often known as the board operator, a legacy from the days when control was accomplished using individual electronic control devices mounted on boards). The operator investigates the cause of the alarm and may take corrective action if required to insure that the process is operating correctly.

It is not uncommon for a process upset to cause multiple measurement indicators and controllers to alarm. Process control system software usually includes a feature that allows the operator to quickly (with the push of a single function tab) locate the first alarm that sounded so that the root cause of the upset can be identified.

Control System Fundamentals

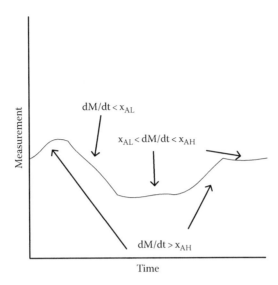

FIGURE 2.24
Instead of using the absolute value of the measurement, a Rate of Deviation alarm uses the Rate of Change of the Deviation, the first derivative of the measurement value. The low deviation alarm would be activated if $dM/dt < x_{AL}$ and the high deviation alarm would be activated if $dM/dt > x_{AH}$.

One of the weaknesses of this approach is the reliance on a person to correctly identify the cause of the alarm or alarms and to take the correct action. If the control operator makes an error, the situation may be made worse, leading to a more significant incident. Thus, as much as possible, we would like to have the plant automation system respond and react to upsets automatically. To mitigate the effect of control operator mistakes and to respond to severe upsets in a timely fashion, a well-designed plant automation system includes a safety automation system that operates independently from the process control system. In the safety automation system, separate measurement devices are connected to a separate control computer where the status of the measurement is evaluated. When a measurement is so far out of normal that it will cause a severe upset, the safety controller will initiate an output to cause an action that is aimed at preventing such an upset. The safety automation system represents a last resort when the process control system cannot bring the process into control and the operator has not taken action (usually because this operator cannot react quickly enough or has too many simultaneous alarms to deal with) or has taken the wrong action.

To see how this works, let's return to our liquid knockout drum example. As shown in Figure 2.4, the drum includes a pressure control loop on the overhead outlet stream and a level control loop on the bottoms outlet stream. In a well-designed process control system, we can assume that both controllers are configured with high and low measurement alarms. Let's consider the level control loop first. If the level of the liquid gets too high, it could get reentrained into the gas stream, which could damage downstream equipment, such as a gas compressor. When the level gets to a specific height, the high level alarm on the level control loop will be activated. At this point, the control operator may decide to override the controller output and partially or totally open the level control valve to lower the level. This is known as taking manual control of the loop or putting the loop into *manual mode*. Once the level returns to a normal level, the operator can return the loop to automatic operation.

But what happens if the operator does not respond to the high-level alarm or if there is an error in the level measurement that prevents the correct level from being determined? If the liquid continues to rise, eventually it will start leaving with the gas, which could cause severe problems in downstream equipment and systems. To cover this possibility, a secondary level measurement device should be installed with an alarm point set above the high-level alarm of the process level device. This alarm is known as a high-high alarm, level or *level alarm, high-high* (LAHH). If the high-high level alarm is activated, the safety level controller could initiate an output that would close off a block valve on the pneumatic line of the control valve. This remotely operated block valve is most commonly known as a solenoid valve and is designated as a "ZV" on the control scheme diagram.

By blocking off the air to the control valve, the pressure on the membrane will be removed, and the spring pushing on the other side of the membrane will move to the maximum position. In a forward acting valve, this will result in the valve opening to its maximum. This is known as an *air-fail-open* (AFO) valve. For a reverse acting valve, loss of air pressure will result in the valve closing completely. This is known as an *air-fail-close* (AFC) valve (see Chapter 10 for the details of how pneumatic control valves work).

For our high-high liquid level, we want the bottoms valve to open fully to drain as much liquid out of the drum as quickly as possible. So in this case, we specify that the level control valve be an AFO valve. If the control valve gets stuck, this action will often allow it to reset and operate correctly.

The safety high-high level controller will maintain a "close" signal to the solenoid valve until the level in the knockout drum drops below a specified point. Sometimes this is specified to be the high-high level measurement value. In other cases, the specified reset value is the high-level alarm value of the process liquid level control loop.

A separate level measurement device is needed because a malfunction or failure of the process level measurement device may occur. If this happens, the level control loop may be receiving a normal liquid level reading even though the liquid level has risen.

The high-high level measurement device can either be a full indicator-type measurement or a switch-type indicator. The former provides more information since this same level measurement can be used to verify the measurement of the process level measurement. However, a level switch is often

Control System Fundamentals

used because it is usually much less expensive than an indicator. When a switch-type measurement device is used, the LE (level element) symbol on the control scheme diagram is changed to an LS (level switch) symbol. The level switch only has two states: 0 (normal) and 1 (alarm).

In addition to concerns about too much liquid in the knockout drum, we are also concerned about having too low of a liquid level. If the liquid completely leaves the vessel, some of the process gas will be sent down the liquid outlet line. This will contaminate this stream, result in loss of the primary process product (the gas stream), and may result in the release of the gas into the environment (e.g., if the liquid were drained to a sewer). To prevent the unintentional release of gas through the bottoms outlet line, another separate liquid level measurement device can be installed in the liquid knockout drum. The alarm point for this secondary level device would be set below the low alarm point for the primary process level measurement device. This is known as a low-low alarm, level or *level alarm, low-low* (*LALL*).

Now, if we hadn't already specified an LAHH safety system, we could use the low-low level alarm to activate the solenoid valve on the process level control valve. In this case, the valve would need to be an AFC valve, so that the outlet is closed upon alarm. But we can't do that in this case because we are using the level control valve failure action to handle a high-high level alarm. Instead, we need to install a separate, remotely operated block valve on the outlet line. This block valve can be located upstream or downstream of the level control valve. The remote operator is usually an electric motor operated valve (MV) but may be an air operated valve (AV). When the low-low alarm is activated, the safety controller will send an output to the MV or AV to close. When the reset point is reached, the controller will send an output to the MV or AV to open.

For the overhead outlet line, a different safety system is used. Safety codes require that all pressure vessels be protected from overpressure failure through the use of a dedicated *relief valve*. See Chapter 10 for a description of how these valves work. If the pressure in the knockout drum gets too high, the pressure control loop high alarm will activate. If the operator does not or cannot correct this situation and the pressure continues to rise, the relief valve will automatically open, sending gas through a designated alternative routing, relieving pressure in the vessel. An installed backup relief valve is provided so that the operating valve can be removed and tested periodically without shutdown of the process.

There is no safety risk if the pressure in the vessel drops below the low-pressure alarm point (unless the vessel is operating at a vacuum, in which case vacuum breakers—reverse relief valves—are installed); therefore, no safety automation control is provided for this scenario. The complete control system for the liquid knockout drum, including the safety control loops, is shown in Figure 2.25.

We will discuss typical safety automation control loops for the most common unit operations in Chapters 3 through 6.

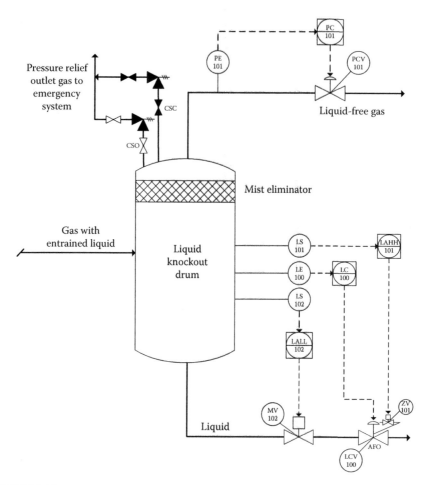

FIGURE 2.25
The independent safety system instrumentation and controls to protect a liquid knockout drum from potentially hazardous or damaging scenarios. The labels CSO and CSC at the relief valves mean that the valves are *Car Sealed Open* or *Car Sealed Closed* (There is a breakable band in the handle of the block valve that isolates the relief valve; to change the valve position the band must be broken. This insures that the state of the valve is always known). The AFO label below LCV-100 indicates that when ZV-101 is energized and blocks off the supply of air to the valve, the valve will fail to the fully open position. No label is needed on MV-102 because the unshaded valve symbol indicates that the valve is normally open, which means the valve will close when LALL-102 is activated.

2.11 Sequential Logic Control

All of the feedback control loop and other control techniques that we have introduced so far in this chapter have been applicable to steady-state continuous operation. However, as we saw in Section 1.2, some processes are operated

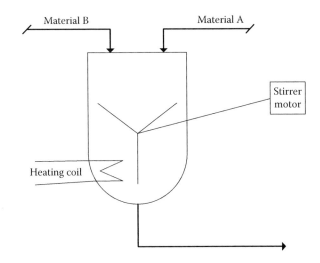

FIGURE 2.26
A simple batch process to produce component C by adding a series of small increments of component B to a reactor containing component A.

in batch mode. More commonly, almost all continuous processes include semi-batch unit operations (Section 1.3). In these systems, operation occurs at one set of conditions until some termination criterion is reached. Then, the operation changes to a second set of conditions, and so on. The process progresses through a sequence of different conditions until the batch process is completed.

An entirely different control paradigm is needed for sequential processes. This paradigm is known as sequential logic control. For each step in a batch sequence, there is a *lineup* or piping configuration. The routing of fluids through the configuration is accomplished using remotely operated valves (MVs or AVs; for simplicity, we'll only refer to MVs since they are more commonly employed). Some of the sequences may require that motors are started or stopped for devices used to move materials (pumps, compressors, conveyors, etc.) or for mixers to blend materials together.

A sequential logic controller is software or firmware that can be configured to change multiple valve positions (open or closed) and/or motor conditions (on or off) simultaneously to change from one step in a process to another based on one or more termination criteria.

Let's consider the case of the simple batch process (Figure 2.26). The goal of this process is to add small increments of component B to a drum or tank containing a stoichiometric excess of component A, then heat and mix the two components together until all of component B has been consumed in the reaction of A + B ==> C. Once all of B has been consumed, additional component B is added to the reactor and the process is repeated until all of component A has been consumed. Now that the drum/tank contains virtually all component C, the fluid is drained out of the drum/tank.

To develop the control scheme, we can walk through the process step by step and add the dependent and independent variables for each step in the sequence. Let's start with a completely empty mixing vessel. Because this process is sequential in nature, we need to install remotely operated block valves (MVs) on all the inlet and outlet lines and label them. The first step in the process is to fill the vessel with a known quantity of component A. To do this, we open the MV on the component A inlet line while keeping the MVs on the component B inlet and the outlet lines closed. To measure the quantity of component A, we can install a flowmeter on the component A line and then calculate the total mass passing through the flowmeter over time (this is known as totalizing the flow). However, there is an easier way to monitor the amount of component A that enters the mixing vessel: we measure the liquid level in the vessel. When the liquid level reaches a given height in the vessel, we know that the correct amount of A is present, so we then shut the MV on the component A inlet line. To do this, we need to add a level measuring element. Note, if the component A fluid is very viscous or tends to foam, then the totalizing flow method is preferred.

The next step is to turn on the mixer motor and open the MV on the component B line. So we need to add a remote start/stop function to the mixer motor. This will insure that as B enters the vessel it is well mixed with A. Because we are mixing the solution, level measurement is no longer accurate, so we need to use the totalizing flow method to insure that the correct quantity of B enters the mixing vessel. To do this, we add a flow-measuring element to the component B inlet line. An alternative way to do this would be to add B before turning on the mixer motor using the incremental level increase to determine the amount of B added and then turning on the mixer to stir up the solution. Once the correct quantity of B is added to the reactor, the MV on the component B inlet line is closed.

Now that the reactor contains a mixture of A and B, the heater can be energized. So we need to add a remote energize/deenergize function to the heater. This is usually accomplished using a relay. When the relay is closed, electricity enters the heating coil and heats the fluid. When the relay is open, no electricity enters the heating coil and no additional heating occurs. As a result, it is more common to depict this function as *close/open* rather than *energize/deenergize* and we will follow that convention in this book.

Once the solution reaches the desired reaction temperature, a feedback temperature control loop should be included to maintain the temperature at its setpoint. So we need to add a temperature measurement element and a temperature controller. The temperature controller will send a close/open signal to the heater to control the temperature in the fluid. We only want the TC output signal to go to the heater when we are at the proper step in the process. To accomplish this, we add a *decision block* in the control system. This decision block, denoted as an XC (miscellaneous control) block, receives a signal from the sequential logic controller (SLC) as well as the output signal from the TC. In the decision block, the two signals are multiplied together.

Control System Fundamentals

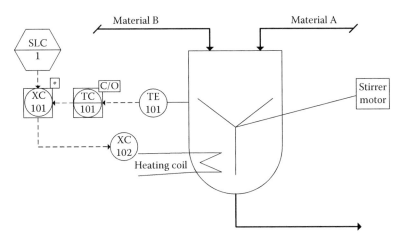

FIGURE 2.27
Detail of the control scheme used to insure that the heater operation is only regulated by the continuous temperature control loop during the correct step in the batch process. The sequential logic controller, SLC-1, sends a "1" to XC-101 during the heating step and a "0" during all other steps in the batch process. The * in the information box of XC-101 indicates that the signal from the continuous temperature controller, TC-101, is multiplied by the signal from SLC-1 with the product sent to XC-102, which is an open/close type relay on the electrical supply to the heating coil.

When the process is at the correct process step for heating, the SLC sends a signal of "1," and when the process is not at the correct process step, the SLC sends a signal of "0." Thus, the heater only receives the TC command at the proper time. This is shown in Figure 2.27.

We maintain the process in this condition until the entire available component B reacts with A to form C. To determine when this is accomplished, we add an online analyzer to measure the concentration of B in the reactor. When the concentration of B falls below a certain threshold (you'll rarely get 100% conversion), this step is completed.

We have now reached a decision point. For the first sequence, and several sequences afterward, we wish to repeat the process of adding B to the reactor and then having the heating/reaction proceed. But at some point, most or all of component A will be consumed. At this point, we do not want to repeat the sequence. We determine this point by adding an online analyzer to measure the concentration of A in the reactor. This may be the same physical analyzer that is measuring B (i.e., a multicomponent analyzer) or a separate analyzer. In either case, at this point we just need to specify that there are two functional measurements needed: the concentration of A and the concentration of B.

If additional A remains, then more B is added to the mixing vessel and the reaction is allowed to proceed. Once all of A is consumed, we stop the stirrer motor and then open the MV on the mixing vessel outlet line so that the contents, essentially all C, can be removed from the vessel. This material will either be routed to other unit operations for further processing, to tankage for storage, or to a packaging area for final preparation prior to sales.

At this point, we have specified all the instrumentation and controls necessary for this process. The complete system is shown in Figure 1.12. The logic flow diagram associated with the process is shown in Figure 1.13.

Sometimes multiple termination criteria must be met before the step is finished. Sequential logic software can easily handle this situation. Another variation is to have a decision point as a termination criterion. For example, the termination criterion might be: "If x is reached before y, go to step 3, otherwise go to step 4." This type of description is very similar to many statements in computer languages such as FORTRAN, Basic, or C. That is not a coincidence. Computer programs are inherently sequential, so the same types of logic used to develop computer programs can be used to design sequential logic for batch and semi-batch control strategies.

Note that within a given step in the batch sequence, continuous control loops are usually required for proper operation. However, a valve that is used for throttling should never be relied upon to block off flow completely. The leakage rate of a throttling control valve is too high and may lead to contamination or upsets if used for block valve service.

A clear, compact method to depict the logic of a batch/semi-batch process on PFDs and P&IDs is the sequential events table. Examples of these tables for the Figure 1.12 system are shown in Tables 1.1 and 1.2 in Chapter 1. A sequential events table is often placed on the bottom of the PFD and P&ID sheet(s) depicting the batch unit operation as shown in Figures 1.14 and 1.15. These tables were described in Section 1.9.

Problems

2.1 Define the following terms: measurement variable, control variable, dependent variable, valve position, operational objective, control objective, controller, setpoint, measurement error, controller output, manual mode, high-high level alarm, low-low level alarm, air-fail-open, and air-fail-closed.

2.2 Describe each of the types of independent variables that are typically available for use in process control.

2.3 Describe a coupled control system and why such systems should be avoided whenever possible.

Control System Fundamentals

2.4 For the following unit operations, (a) identify the governing mass balance equations, (b) determine the number of independent variables, (c) determine the number of control variables, and (d) design simple feedback control loops for each control variable:

    2.4.1 A single effect evaporator

    2.4.2 A crystallizer

    2.4.3 A polymer extruder

    2.4.4 A gas phase endothermic reactor with a steam heating jacket

2.5 For the following unit operations, (a) identify the governing mass balance equations, (b) determine the number of independent variables, (c) determine the number of control variables, and (d) design at least one combined feed forward/feedback control loop for each unit operation.

    2.5.1 Three liquid streams are combined and fed to a pump that includes a low flow recycle line. The flow of the largest of these streams varies greatly and results in pump cavitation if the low flow recycle system does not react in time.

    2.5.2 A three-effect evaporator with countercurrent flow is used to concentrate an aqueous liquid stream.

    2.5.3 A two-stage crystallizer system is used to remove a contaminant to a low level via precipitation from an aqueous liquid stream.

    2.5.4 A three-step distillation system to separate a mixed feed stream into propane, butane, pentane, and hexanes.

2.6 For the following unit operations, (a) identify the governing mass balance equations, (b) determine the number of independent variables, (c) determine the number of control variables, and (d) design at least one ratio control loop for each unit operation.

    2.6.1 An endothermic tubular reactor, with multiple reaction tubes, that uses a fired heater section to provide energy at a high temperature in the vessel space surrounding the tubes.

    2.6.2 A neutralization tank that adds acid to a high pH waste stream.

    2.6.3 A process stream and a nutrient stream are added to a photo-bioreactor. In the photoreactor, light banks are cycled on and off to add the correct amount of energy (which can be correlated to the temperature of the outlet stream) to a water-based bioreactor.

2.7 For the following unit operations, (a) identify the governing mass balance equations, (b) determine the number of independent variables, (c) determine the number of control variables, and (d) design at least one cascade control loop for each unit operation.

2.7.1 A neutralization tank that adds acid to a high pH waste stream.

2.7.2 Three process streams of varying flow rates are mixed together in a pressure vessel that is used to smooth out the combined flow rate into the next unit operation.

2.7.3 A filtration system that adjusts the recycle rate of a waste stream back through the filter, along with new waste liquid based on the particle concentration in the outlet waste stream.

2.8 Another example of the use of a feedback trim cascade control loop is in the operation of a Claus sulfur process. In this process, a feed gas containing H_2S is partially combusted with air to yield a mixture of H_2S, SO_2, and S via the reactions:

$$2H_2S + O_2 \rightarrow 2H_2O + 2S \qquad (2.2)$$

$$2H_2S + 3O_2 \rightarrow 2SO_2 + 2H_2O \qquad (2.3)$$

The flue gas is cooled to condense out the sulfur. It is then sent through a series of reheaters, catalytic reactors, and condensers to produce additional sulfur via the reaction:

$$2H_2S + SO_2 \rightarrow 2H_2O + 3S \qquad (2.4)$$

The key to maximizing sulfur recovery is to have the correct H_2S-to-oxygen ratio in the initial thermal oxidizer. Theoretically, this is a simple 2:1 ratio. However, due to variations in the concentration and flow rate of H_2S in the process gas stream plus variations in the reaction efficiency of the thermal oxidizer and the three catalyst beds, the true optimum can only be found experimentally. As a result, the composition of either H_2S or SO_2 is measured in the final stream and this measurement is used in a feedback trim cascade controller to the inlet air ratio controller.

Duplicate the drawing shown in Figure 2.28 and add a ratio with feedback trim control scheme to meet the operational objectives described above.

2.9 Consider the control scheme described in Figure 2.18 and the accompanying text. A more stable and responsive scheme uses the inlet cooling water flow and temperature of the outlet gas in a feed forward/feedback control as the slave in a cascade control scheme with an analyzer on the outlet stream as the master control loop. Modify Figure 2.18 to show this improved control scheme.

2.10 Consider the system shown in Figure 2.21. Many local regulatory agencies require that the opacity of the flue gas exiting the furnace or boiler stack be measured. Opacity is a measure of the quality of the combustion. If combustion is inefficient, then the flue gas will include soot

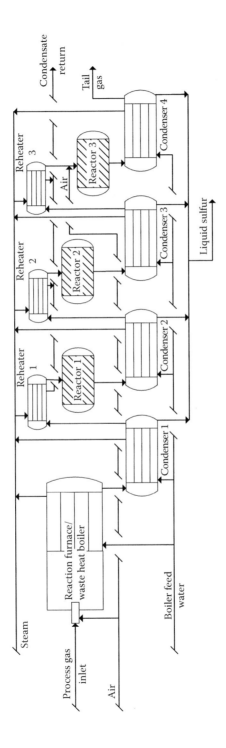

FIGURE 2.28
Simplified flow sheet for a Claus sulfur plant.

and/or unburned fuel, which will decrease the opacity of the existing gas. If the opacity exceeds a certain point, action must be taken to increase the air-to-fuel ratio in order to insure that the discharge is within acceptable limits. Duplicate the scheme shown in Figure 2.21 and add a safety automation system control scheme that will take "last resort" action to add air into the burner if the opacity meter reaches its low-low alarm, analysis point.

2.11 Consider the neutralization system shown in Figure 2.14. This scheme can be improved by adding a cascade control loop that uses the outlet pH reading as the master dependent variable. Duplicate the scheme shown in Figure 2.14 and show how this improved scheme would be depicted.

3

Motive Force Unit Operations Control

The goal of this chapter is to describe typical control designs for the most common types of equipment used to move material through processes. In a real-time process, material moves through the various unit operations where it is transformed from its original state into usable products, by-products, and wastes.

For fluids—gases and liquids—movement through the process occurs due to a pressure difference between the upstream and downstream unit operations. Fluids will always flow from an area of high pressure to an area of low pressure. The pressure of the fluid is increased using pumps (liquids), compressors (gas, high pressure), blowers (gas, moderate pressures), and fans (gas, low pressure) at strategic points in the process to facilitate this movement.

Solids are more challenging. Usually they must be physically moved, such as by using a conveyor or an extruder. Sometimes gravity is used to move a solid, while at other times the solid is entrained in a gas (fluidization) or liquid (slurry) so that it can be moved using the entraining fluid.

3.1 Incompressible Fluids: Pumps

A pump is used to increase the energy of an incompressible fluid. According to the Bernoulli equation, the energy contained in a fluid can be described by three types of "energies": flow, pressure, and potential energy. For a given level of energy, these are related by the following ideal equation:

$$\frac{P}{\rho} + \frac{v^2}{2} + gh = \text{Constant} \tag{3.1}$$

where
  P is the fluid pressure
  ρ is the fluid density
  v is the fluid velocity
  g is gravity
  h is the fluid's vertical position relative to a reference point

The change in potential energy is related to the piping configuration around the pump, but in general, the change in fluid position due to vertical changes in piping is very small compared to the pressure and flow rate of the fluid with certain important exceptions such as when a reflux pump at ground level must supply liquid to the top of a 30 m high distillation column. So usually we can ignore the potential energy term and consider variations in flow and pressure changes to be interrelated variables for a pump.

The increase in pressure that occurs due to the actions of the pump is known as the *pump head*. For each pump, there is a specific empirically determined function that defines the relationship between the flow rate through the pump (usually given in terms of actual volumetric flow units) and the pump head generated by the pump. This function is known as the pump curve (see Figure 3.1 for an example of a pump curve for a fixed speed pump). When a variable speed motor is used to drive the pump, there are pump curves associated with each speed, creating a control surface (Figure 3.2). Since the fluid is incompressible, any increase in flow rate caused by the pump will also increase the flow rate of the fluid before the pump.

Fortunately, we do not need to reference the pump curve to efficiently control the operation of a pump. With only one degree of freedom, we only specify one independent (control) variable to control pump operation. We can use any of the following:

1. A throttling control valve on the outlet piping after the pump
2. A throttling control valve on the inlet piping before the pump
3. The fraction of time the pump motor is operating
4. The resistance to current in the motor speed control circuit (motor/pump speed)

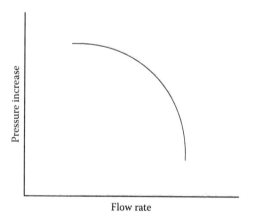

FIGURE 3.1
A typical pump curve for a fixed speed pump.

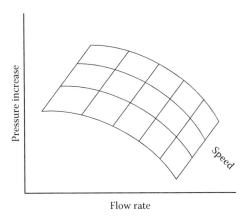

FIGURE 3.2
A more advanced pump curve for a variable speed pump.

The choice is based on circumstances, some of which we will now briefly explore.

For a fixed speed pump, the most common choice is the first, a control valve installed on the outlet piping of the pump. The reason the outlet is normally used instead of the inlet has to do with the nature of throttling control valves. These valves work by changing the fraction of the total cross-sectional area that is open. When the fraction is decreased, the fluid is choked through a smaller opening that increases the resistance to flow. This resistance will result in decreases in both the flow rate and pressure of the fluid, which for incompressible fluids are interrelated. Smoother and finer control can be accomplished in the outlet piping because there is more pressure available after the pump. In addition, for many types of pumps, such as the most common type, centrifugal, a minimum pressure must be maintained in the inlet line of the pump for proper operation. Either flow or pressure can be used as the dependent variable, as shown in Figure 3.3. The choice is usually determined by the primary purpose of the pump: to increase fluid flow or increase fluid pressure.

There is a lower limit to the amount of fluid that must flow through the pump to insure proper operation. To extend the range of a fixed speed pump below its lower operating level, a minimum flow recycle line containing a control valve can be installed. For the most efficient operation, this control valve should be 100% closed if the flow meets or exceeds the lower flow limit of the pump (Figure 3.4). If the flow drops below this point, the control valve is opened to insure that the total flow through the pump matches the lower limit, regardless of the actual flow of fluid required by the process. There are several of other ways to implement this control. The most common is to use an override controller. We will look at how these controllers work and how to apply them to this specific application in Chapter 7.

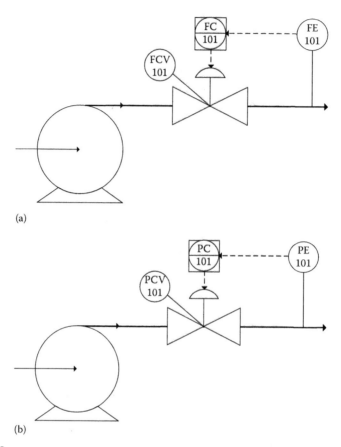

FIGURE 3.3
A basic fixed speed pump control using: (a) flow or (b) pressure as the dependent (measurement) variable.

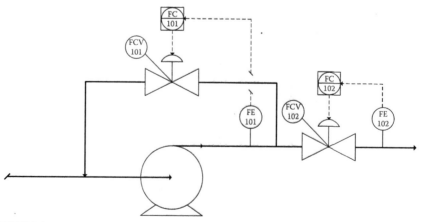

FIGURE 3.4
A fixed speed pump control with a simple minimum flow recycle.

Sometimes, the minimum flow recycle is always used. This is less energy efficient but may be justified in certain cases. For example, some types of pumps introduce a pulse into the fluid (e.g., reciprocating). The minimum flow recycle can be used to dampen this pulse. For systems where the flow recycle is always used, the control valve on the pump outlet line can be removed and the recycle flow can be used as the one independent variable in the system.

While much less common, there are some systems where it is better to install the control valve on the inlet to the pump. This configuration is limited to those installations where the net positive suction head is guaranteed under all potential operating conditions. The most common time such a configuration is used is when there is a suite of pumps configured in a complicated piping scheme. Consider the process arrangement shown in Figure 3.5. An inlet manifold routes fluid to three pumps: two are in operation and one is a standby that is used if either of the two pumps requires repair. The two in operation are not connected to a common outlet manifold. Instead, they send fluid along two separate outlet pathways.

Let's analyze this system. The total inlet flow rate is equal to the sum of the two outlet flows. The inlet pressure at the manifold is the same, regardless of the downstream pathway (since pressure is a state variable). The outlet pressures are completely independent from each other but they are related to the flow rate through each manifold via the pump curve. So, there are two degrees of freedom. We could install control valves on the outlet of each

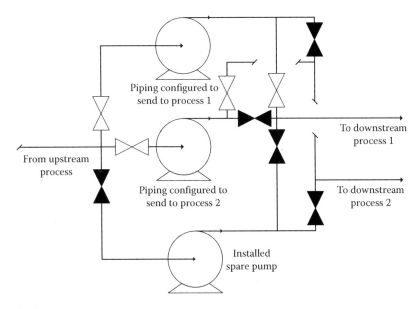

FIGURE 3.5
A pump and piping manifold to send liquid from the upstream process steps to two different downstream process steps at different flow rates and/or pressures.

pump. However, a change in the condition of one fluid will serve as a disturbance variable for the other. If we continuously adjust the condition of both fluids, the controls will be difficult because each will act as a disturbance variable to the other.

A smoother, easier control can be accomplished by moving one of the control valves to the common inlet manifold. A change in the condition of the total inlet fluid will still act as a disturbance variable for the outlet line control valve, but the outlet control valve is not a disturbance variable for the inlet line control valve. Thus, the system is more stable and easier to control. This scheme is shown in Figure 3.6.

In this example, the pumping manifold is an independent unit operation used to divide a fluid into two streams and to give them sufficient energy to move through the process, as defined by the requirements of the system. If the fluid condition of one stream is critical and the other is not as important, the pressure of the important stream could be paired with a control valve on the outlet of that stream and the flow rate of that same stream paired with the inlet control valve. This would insure that the important stream has the exact pressure and flow required. However, the pressure and flow of the other stream could not be specified and would vary as necessary to keep the important stream at its proper conditions.

Another typical stand-alone pump application is when a pump is used at the inlet of a process. In this case, we might want to insure that there is sufficient pressure in the fluid to drive it through a series of process steps or to meet the specific requirements of the first unit operation. For these types of

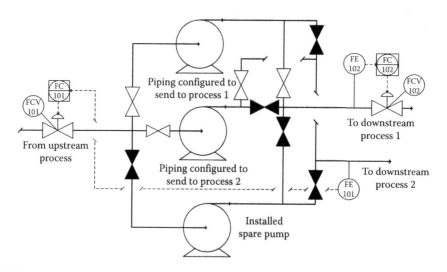

FIGURE 3.6
A control scheme for a pump and piping manifold to send liquid from the upstream process steps to two different downstream process steps at different flow rates and/or pressures with fixed speed pumps.

applications, we can pair the outlet pressure with a control valve installed on the outlet line of the pump. Similarly, at the end of the process, we may want to transfer a product to a storage area. The same control would be used: outlet pressure paired with a control valve installed on the outlet line of the pump.

Another way to control a pump is to turn the motor on and off at intervals. The pump usually operates at 100% of its normal operation while running. This is known as on/off control. The most common application for this type of system is when a pump is used in conjunction with a unit operation containing a liquid volume. The dependent variable paired to the motor on/off control variable is the level of the liquid in that unit operation (e.g., pressure vessel, tank, sump). When the liquid level rises to a specified point, the motor is energized and the pump transfers the fluid out of the unit operation. When the liquid level drops to a specified point, the motor is deenergized to stop fluid transfer.

For important unit operations, a normal level measuring device is used by the on/off controller to specify the high- and low-level limits (see Chapter 8 for a description of on/off controllers). For less critical unit operations, a level switch may be employed. The switch activates at the high limit but doesn't deactivate until the low limit is reached (this is known as a setpoint offset). A typical example of this second case is a system to transfer the liquid from a drainage sump to a wastewater treating system (see Figure 3.7).

Variable speed pump/motor systems can reduce the energy consumption of the process but are more expensive to build and require more maintenance over their lifetimes. The decision to specify a variable speed device should be based on an economic analysis of the total lifetime costs of the pump. However, there is a clear trend toward using variable speed systems as energy costs increase and as comprehensive life cycle sustainability becomes more important.

There is still only one degree of freedom for the pump. At any given speed, there will be an interrelationship between the fluid's pressure head and flow rate. But instead of adding internal energy from the pump motor and then removing that energy immediately in the form of friction across an outlet or inlet control valve, we can adjust the energy input instead. The pumping system is controlled by varying the resistance to current in the motor speed control circuit. This directly changes the speed at which the motor turns and thus changes the motor energy output used to drive the pump.

Consider our pumping scheme example in Figure 3.5. If we replace the fixed speed motors by variable speed motors, then the two independent variables will be the speeds of the two motors. We will still have some control issues because the change in motor speed of one pump will be a disturbance variable to the control of the other pump. There are a number of more advanced techniques that can be used to handle this, which we will explore in Chapter 7. But to give some insight now, let's assume that the flow rates through the two pumps are similar and that the desired outlet pressures are

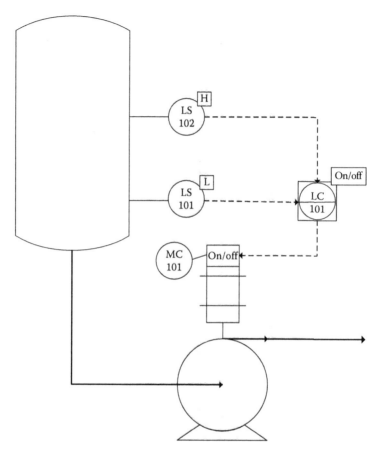

FIGURE 3.7
An on/off control scheme for a fixed speed pump used to remove liquid from a drum based on the activation of low- and high-level switches.

also similar. In this case, we can force the two pumps to operate at the same speed without much loss of efficiency. So we create an additional dependency: speed of motor B must equal the speed of motor A. This frees up one independent variable that we can then satisfy by installing a control valve on the common inlet manifold.

Most of the time, pumps are tightly integrated with other unit operations as part of unit operation systems. In these cases, the overall objectives of the entire unit operation system would be analyzed to determine the best dependent variable to pair to the pump's independent variable. We saw an example of this in Figure 3.7. As another example, if Figure 3.5 pump scheme were directly downstream of a pressure vessel, it might be desirable to pair the liquid level in the pressure vessel to an inlet control valve, while pairing flow (Figure 3.8) or a ratio of the two flows with the other control valve.

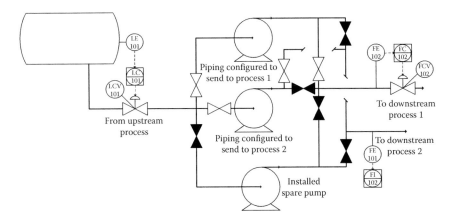

FIGURE 3.8
A control scheme for a pump and piping manifold integrated with an upstream pressure vessel that sends the vessel outlet liquid to two different downstream process steps at different flow rates with fixed speed pumps while maintaining the liquid level in the vessel.

The choice of dependent variable to pair with the outlet or inlet control valve is typically based on the operating objectives of the overall unit operation system that includes the pump. For example, if a pump is supplying cooling water to a heat exchanger, then the temperature of the process fluid leaving the heat exchanger would be the dependent variable paired with the pump's independent variable (most commonly speed for a variable speed pump, Figure 3.9, or

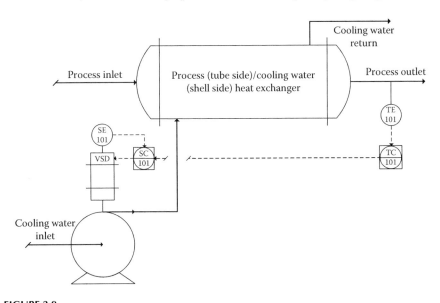

FIGURE 3.9
A control scheme for a variable speed cooling water supply pump integrated with a heat exchanger used to cool a process fluid.

a throttling control valve on the cooling water outlet leaving the heat exchanger for a fixed speed pump).

We will cover systems where pumps are integrated with other equipment into unit operation systems for the most common types of systems in Chapters 4 through 6.

3.2 Compressible Fluids: Compressors, Blowers, and Fans

Compressors, blowers, and fans are used to increase the energy of gases. These equipment types are all functionally identical. The use of one label over another is typically based on the fraction of the energy added to the device that goes into increasing flow rate versus the fraction that goes into increasing pressure. In fans, the ratio of flow energy to pressure energy is very high. In fact, a fan typically can only add a few kPa of pressure to the fluid. By contrast, compressors have a much lower ratio. Reciprocating compressors can be designed and configured to increase the pressure head by 1000 kPa or more. Blowers are intermediate devices where the ratio is typically still biased toward flow, but these machines can generate pressure heads of a few hundred kPa.

To add further confusion, compressible fluid motive force devices that operate below atmospheric pressure are often labeled as *vacuum pumps*. But they aren't pumps at all; they are compressors, blowers, or fans that happen to have suction pressures that are below atmospheric pressure! For the sake of simplicity, we will refer to all of these devices—compressors, blowers, fans, and vacuum pumps—as compressors.

Because the fluid is compressible, the density term in the Bernoulli equation (3.1) is no longer a constant. Therefore, even ignoring the potential energy term, there are two degrees of freedom in compressor systems. We have the same four options as for pumps:

1. A throttling control valve on the outlet piping after the compressor
2. A throttling control valve on the inlet piping before the compressor
3. The fraction of time the compressor motor is operating
4. Varying the motor/compressor speed

As with pumps, the choice is based on circumstances, some of which we will now briefly explore.

For fixed speed compressors, the most common configuration is to install control valves on both the inlet and outlet lines for the compressor. Motor on/off is only used for small devices and is usually limited to fans. Most large compressors have variable speed motors. For these machines, the most

Motive Force Unit Operations Control

common configuration is a control valve on the outlet line plus a speed controller on the motor.

Some very large compressors are not driven using an electric motor. Instead, a turbine is used. The most common turbines are either steam or gas driven devices. For these systems a control valve is installed on the inlet line for the steam or gas into the turbine. By controlling the turbine driver fluid, the speed of the driver and thus the compressor can be varied.

When compressors are independent unit operations, the most common dependent variables paired with the two independent variables are outlet pressure and outlet flow rate. Both variables are responsive. They are also related to each other and thus act as disturbance variables for the other control loop. Figure 3.10 shows this configuration for a variable speed compressor system with a steam turbine driver.

As with pumps, for any given speed, there is one specific pressure head and fluid flow rate through the compressor. But there is one important difference compared to pumps, the compressibility of the fluid. In a pump, if a given pressure head is achieved, the outlet pressure is fixed. But for a compressor, the outlet pressure is related, but not dependent solely upon the pressure head because the fluid's density is not a constant. As such, there is some interaction between the two control loops no matter which dependent variable is paired with which independent variable, but this interaction is usually manageable.

Following the second law of thermodynamics, when the compressor imparts energy into a compressible fluid, a portion of that energy is transformed into internal energy and the fluid heats up. There is a common misconception that compression of a gas that is near its saturation temperature

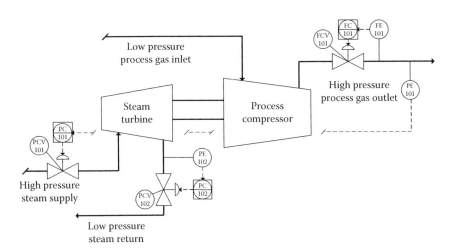

FIGURE 3.10
A control scheme using outlet flow and outlet pressure for a compressor with a steam turbine driver. By controlling the flow rate of steam into the turbine, the speed of the turbine is controlled, which controls the energy input to the fluid in the compressor.

at the suction of the compressor may lead to partial condensation within the compressor. However, unless you have a highly unusual fluid, the increase in internal energy of the fluid will always be sufficient to keep the entire fluid in the vapor phase.

Unfortunately, the increase in internal energy resists further reductions in the density of the fluid by the compressor. As the temperature increases, it takes increasingly more energy to achieve the same pressure increase, and this leads to requirements for substantially larger drivers and substantially greater energy consumption to achieve the objectives of the compressor. To overcome this, it is common to provide interstage cooling whenever the ratio of the outlet pressure (in absolute pressure units) to the inlet pressure is greater than around four.

Consider an example where a compressor is specified to increase the pressure of a gas from atmospheric pressure (101 kPa) to 9 atm (909 kPa), a compression ratio of 9:1. Using the rule of thumb of adding interstage cooling whenever the compression ratio exceeds 4:1, we would add an interstage cooler in the system. The first stage will compress the gas from 1 atm (101 kPa) to around 3 atm (303 kPa). The gas is then routed through a heat exchanger to remove the internal energy added during the first stage of compression. A knockout drum is often added after the heat exchanger to remove any fluid that condensed while it was cooling in the heat exchanger. The fluid from the first stage knockout drum is then routed to the second compressor stage where it is compressed from around 3 atm (303 kPa)* to 9 atm (909 kPa).

The cooling fluid flow rate to the interstage cooler is controlled using the gas outlet temperature from the knockout drum (better to measure after the knockout drum to avoid liquid drops impacting the temperature measurement probe). A simple liquid level control can be used on the bottom of the knockout drum. The material can be sent back to a spot in the process where it can be recovered. The entire control configuration using a variable speed compressor is shown in Figure 3.11.

Many types of compressors, including the most common type, centrifugal, can experience a phenomenon known as *surge*. Surge occurs when the volumetric flow of fluid is too low for the change in fluid density caused by the pressure increase in a portion of the compressor (Figure 3.12). In centrifugal compressors, this occurs when the quantity of gas entering a compressor wheel cannot fill the entire space inside of the wheel as its density is increased. When this occurs, fluid from downstream of the surge point will reverse direction and flow backward to fill the void until the pressure and volume are stabilized. After sufficient fluid has returned into the surge point, the flow direction returns to normal. But, immediately the surge condition occurs again, causing flow reversal. It is this back-and-forth flow of the fluid that gives this condition its name; the fluid "surges" through the machine.

---

* *Note*: the actual inlet pressure to the second stage will be 20–30 kPA lower due to the pressure drop the fluid experiences in the interstage cooling loop.

Motive Force Unit Operations Control

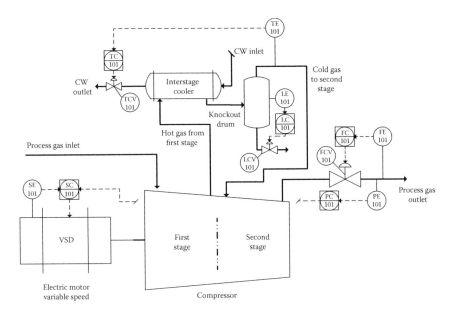

FIGURE 3.11
A control scheme for a two-stage compressor with an electric motor. An interstage cooler along with a knockout drum is installed to reduce the fluid temperature half way through the compressor operation in order to reduce the overall power requirement of the compressor. Note that the two stages should consume roughly the same power in order to minimize stress on the compressor shaft. If the two stages have very different power requirements, then each stage is provided with its own driver.

Surge not only upsets the process, it can cause substantial damage to the compressor if allowed to continue. The machine can literally shake itself to pieces! The solution is to always keep the flow rate through the compressor high enough to avoid the surge point. To accomplish this, a minimum flow recycle line is installed. A fast-acting control valve is installed on the recycle line to insure that the flow through the machine can be quickly increased if surge is detected. To detect surge, the inlet and outlet pressures are both measured and their values are compared (usually by calculating the pressure head). If the rate of decrease in the pressure exceeds a certain value, the surge valve is sent to the open position. The rate of decrease is used rather than the absolute differential pressure (head) value because it provides a faster response and may allow the control action to avoid surge altogether. This situation, when the pressure head is rapidly decreasing, is known as *incipient surge* because the surge condition is corrected before compressor actually gets to the point where the flow reverses.

More sophisticated controls are usually overlaid on top of the emergency surge control just described. These controls, depicted graphically in Figure 3.12, calculate the operating point of the compressor and compare it to a function known as the surge control line. If the operating point approaches the surge

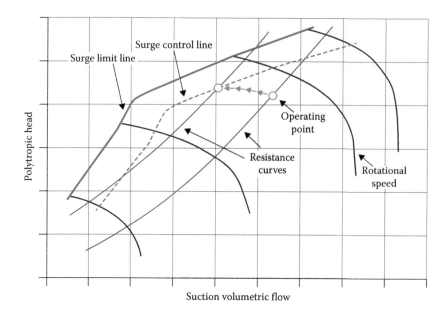

FIGURE 3.12
A compressor map. Each manufacturer will provide an ideal map, but usually an updated map is determined during commissioning. It should be updated by actual tests every couple of years or when there is a substantial change in inlet gas composition. The map plots the compressor pressure increase (ratio of outlet to the inlet pressure; denoted here as the polytropic head) versus flow for various compressor rotational speeds (denoted by the solid black curves). On the left is the surge line. Surge is expected if the conditions are to the left of this curve. A control line is usually calculated to be around 10% to the right of this line to insure avoidance of surge even if the conditions change. The operating line (denoted as the peak efficiency line in this figure) is the desired condition and is slightly to the right of the control line. The regulatory control system setpoints are specified to correspond to the operating line. When operation is to the left of the control line, the control system begins to open the recycle valve to move operation back toward the right to avoid entering the region where surge is predicted. (Courtesy of the Compressor Controls Corporation.)

control line, the recycle control valve begins opening to move the operation to a safer condition. The goal is to operate as close to the surge point as possible while minimizing the possibility of experiencing surge. This type of advanced control system is known as an antisurge controller.

Another consideration for the compressor recycle line that is different than that for a pump is the heat of compression factor. If a substantial quantity of the flow entering the compressor comes from an uncooled recycle line, the performance of the compressor will be affected. The increased internal energy of the fluid resists compression. Therefore, it is important to include a heat exchanger that cools the fluid in the recycle line. The best solution is to find a place in the process upstream or downstream of the compressor (but within a reasonable time lag from the compressor) where the heat can be removed and to incorporate that heat transfer into the recycle loop.

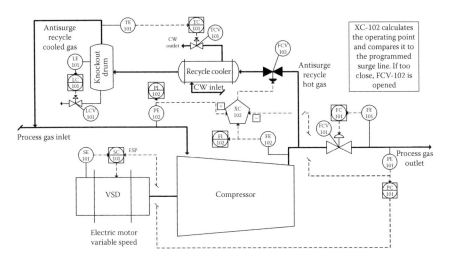

FIGURE 3.13
Antisurge control with dedicated recycle cooler for a compressor with an electric motor. Note that we have used a Pentagon symbol, as defined in Section 1.10, to indicate that this device is part of the unit's safety system.

Where this isn't possible, a dedicated heat exchanger can be added. A compressor control scheme incorporating an antisurge recycle line with a dedicated recycle cooler is shown in Figure 3.13.

If the gas is not valuable and not toxic (e.g., air), a vent line can be used instead of a recycle line. In this case, excess fluid is vented to the atmosphere rather than recycled through the process. A compressor control scheme incorporating an antisurge vent line is shown in Figure 3.14.

3.2.1 Other Compressible Fluid Devices: Expanders and Turbines

While used more rarely, two other important unit operations for compressible fluid transport are expanders and turbines. In these devices, energy is extracted from the fluid in the form of shaft work. Expanders and turbines can be used to recover energy stored in the form of pressure, fluid momentum, or reactive potential energy (i.e., heat of reaction). Fluid flows through a device similar to an axial compressor/blower, but in the opposite direction. In each section of the expander, the volume available to the gas is larger. As the gas expands to fill the space, its pressure is reduced. This provides a pressure driving force that encourages flow from the high-pressure side of the chamber to the low-pressure side. Vanes or other channels are used to extract some of the momentum out of the fluid and use that force to turn the shaft of the expander. With this energy extracted, the gas expansion will cool the gas.

Functionally, there is no difference between the two devices. We will classify them by application. An expander is a device where the primary

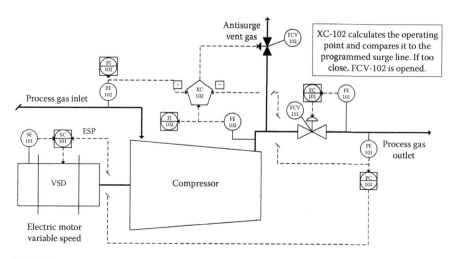

FIGURE 3.14
Antisurge control for a nontoxic gas with insufficient value to warrant a recycle; a gas that can be vented to atmosphere. The most common is a process or utility compressed air stream.

purpose is the reduction of temperature and/or pressure of the fluid and a turbine is a unit operation where the primary purpose is to extract shaft work from the fluid.

A common use of expanders is to provide very low temperature fluids for cryogenic separation operations. For instance, sometimes the most efficient way to separate two or more components in a mixture is a very low temperature distillation. An example of this is the large-scale separation of oxygen and nitrogen contained in air. The most practical way to get the fluid temperature cold enough for this application is to compress the air to high pressure, remove the heat of compression using a heat exchanger, and then expand the air to low pressure in an expander, as shown in Figure 3.15.

The most common types of turbines use either steam or a combustible gas as their fluid. High-speed turbines extract pressure energy from high-pressure steam and use the steam to drive the shaft of the turbine. Condensing steam turbines also extract some of the latent heat released as the steam condenses to liquid water. Current technology limits these turbines to lower speeds due to the stresses added to the turbine due to the extreme density change as the water changes from vapor to liquid phase.

In a gas turbine, a combustible gas is mixed with an oxidant (usually air) and combusted to CO_2 in a burner chamber. The combustion process generates a very hot, pressurized gas. This energy is extracted as shaft work as the combustion product gas passes through the turbine blades.

Turbines are often used instead of electric motors. In this application, the turbine shaft is connected to the shaft of the device being driven through a coupling. Gears can be used to change shaft speeds between the turbine and the user device. We saw an example of this in Figure 3.10 where a steam

Motive Force Unit Operations Control

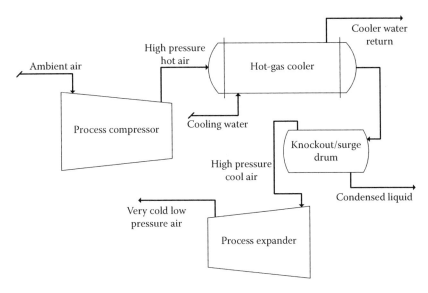

FIGURE 3.15
A compression/expansion system to generate very cold air.

turbine was used to power a compressor. Common uses are to drive large compressors, electric generators (steam turbines are used in power plants to generate electricity), and pumps.

For expander/turbine systems, there are three possible independent variables: the fluid inlet stream, the fluid outlet stream, and the speed of the expander/steam turbine. Since a compressible fluid is processed, there are two independent variables, with the third possibility related to the other two by material balance. The choice depends upon the application but most commonly involves one control valve, either on the inlet or outlet lines plus a speed control for the expander/turbine. However, there are applications where the shaft speed is not important (as long as it is within the expander/turbine's normal operating range) and two control valves are used.

For the cryogenic cooling application described earlier, outlet fluid temperature and outlet fluid flow rate are often used as the dependent variables. This insures that the expander provides adequate cooling for the process. In a steam turbine attached to an electric generator, the electric output of the generator and the outlet pressure of the steam are often used.

For a gas turbine, the unit has an additional independent variable, the amount of oxidant blended into the combustible gas. A control valve on the oxidant inlet is typically used for this additional degree of freedom. The demand from the user device is usually used for one of the independent variables. The combustion product gas outlet temperature or pressure is often used as a second dependent variable, while the oxidant-to-fuel ratio is often used (via a ratio controller) as the third dependent variable. To validate efficient combustion, the outlet concentration of oxidant in the combustion product gas exiting the turbine can be measured and used in a cascade as a feedback trim to the ratio controller (see Section 2.8).

3.3 Solids Handling Devices

Solid materials are not fluids. In the vast majority of cases, we can't simply put solids in a pipe, apply a pressure, and expect it to move through the process. Solids transport systems can be divided into three basic categories:

1. Mix the solid with a gas (entrainment, fluidization) or liquid (slurry, fluidization) such that the mixture behaves like a fluid.
2. Place the solid on a physical surface that moves.
3. Use physical pressure or gravity directly on the solid for short distance transport.

We will now look at common examples for each of these categories and look at ways to design the associated control strategies.

3.3.1 Solids Mixed with Gases and Liquids

Solids can be mixed with a sufficient quantity of gas or liquid so that the mixture behaves like a fluid. Liquids can be moved with pumps and gases with compressors. In general, the controls for such systems are identical to those described in Sections 3.1 and 3.2. If both the fluid and solid sources are defined by upstream process units, then there are no additional degrees of freedom and the controls described above are sufficient. However, in many cases, either the amount of fluid blended with the solid or the amount of solid added to the fluid is an independent variable.

If the fluid rate is an independent variable, a control valve can be installed on the fluid inlet line. If a pump or compressor is used to pressure the fluid into the blending unit operation, the independent variable is usually integrated with the controls for the pump or compressor. For example, a pump or compressor speed control for a variable speed machine might be used as the independent variable.

If the solids feed rate is an independent variable, then the controls will be integrated with the controls of the device that provides the solids to the mixing point, which we will discuss in the next subsections.

If the fluid and the solid are continuously mixed together and the flow rate of each can be measured, then a good dependent variable to pair to the additional independent variable is the ratio of solids to fluid. This would be accomplished using a ratio controller as described in Section 2.7.

Unfortunately, it may not be possible to accurately measure the solids flow rate. In this case, an indirect measure needs to be used. For example, the density or viscosity of a solid-liquid slurry could be measured and paired to the independent variable. These measurements are not very responsive, so incorporating a cascade control that uses the density or viscosity to adjust the setpoint of a ratio or flow controller might be useful (Section 2.7). An example of a control system to slurry a solid with a liquid to a specific solids concentration using density as the indirect measurement is shown Figure 3.16.

When a gas is used to fluidize and continuously transport a solid, the complexity of the gas-solid system is too great and there are no indirect measures that can be used to automatically optimize the gas-solid ratio. In this case, manual control may be necessary.

Often solids/fluids mixing occurs in a batch process step. For example, a fixed quantity of solids is added to a mixing vessel. Then, the mixing fluid is added. For slurries, a measure of the density or viscosity of the mixture can be used to determine when sufficient liquid has been added. For gases, the fluidization is typically a continuous step. An empirically calculated quantity

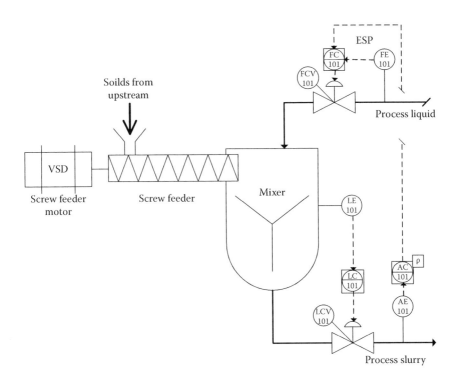

FIGURE 3.16
A typical control scheme to blend a solid into a liquid to obtain a slurry in a mixing vessel, where the solids flow rate is set by an upstream unit operation. Density (ρ) is used as an indirect measure of the solids content in the blended slurry product.

of gas is routed over the top of a solid or through a bed of the solid. A portion of the solid will become entrained in the gas and thus be made mobile.

3.3.2 Physical Motive Force Devices

The most common way to transport solids to, within, and from a process is with a motor-driven conveyor. Each conveyor has a fixed width, a maximum load capacity, and a range of travel speeds. There is one independent variable, and it is virtually always satisfied by travel speed control. For conveyors that are independent unit operations, the dependent variable is usually some measure of flow rate with the most common being mass flow rate measured using a weigh scale embedded with the conveyor. Like pumps and compressors, conveyors are frequently integrated with other equipment as a unit operation system. The dependent variable is usually determined by the performance objectives of the overall unit operation system.

For example, consider a process where a solid is continuously added to a mixer where it is dissolved into a liquid. There are three potential independent variables: inlet solids flow rate, inlet liquid flow rate, and outlet mixture flow rate. Only two of the three are truly independent. The flow rate of one of the two inlet streams, the solid or the liquid, may be determined by the upstream process. Alternately, the outlet flow rate might be specified by the production demand of the unit.

Let's assume that our process requires that we generate a fixed quantity of mixed solution at a specified concentration. This is common when a solid chemical must be added into the process, for example as a solvent, to adjust pH, or as a reactant. The goal of the unit operation is to provide the required quantity of the solid to the process by dissolving it into a liquid to a specific concentration or by creating a slurry that can be pumped into the process stream.

The outlet flow rate of the mixture is set by the process. If the amount required is manually set, the control valve on the mixing vessel outlet line will be paired with a measurement of the outlet mixture flow rate. Similarly, if the true setpoint is a measurement well downstream in the process, so that there is a large time lag, a cascade loop may be added to the local flow control loop. The liquid level in the mixer is used to control the quantity of liquid added to the vessel. As downstream demand changes, which will change the flow rate through the outlet line control valve, the level controller will automatically adjust the liquid feed to the mixer to accommodate the change in outlet flow rate.

A direct measure of a key chemical element of the solid feed stream or an indirect measure of a key property such as pH, viscosity, or density in the outlet mixture is then paired with the speed control of the conveyor providing solids to the mixing vessel. This variable insures that the correct ratio of solids to liquid is mixed together in the mixer. The entire control scheme is shown in Figure 3.17 using viscosity as the indirect mixing dependent variable. If this measure is difficult or unreliable, the flow of solids can be

Motive Force Unit Operations Control

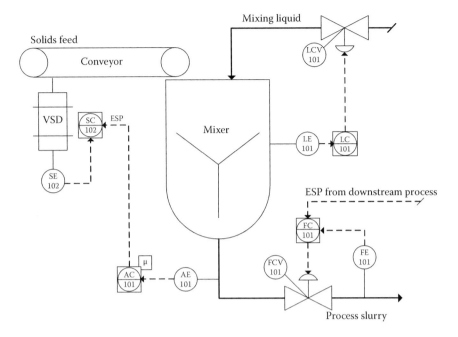

FIGURE 3.17
A typical control scheme to add a solid to a liquid process stream in a mixing vessel, where the flow rate is set by a downstream process and viscosity (μ) is used as an indirect measure of the quality of the blended product. Note that the outlet may be either a liquid (if the solid dissolves into the liquid) or a slurry.

measured and a ratio controller can be used to change the conveyor speed to maintain the solids-to-liquid ratio.

To move solids into a pressurized unit operation, a screw conveyor or extruder may be used. The turning speed of the screw controls the rate of flow of the solid. Controls are analogous to conveyor controls.

3.3.3 Using Physical Pressure or Gravity Directly on the Solid for Short Distance Transport

A common way to move solids into or out of a pressure vessel is to use pressurized hoppers (also known as lock hoppers). Hoppers are a semi-batch operation typically integrated with an associated conveyor and the equipment that processes the solid. Let's look at an example where a solid is added to a reaction vessel and mixed with a reacting liquid solution to form a solid reaction product. The overall system is shown in Figure 3.18.

The solid raw material is brought by a conveyor to the overhead hopper. The material drops into the hopper until a given amount has been added. This can be measured using totalized flow or by level measurement. Once the hopper is full, the inlet conveyor is stopped and a gate is slid over the

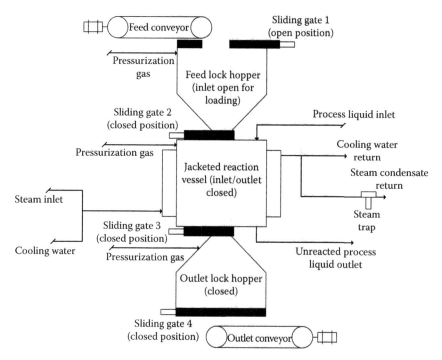

FIGURE 3.18
Using pressurized lock hoppers to load and unload a solid into a reaction vessel.

opening of the hopper, providing a seal. Then, nitrogen (or another gas) is used to pressurize the hopper via a nitrogen fill line. When the pressure inside of the hopper reaches the desired level, a gate isolating the hopper from the reactor is opened and the solid flows into the reactor. When all or part of the solids has transferred from the hopper into the reactor, the isolation gate is closed. Next, the reacting liquid is added to the reactor. The liquid's temperature may be increased or decreased prior to entry into the reactor in order to meet optimum reaction requirements.

Additional heating or cooling may be required to bring the reactor to the correct reaction temperature. The most common method is to jacket the reactor with a shell filled with a heating or cooling fluid (Figure 3.18). Once the reactor is at temperature, the reaction will be carried out. Most reactors operate isothermally or near isothermally but the reactions are either exothermic or endothermic. So, the jacket may be used to keep the reaction at or near the optimum reaction temperature. There are cases where the jacket is filled with steam initially to heat the mixture to the reaction temperature and then switched to cooling water to remove the heat of reaction for an exothermic reaction.

After the reaction reaches completion, any remaining liquid is drained from the reactor. The reactor might then be pressurized with nitrogen to help get the solids out of the reactor. A gate separating the reactor from the outlet hopper is opened and the solids are pressurized out of the reactor. Sometimes just a chute is needed to direct the solids onto a conveyor for transport to the next unit operation in the process.

Recall from Section 2.9 that sequencing between steps in a batch unit operation requires remotely operated block isolation valves (most commonly motor operated valves, MVs). The P&ID representation of this batch process is shown in Figure 3.19 (using both steam for heating and water for cooling). The sequential events table is provided in Table 3.1.

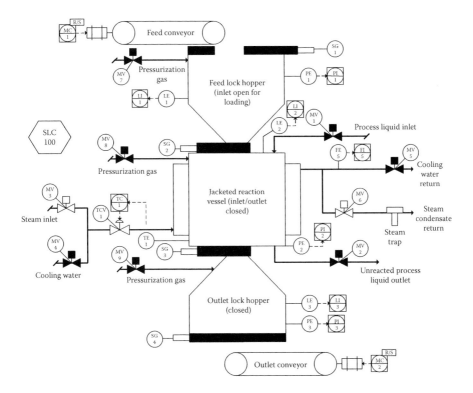

FIGURE 3.19
A typical control scheme for a batch process system that uses pressurized lock hoppers to load and unload a solid into a reaction vessel. The first step from the sequential events table (Table 3.1) is depicted. All of the instruments and control blocks are linked to the sequential logic controller, SLC-100. SG stands for "Slide Gate" and is analogous to MV for a remotely operated block valve.

TABLE 3.1

Sequential Event Table Associated with Figure 3.19

Step	Description	MV1	MV2	MV3	MV4	MV5	MV6	MV7	MV8	MV9	SG1	SG2	SG3	SG4	MC1	MC2	Termination Criteria
1	Fill feed hopper with solids/preheat reactor	C	C	O	C	C	O	C	C	C	O	C	C	C	R	S	LE1 = L1 max
2	Pressurize feed hopper	C	C	O	C	C	O	O	C	C	C	C	C	C	S	S	PE1 = P1 max
3	Fill reactor with solids	C	C	O	C	C	O	O	C	C	C	O	C	C	S	S	LE1 = L1 min
4	Add process liquid to reactor	O	C	O	C	C	O	C	C	C	C	C	C	C	S	S	LE2 = L2 max
5	Reaction with heat removed by cooling water	C	C	C	O	O	C	C	C	C	C	C	C	C	S	S	FE5 = FE5 min (when reaction is complete, excess heat is no longer generated)
6	Empty unreacted liquid out of reactor	C	O	C	C	C	C	C	C	C	C	C	C	C	S	S	LE2 = L2 min
7	Empty reactor to outlet hopper using gas pressure to assist	C	C	C	C	C	C	C	O	C	C	C	O	C	S	S	LE3 = LE3 max
8	Pressurize outlet hopper	C	C	C	C	C	C	C	C	O	C	C	C	C	S	S	PE3 = P3 max
9	Empty outlet hopper	C	C	C	C	C	C	C	C	O	C	C	C	O	S	R	LE3 = LE3 min

3.4 Startup and Shutdown for Large Motive Force Systems

Large pumps and major compressors require special handling during startup and shutdown events to avoid permanent damage. Elevated fluid temperatures due to process conditions, heat of compression (for compressors), and shaft rotational friction will expand the metallic elements of the machine. Yet the device must be able to accommodate this expansion while still maintaining the precise tolerances required to operate efficiently.

These devices often contain a rotating shaft that supports the blades, chambers, and other components of the pump/compressor. These components result in stresses to the shaft during operation (these will be discussed more fully in Section 3.5) and exert a downward pressure when the machine is stopped. In fact, for major compressors, we try to avoid complete stoppage of shaft rotation unless the shaft is going to be removed from the compressor shell.

For these and other reasons, a sequence of steps is typically used for the startup and shutdown of these machines. Sequential logic, typically programmed into a programmable logic controller, is used. Almost all vendors supply startup and shutdown sequencing and all necessary associated instrumentation plus a dedicated PLC when the device is purchased.

However, it is important to know when these systems are used. They should have interfaces to the process safety automation system (usually through the equipment protection system, Section 3.5). Further, over time the hardware and software employed may require replacements if they become unreliable. The original equipment is unlikely to be available and more modern systems may need to be employed.

The specific sequences are device specific and beyond the scope of this text.

3.5 Equipment Protection Systems

High-speed pumps and major gas compressors usually include equipment protection systems. These systems protect the equipment from damage due to upsets in operation. The key parameters for these systems are excess speed, vibration, surge, and damage due to uncontrolled shutdowns. Other parameters may be important depending upon the actual physical device employed.

For centrifugal pumps, compressors, expanders, and turbines, there is a great deal of momentum built up in a high-speed rotational shaft. Sudden braking will likely result in substantial damage to the machine with possible collateral damage to other equipment. If possible, we'd like to avoid adding stresses to the shaft by cooling it down unnecessarily. Even if the unit must be cooled down, we'd still like to keep the shaft rotating at a minimum level to keep its weight from causing permanent bending. Thus, there are typically multiple types of shutdowns that can be initiated, depending upon the severity of the event. These often include minimum flow recycle, hot standby, cold roll, and emergency or complete stop.

When possible, the device should be isolated from the process and placed in a 100% minimum flow recycle mode. From this condition, the unit can be placed back in service quickly and safely after the upset event has been resolved. If the process will be out of service for a more extended period, the device can be placed in *hot standby*, which is also commonly known as *hot slow roll*. Under this condition, the device is isolated from the process and the speed of rotation is reduced to a slow roll. Steam or another hot fluid is bled into the inlet flow line and moves through the device to keep the surfaces hot. This avoids thermal contraction of the metal and allows the unit to be restarted much faster than if the machine cools down to room temperature.

When the device will be out of service for weeks, the unit may be placed in a *cold roll* (slow roll) condition. In this case, the device is isolated from the process and the speed of rotation is reduced to a very slow rotation. Process gas is purged out of the compressor and the device cools down to room temperature. The slow rotation keeps the weight from bowing the shaft or damaging the bearings. Emergency or complete stop means the machine is completely stopped and deenergized. In this case, the shaft should be physically removed from the device if the shutdown lasts for more than a few days.

When possible, it is best to start with the least severe shutdown condition and then move to the more rigorous as required. Shutdown logic is usually provided by the equipment vendor and included as sequential logic in a PLC.

When high-speed rotational equipment operates, vibration is induced due to imperfections in the balance of the shaft. In cylindrical coordinates, this vibration can occur in the axial, rotational, or (less commonly) radial directions. For major compressors, vibration monitors are typically installed for axial and rotational vibration. If the vibration exceeds a high limit, a shutdown action is initiated. Usually, a separate vibration monitoring safety system is provided. These are typically provided in a PLC. However, this system does not directly shut down the machine. Instead, it must be interfaced to the equipment shutdown or equipment protection system. The vibration system notifies the appropriate system that a high limit action should be initiated.

3.6 Switching Controls for Parallel Motive Force Units

Motive force units are mechanical devices with rotating parts. Because of this, they typically required more off-line service than most other unit operations. Fortunately, pumps and conveyors are relatively inexpensive. Yet if one of

Motive Force Unit Operations Control

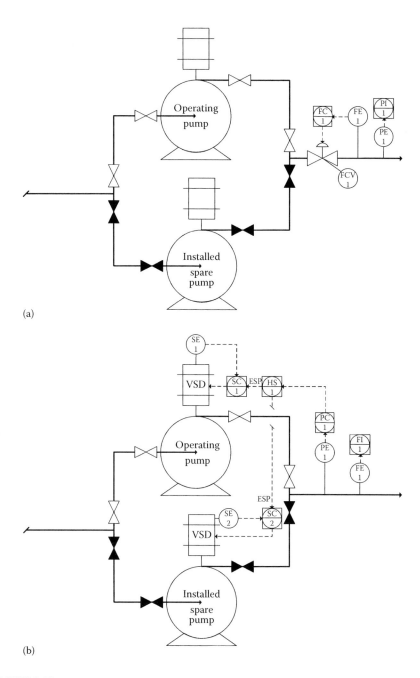

FIGURE 3.20
Typical piping and control configurations for a pump set where one pump is in operation and one is an installed spare: (a) fixed speed pumps with flow control and (b) variable speed pumps with pressure control (the common pressure signal is sent through a hand switch so the control operator can manually select the pump that is in operation).

these does need to be taken out of service, it could result in millions of dollars in production losses. To avoid this, when pumps and conveyors are used in continuous processes, a 100% capacity installed spare is typically provided. There are exceptions of course, but in general, we can assume that if a pump or conveyor is installed, then an installed spare will also be installed.

For compressors/blowers/fans, the installation of a 100% capacity spare is less universal. The governing factors are the capital cost for the unit and whether the unit can remain at shutdown without damage to the shaft or other components for an extended period. Large industrial compressors can cost several million dollars. For these high-cost installations, an installed spare is not provided. By contrast, for a relative low-cost fan, the spare will be included.

Figure 3.20 shows a typical installation for a pump with an installed spare. The spare pump is isolated from the process using manually operated block valves. A valve should be placed as close to the tee where the process stream splits into two to avoid creating a stagnant zone where accelerated corrosion can occur. To put the spare in service, the isolating block valves are opened on the spare pump, and then the pump motor is started. After the spare pump is in operation, the operating pump motor is deenergized and the isolating block valves on the operating pump are closed.

Problems

3.1 Consider the pump arrangement and control scheme shown in Figure 3.6. Modify this scheme such that the ratio of the flows is paired with the control valve on the inlet line.

3.2 Consider the pump arrangement shown in Figure 3.6. If the fluid condition of one stream is critical and the other is not as important, the pressure of the important stream could be paired with a control valve on the outlet of that stream and the flow rate of that same stream paired with the inlet control valve. This would insure that the important stream has the exact pressure and flow required. However, the pressure and flow of the other stream could not be specified and would vary as required to keep the important stream at its proper conditions. Modify Figure 3.6 control scheme to reflect this situation.

3.3 Modify Figure 3.6 control scheme to reflect variable speed motors on the pumps.

3.4 Modify Figure 3.6 control scheme to reflect variable speed motors on the pumps where the motors are maintained at the same speed.

Motive Force Unit Operations Control

3.5 For slurries, it is common to install three 60% capacity pumps. Under full throughput operation, two of the three pumps will be in operation and the third will be an installed spare. Three pumps are often used because slurry pumps typically require more maintenance than pure liquid pumps. So in the rare cases when two pumps are off-line undergoing maintenance, the process can still operate at 60% of its rated capacity. Consider the case where either one or two fixed speed pumps are in operation with two or one pumps out of service to increase the momentum of a slurry based on the downstream process requirements. If two pumps are in service, the flow through the two pumps should be equal. Draw a control scheme that will accomplish these operational objectives.

3.6 Repeat Problem 3.5 for the case where the pumps are variable speed pumps.

3.7 Modify the control scheme shown in Figure 3.7 for the case where a level element (level measurement) is used instead of two level switches.

3.8 Modify the control scheme shown in Figure 3.9 if a fixed speed pump is used instead of a variable speed pump.

3.9 Modify the process and control scheme shown in Figure 3.10 if the steam turbine driver is replaced with a variable speed electric motor.

3.10 While not common, some compressors are configured to use the pressure increase across the unit operation as one of the dependent variables. Modify the control scheme of Figure 3.10 to show how this would be accomplished.

3.11 Develop a control scheme for a compression system used to raise the pressure of a gas from atmospheric pressure to 4000 kPa. Assume there is a 25 kPa pressure loss for any interstage cooler/KO drum units employed (*hint:* your solution should include a three-stage compressor: 1, 101–355 kPa; 2, 330–1162 kPa; 3, 1137–4000 kPa).

3.12 Develop a control scheme for a variable speed expander that must reduce the temperature of a specified quantity of gas to a specific temperature.

3.13 Consider the case of a combustion gas turbine connected to an electrical power generator. Develop a control scheme that efficiently generates the required quantity of electrical power. Include a ratio with feedback trim control loop for the air-to-fuel ratio based on the CO content of the turbine exhaust gas.

4

Heat Transfer Unit Operations Control

The goal of this chapter is to understand typical control designs for the most common types of equipment used to transfer heat within processes. There are a lot of reasons to transfer energy in processes. We saw one example in Chapter 3 for compressible fluids. During compression, some of the shaft work added to the fluid increases the internal energy of the gas instead of going into increasing the fluid's pressure and/or flow rate. As the temperature of the fluid rises, it gets increasingly harder to compress the fluid against the internal energy that has increased in the fluid. When the ratio of the outlet to inlet pressure exceeds 3-4, the unit operation can be improved by taking the fluid out of the compressor partway through the compression process, cooling it, and then putting it back into the compressor. The cooling is a heat transfer operation and is performed in an indirect contact heat exchanger.

Consider a process that produces a food or pharmaceutical product. Usually, one of the final steps is to dry the solid product prior to packaging. For small-scale processes, the solid may flow through an oven. An oven is another type of indirect contact heat exchanger where the heat is usually provided by the heat of resistance as electric current flows through wires. For larger processes, this drying step may be by direct contact. A hot inert gas stream (usually air) can be blown through or over the solid. As the solid heats, water evaporates out of the solid, leaving a dryer solid that meets the product specification.

Here is another example. Most reactions are conducted isothermally at a near-optimum reaction temperature. But reactions are either exothermic or endothermic, so energy must either be removed or added to the reactor to keep the temperature constant. The controls for this type of heat transfer will be discussed in Chapter 6.

4.1 Fluid-Fluid Heat Transfer

The most common heat transfer application is the indirect transfer of energy between two fluids in a heat exchanger. There are a number of types of heat exchangers, but they are all functionally equivalent. We will use the most common type, the shell and tube heat exchanger, in this discussion, but the information should be readily adaptable to any of the other common indirect fluid-fluid heat exchangers.

To properly design the controls for a heat exchanger, we must know the primary process objective of the unit operation. This might seem unusual. After all, isn't the objective always to get an outlet process fluid to a specified temperature? Surprisingly, the answer is no. There are many instances where a different objective is desired. We will look at various commonly occurring situations in the subsections below.

4.1.1 Heating or Cooling a Process Stream to a Specified Temperature Using a Utility Stream

A simple heat transfer application is to use a cold utility fluid (most commonly water) to cool a process stream to a desired temperature or to use a hot utility fluid (most commonly steam) to heat a process stream to a desired temperature. Usually, the flow rate of the process stream through the heat exchanger is specified by an upstream and/or downstream unit operation. Further, there is usually no control over the temperature/energy content of the utility fluid. This leaves one independent variable for an incompressible utility fluid (inlet or outlet flow rate) or two independent variables for a compressible utility fluid (inlet and outlet flow rate or pressure).

To address the case of one independent or control variable when the utility fluid is a liquid (incompressible), a control valve is usually installed on the outlet of the utility stream leaving the heat exchanger. However, if there is a companion pump to move the utility fluid through the heat exchanger, speed control is sometimes used instead. The temperature of the process stream leaving the heat exchanger is measured and this measurement variable is used in a feedback control loop with the control valve. Figure 4.1 shows the case without an integrated pump where the control variable is a control valve located on the utility fluid outlet line.

If a very hot process fluid is being cooled, the size of the heat exchanger can be reduced by vaporizing the utility stream liquid. The most common application is to feed boiler-quality water (commonly known as boiler feed water or BFW) into the utility side of the heat exchanger and use the process fluid's energy to vaporize the water to steam. This also has the advantage of capturing the energy we removed from the process stream in a usable form. The steam produced can often be used as a heating utility fluid in another part of the process.

When we vaporize the utility liquid, we introduce an additional independent variable because the vapor is a compressible fluid. For example, for the configuration shown in Figure 4.1, we usually add another control valve on the inlet of the utility stream. Under normal operation, we want to keep as much of the heat transfer surface contacting utility liquid as possible so that the heat transfer is the most efficient. However, if we are operating below capacity, one way to control amount of heat transfer is to vary the fraction

Heat Transfer Unit Operations Control

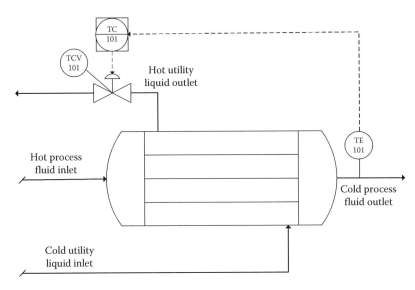

FIGURE 4.1
Using an incompressible utility liquid to cool a process stream to a specific temperature.

of the tubes that are covered by the utility. Both cases demonstrate that an important parameter is utility fluid liquid level in the exchanger.

If we are making steam and want to use it in another application in the plant, we have to insure that the steam has a temperature/pressure that can be used. Usually, the steam will be routed into a common steam header rather than directly to another heat exchanger. So the pressure of the steam must be slightly higher than the pressure in the steam header. Thus, the outlet utility steam pressure is an important parameter.

Based on the above, we have two independent variables (inlet and outlet line control valves) and three important dependent variables (process stream temperature, utility fluid liquid level, and utility stream outlet pressure). We normally choose the two most responsive parameters: temperature and pressure. The more stable pairing is to match the process stream temperature to the inlet control valve and the utility stream outlet pressure to the outlet control valve. As long as operation is normal, the liquid level will automatically adjust and stay within a reasonable level for good operation.

However, it is useful to put a low-low level safety system on the exchanger to insure that the heat exchanger retains liquid. If not, the duty will fall quickly and drastically because all the energy from the process fluid will be transferred into sensible heat (temperature increase of the utility vapor) instead of into latent heat (vaporization of the utility liquid into vapor). There are two circumstances when this situation would occur. The first is if the utility outlet control valve is stuck in a full or partially open position. For this case, pressure in the outlet line will begin to drop and eventually the pressure will build back up since the steam being produced will not be able to enter the header unless the pressure is high enough. This, in turn, will slow the flow of vapor out of the exchanger and allow the liquid level to be restored in the heat exchanger. In essence, the process will self-correct if this situation were to occur.

The second circumstance is if the inlet control valve sticks in a full or partially closed position. In this case, the process will not self-correct and the process stream will not be cooled to the desired temperature. To protect against this situation, a low–low liquid level measurement (element or switch), a low–low level alarm control block, and a solenoid valve installed on the utility inlet control valve are used. The utility inlet control valve should be an air-fail-open valve so that liquid will enter the heat exchanger until the level is restored to a normal level. The full control scheme is shown in Figure 4.2.

For an incompressible fluid with no phase change, steam is often used as a utility stream when heating the process fluid. Most commonly, the steam is condensed in the heat exchanger, allowing latent heat to be released and used to heat the process fluid. Because there is a gaseous utility stream, there are two independent variables—inlet steam and outlet steam condensate.

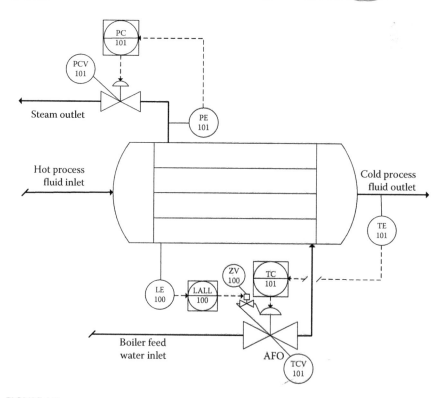

FIGURE 4.2
Control scheme for a heat exchanger that is generating steam to cool a very hot process stream.

A common configuration is to install a steam trap on the outlet line and a control valve on the inlet line. A steam trap is a mechanical device that allows condensate to flow through the trap while holding back any vapor (steam). Functionally, it acts as a mechanical level control loop.

For more precise control, a condensate drum or pot is installed on the steam condensate outlet line. The condensate drum provides a spot where the liquid condensate can build up and separate from the steam vapor. The liquid level in the drum is measured and used in a feedback control loop with a control valve on the condensate drum's liquid outlet line.

In addition to the condensate outlet, a control valve is usually installed on the inlet steam line. The process fluid temperature is still our primary parameter, so the process fluid outlet temperature is paired with the inlet control valve.

A key to this type of heat exchanger is to insure that the steam is condensing in the heat exchanger. If the steam trap "sticks open" or the condensate pot outlet control valve sticks in a partially or fully open position, most of the heat that will be removed from the steam will be sensible heat. When this happens, the process fluid temperature will drop and the primary control loop will open the inlet steam valve more to push more steam through the heat exchanger. But this is exactly the opposite of what is needed. Instead, we need to shut off the steam flow so that steam will condense in the heat exchanger, reestablishing a condensate liquid level. To accomplish this, a low–low condensate level measurement element should be installed on the utility side of the heat exchanger (when using a steam trap) or in the condensate drum. When the low–low level is reached, the inlet steam valve should be failed to the closed position. Once the level builds back up, the steam valve can begin opening again.

Sometimes this low–low level action will reset a stuck steam trap or control valve and solve the problem automatically. At other times, a cycle of low–low action and reset will occur, which should alert the operators that the steam trap/control valve is malfunctioning so that manual intervention can be performed. The complete configuration when using a steam trap is shown in Figure 4.3.

4.1.2 Vaporizing a Liquid Process Stream Using a Utility Stream

Rather than increasing the temperature, heat exchangers are often used to change the phase of a process stream—liquid to vapor or vapor to liquid. Let us look at the liquid-to-vapor case first. The process scheme is shown in Figure 4.4. In this application, a hot utility stream is used to vaporize a liquid.

On the process side, there are two degrees of freedom since the fluid leaving the heat exchanger is compressible. Usually the inlet flow rate is set by the process, leaving one degree of freedom, the outlet vaporized fluid. A control valve is usually used on this outlet stream to serve as the independent variable.

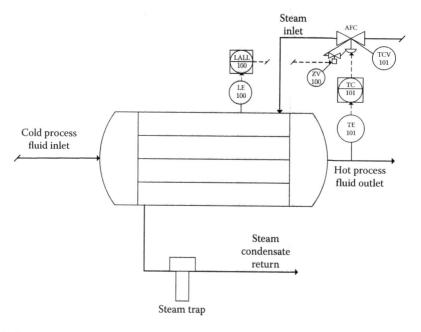

FIGURE 4.3
Control scheme for a heat exchanger that is condensing utility steam to heat a process stream to a specific temperature and employing a steam trap.

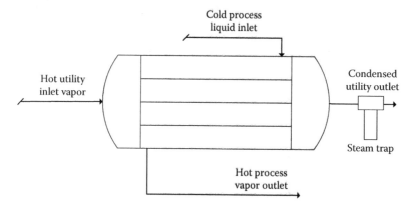

FIGURE 4.4
Process scheme for a heat exchanger that is condensing a utility steam to vaporize a process stream.

If the utility fluid is condensing (e.g., utility steam is condensing to liquid water), there are two degrees of freedom on the utility side of the heat exchanger. A steam trap or the control valve on the liquid outlet of a condensate pot is used for one control variable, while a control valve is usually installed on the utility inlet stream.

The primary objective is to vaporize the process fluid. As long as there is some process liquid in the heat exchanger and some vapor in the heat exchanger, this objective is satisfied. We can measure the level of the process liquid fluid in the exchanger to insure this is occurring. This is true whether the process fluid is on the shell side or the tube side of the exchanger.

A second objective is to optimize the heat transfer area utilized to insure efficient heat transfer. This can be accomplished by varying the level of the process fluid, which in turn will adjust the fraction of available tube surface area actually used to transfer heat from the utility fluid into liquid process fluid. We rarely operate processes at their design conditions, so this is a useful capability. This is done by matching the process fluid level to a control valve on the utility fluid. If steam is used, the control valve is usually on the steam inlet line. Level is not a responsive dependent variable so it sometimes does not work very well when paired directly to the control variable. A more responsive variable, such as the steam inlet flow rate, is typically used as a slave variable in a cascade loop with the level variable serving as the master.

The steam trap or condensate pot liquid level is used to insure that no utility vapor leaves the heat exchanger before condensing, exchanging its latent heat with the process fluid.

This still leaves the process outlet control valve available for another objective. A third objective is to insure that the process vapor has sufficient pressure to flow to the next unit operation. We can measure the pressure of the process vapor stream in the outlet line for this objective. This measurement can be paired with the process outlet control valve.

The most common utility used for vaporizing a process fluid is steam. If a steam trap is used on the steam condensate outlet line, then when the process fluid liquid level is paired with the steam inlet control valve, no additional control variables remain. As described in Section 4.1.4, a low–low level measurement should be used to air-fail-close the steam inlet valve when the utility condensate level is lost. This configuration is shown in Figure 4.5.

4.1.3 Condensing a Gaseous Process Stream Using a Utility Stream

Now we will look at the other case, condensing a gas. On the process side, there are two degrees of freedom since the fluid is compressible entering the heat exchanger. Usually the inlet flow rate is set by the process, leaving one degree of freedom, the outlet liquid fluid. A control valve is usually used on this outlet stream to serve as the independent variable.

Sometimes, a vapor trap (i.e., a steam trap used for a process fluid rather than for utility steam) is installed on the outlet stream instead of a control valve as shown in Figure 4.6a. This insures that only liquid leaves the heat exchanger. Another common configuration is to use a condensate drum (pot) after the heat exchanger, as shown in Figure 4.6b. In this case, the independent variable shifts from the heat exchanger outlet to the condensate drum

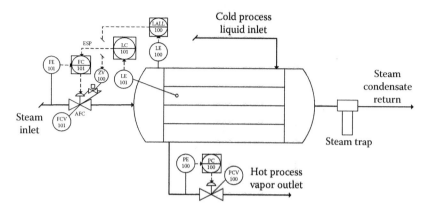

FIGURE 4.5
Control scheme for a heat exchanger that is condensing a utility steam and employing a steam trap to vaporize a process stream.

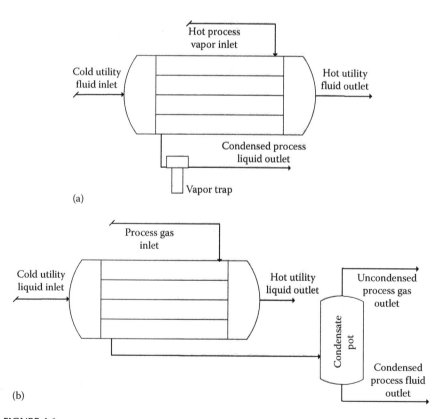

FIGURE 4.6
Outlet configurations when condensing a process fluid in a heat exchanger: (a) using a vapor trap and (b) using a condensate pot.

liquid outlet line. The most common application of this configuration is the overhead system of a distillation system, which we will cover in more detail in Chapter 5. It is much more common to use a condensate drum for condensing process fluids than for condensing utility steam because of the potential impact of poor operation due to problems with the operation of vapor traps on the overall process economics.

The primary objective is to condense the process fluid. As long as there is some process liquid in the heat exchanger and some process vapor in the heat exchanger, this objective is satisfied. We can measure the level in the process fluid either on the shell or tube sides of the exchanger (wherever the process fluid resides) to insure this is occurring.

If there is a control valve on the process liquid outlet line, process fluid liquid level is often paired to it. Another option is to pair outlet liquid flow rate to this control valve. This is possible because the inlet stream is compressible. In a continuous process, where steady state is the goal, this second strategy works only if the specified liquid flow rate is nearly equal to the inlet gas flow rate. If not, eventually there could be too little inlet fluid to satisfy the outlet flow requirement and the liquid level in the heat exchanger will be lost. Conversely, if the outlet flow is smaller than the inlet flow, the heat exchanger could fill completely with liquid. A better solution is to use a cascade control loop with process fluid liquid level in the heat exchanger adjusting the setpoint for the outlet flow rate controller, which is similar to how we controlled the steam in the scheme shown in Figure 4.5. If a condensate drum configuration is used, then the process liquid level measurement is moved to the condensate drum.

A key to this type of heat exchanger is to insure that the process gas is condensing in the heat exchanger. If the vapor trap "sticks open" or the condensate pot outlet control valve sticks in a partially or fully open position, process gas may leave the heat exchanger. A low–low process liquid level measurement element should be installed for the process side of the heat exchanger when a control valve is installed on the heat exchanger process liquid outlet line or when using a vapor trap. When a condensate drum is used, the low–low level measurement is installed on the condensate drum. If the low–low level is reached, the process outlet stream needs to be blocked closed until the level builds back up. This can be accomplished by using an air-fail-close control valve or by installing a separate block valve in the same process outlet line. Figure 4.7 shows this full configuration for a typical system using a condensate drum.

Notice in Figure 4.7 that no temperature measurements are used in the control scheme. This is because the heat exchanger is primarily used to change phase, not to change temperature. Sometimes, in addition to changing phase, the heat exchanger is used to subcool the process liquid or superheat the process gas. In those cases, the controls would be configured like those described in Sections 4.1.1 and 4.1.2, except that the safety automation system would include low–low level action based on the process side of the heat exchanger.

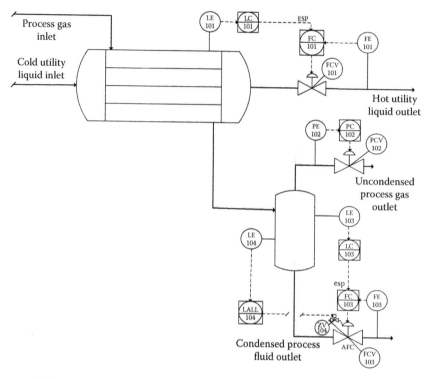

FIGURE 4.7
Control scheme for a heat exchanger that is condensing a process stream using cooling water and employing a condensate drum on the process outlet stream. *Note*: PCV-102 would normally have no flow. Also, the condensate pot has been drawn lower than the heat exchanger. In actual practice, the drum is usually installed such that the level in the drum is approximately equal to the level of condensed process fluid in the heat exchanger.

4.1.4 Process-Process Heat Transfer

Commercial processes typically require many process fluid heating and cooling steps. Sometimes, there is one process fluid that needs to be heated and another fluid (or the same fluid at a different point in the process) that needs to be cooled down at conditions that allow each process fluid to serve as the utility stream for the other process fluid. A common example occurs around chemical reactors. It is very common to adjust the temperature of the inlet reactant streams to a reactor so that the reaction will occur at the optimum temperature. Usually, this involves heating up the process fluid, since reaction kinetics are usually greater at higher temperatures. After the reactor, it may be desirable to cool the reaction product stream down prior to the next unit operation. In this case, the hot outlet reactor product stream and the cold reactor inlet stream may both be routed through the same heat exchanger (Figure 4.8). This type of heat exchanger is often known as a cross exchanger. A more generic representation of a cross exchanger is shown in Figure 4.9.

Heat Transfer Unit Operations Control

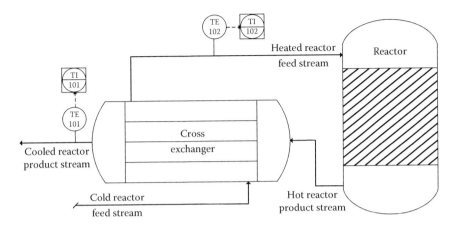

FIGURE 4.8
Exchanging energy between the inlet and outlet streams of a reactor in a cross exchanger.

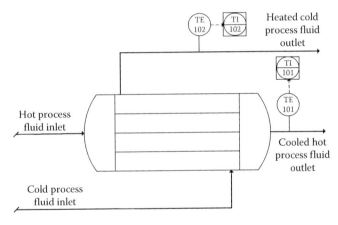

FIGURE 4.9
A more generalized process scheme for exchanging energy between two process streams in a cross exchanger.

The flow rates and inlet temperatures of the two process streams are usually fixed by adjacent unit operations (e.g., the reactor in the example above), so there are no independent variables in the heat exchanger. If the actual heat transfer between the two units is not critical, this may be acceptable, in which case there will be no controls required for the cross exchanger. However, it is more common that one of the two streams will have a requirement to add or lose more duty than the other stream.

To avoid overheating or overcooling the stream that requires less heat transfer, a bypass line is installed around the heat exchanger for the stream having the higher duty requirement. A control valve is then installed in the bypass line to provide one degree of freedom. The key parameter is the outlet

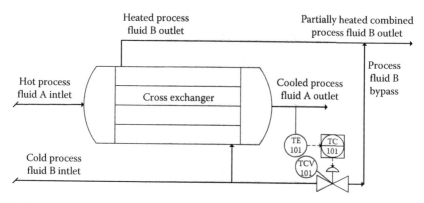

FIGURE 4.10
A cross exchanger where the energy available in process fluid B exceeds the required duty for process fluid A; a bypass is placed on the process fluid B side of the exchanger to avoid overcooling stream A.

temperature of the stream requiring less heat transfer. This temperature is paired with the bypass line control valve, as shown in Figure 4.10. Essentially, the fraction of the process stream with the higher heat transfer duty that flows through the cross exchanger is adjusted so that the duty within the cross exchanger matches the duty required by the other stream.

To completely satisfy the process requirements for the stream with the higher duty requirement, a second heat exchanger, known as a trim exchanger, is installed downstream of the cross exchanger. Usually, a utility fluid is used in the trim exchanger to bring this process stream to the correct condition.

Consider the following example. A stream containing pure hexane needs to be heated from 15°C to 40°C at a flow rate of 1000 kg/h. The energy that must be absorbed by this stream in J/h is:

$$q = \dot{m}\bar{C}_p \Delta T \tag{4.1}$$

where
q is the required heat duty
$\dot{m}$ is the mass flow rate = 1000 kg/h
$\bar{C}_p$ is the average heat capacity, which for hexane is around 2.26 J/(K kg)
ΔT is the absolute temperature difference in the fluid between the inlet and outlet conditions = 25 K

Using these data in Equation 4.1 yields a required heat duty of q_{hex} = 56.5 kJ/h.
Let us assume that we have another process stream, a weak acid stream with an average heat capacity of 3.9 J/(K kg) at 600 kg/h that needs to be cooled from 85°C to 25°C. Using Equation 4.1, the required heat duty is q_{WA} = 140 kJ/h.

There is more duty available in the weak acid stream than is required in the hexane stream. So if we use the configuration shown in Figure 4.10 with the hexane stream as process fluid A and the weak acid stream as process fluid B, fluid A's outlet temperature condition can be met.

What about fluid B, the weak acid stream? If we rearrange Equation 4.1:

$$\Delta T = \frac{q}{\dot{m} C_p} \qquad (4.2)$$

we can calculate the temperature of the combined weak acid stream after the cross exchanger:

$$T_{WA,out} = 353 \text{ K} - \left(\frac{56,500 \text{ J/h}}{(600 \text{ kg/h})(3.9 \text{ (J/K kg)})} \right) = 329 \text{ K} = 56°C$$

This is not the desired temperature of 25°C, so the weak acid will need to be cooled further in another heat exchanger. If another process fluid were available, it might be possible to cool the stream with a second cross exchanger. But a more common scenario is to remove the rest of the heat using a utility fluid in a trim exchanger. In this case, the weak acid could exchange energy with a refrigerated water stream in a trim exchanger to reach the desired condition.

The process scheme when a trim exchanger is used to cool the "hot process fluid" after a cross exchanger is shown in Figure 4.11. Notice we had to use a more expensive utility, refrigerated water, in the trim exchanger because the desired weak acid temperature is too low for most cooling water systems. The cold hexane stream enters the cross exchanger system at a low enough temperature but cannot accept all of the energy the weak acid stream needed to release. But there is still a solution: rearrange the exchangers! If refrigerated water is not available or is too costly, the trim exchanger can be placed before the cross exchanger, as shown in Figure 4.12. This introduces some lag time in the controls, but eliminates the need for the refrigerated water since the hexane inlet temperature is sufficiently low to drive the heat transfer needed to reach the weak acid outlet temperature.

As a general rule of thumb, there should be at least a 5°C difference in temperature between the two fluids at all points in the exchanger. This temperature is known as the approach temperature.

4.1.4.1 Cross Exchanger with Phase Change

Instead of temperature, one of the two fluids in the cross exchanger may be changing phase. If it is the stream with the smaller duty, the configuration is the same as that shown in Figures 4.10 and 4.11. However, instead of outlet

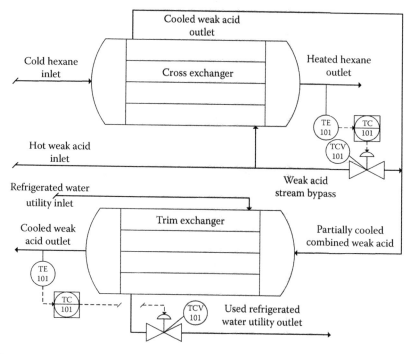

FIGURE 4.11
The complete heat exchanger system for the example of heating a hexane stream and cooling a weak acid stream when the weak acid stream has a higher duty requirement than the hexane stream.

temperature, liquid level (if the process fluid is condensing) or outlet pressure (if the process fluid is vaporizing) may be paired with the bypass line control valve.

Let us assume that in the example described above, 100 kg/h of hexane needs to be vaporized at 69°C. Equation 4.1 must be modified to account for latent rather than sensible heat transfer as follows:

$$q = \dot{m}\lambda \qquad (4.3)$$

where λ is the heat of vaporization, which for hexane at 69°C is around 365 kJ/kg.

Then q_{hex} = 3.7 MJ/h. If there is 20,000 kg/h of weak acid that must be cooled from 145°C to 85°C, then q_{WA} = 4.7 MJ/h. In this case, the process configuration is the same as in Figure 4.10 although the controls are slightly different. Since the goal is to vaporize liquid hexane rather than to heat up a liquid hexane stream, temperature is no longer an effective measurement variable. However, as long as there is some liquid in the heat exchanger as well as some vapor in the heat exchanger, the vaporization is occurring. So, a good control for this system is to measure the liquid level on the hexane side

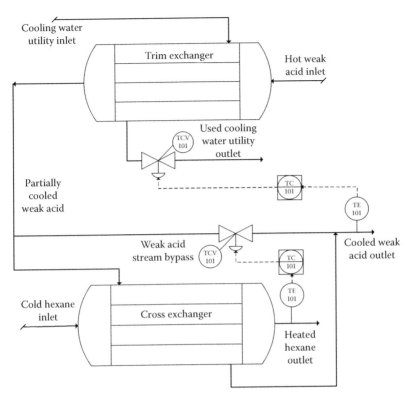

FIGURE 4.12
An alternate configuration for the heat exchanger system for the example of heating a hexane stream and cooling a weak acid stream when refrigerated water is unavailable or too expensive.

of the cross exchanger and use that level measurement to adjust the control valve on the bypass line for the weak acid side of the cross exchanger. This will insure that the correct quantity of weak acid enters the cross exchanger to match the duty requirement for the hexane vaporization. Since the hexane outlet stream is compressible, an additional degree of freedom is introduced into the system. A common control utilizing this degree of freedom is to measure and control the pressure of the hexane vapor leaving the heat exchanger.

As in the example described earlier associated with Figure 4.10, there is insufficient duty in the cross exchanger to meet the cooling requirement for the weak acid stream. If this temperature is important, a trim exchanger can be added, where utility cooling water is used to remove the remaining energy from the weak acid stream. The complete configuration is shown in Figure 4.13.

When the fluid changing phase is the one with the higher duty, then the configuration must be modified. In this case, the trim exchanger should be installed in parallel with the cross exchanger. A control valve is installed on the inlet to one of the two parallel heat exchangers (usually in the line with

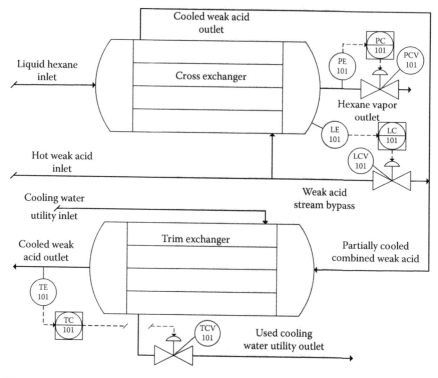

FIGURE 4.13
Process and control configuration for the heat exchanger system for the example of vaporizing a hexane stream and cooling a weak acid stream where the hexane stream has the smaller duty requirement.

the larger flow rate). This control valve can be paired with the outlet temperature of the process stream with the lower duty requirement. The liquid level (if condensing) or outlet pressure (if vaporizing) of the process fluid in the trim exchanger can then be used to control the utility stream in the trim exchanger. This insures maximum utilization of the cross exchanger, which will minimize energy costs.

If the higher duty stream fluid is condensing, then a vapor trap or condensate drum can be added either to the combined condensed stream after the parallel heat exchangers or, more commonly, to the individual condensed streams. Each exchanger should also have a safety system low-level control to insure that fluid is actually condensing in each heat exchanger, not just changing its temperature. Figure 4.14 shows this type of regulatory control scheme for the case where the cross exchanger has the larger flow rate and the larger duty process stream is condensing (note the low level safety system is not shown).

Sometimes, the heat transfer in a cross exchanger is limited by the temperature driving force between the two fluids. In other words, the cold stream

Heat Transfer Unit Operations Control

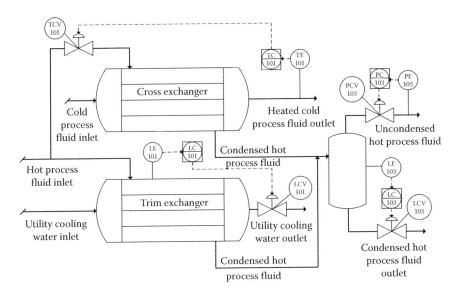

FIGURE 4.14
Exchanging energy between two process streams in a cross exchanger with a supplemental trim exchanger configured in parallel for the stream with the higher duty requirement to insure total condensation of the stream and employing a common condensate drum.

cannot heat to a high enough temperature and the hot stream cannot be cooled to a low enough temperature. For this case, both process streams will need to have a trim exchanger.

For this scenario, shown in Figure 4.15, all of the controls pertain to the trim exchangers and no controls will be needed on the cross exchanger.

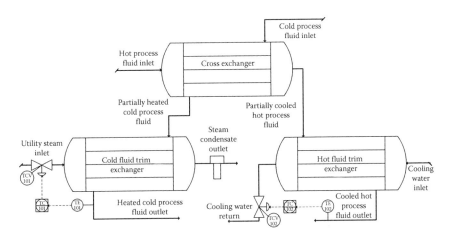

FIGURE 4.15
Exchanging energy between two process streams in a cross exchanger with two supplemental trim exchangers.

Usually, the trim exchanger is placed after the cross exchanger to reduce the signal lag between the final exchanger outlet temperature and the control valve on the utility stream, but there are circumstances where placing the trim exchanger first may be desirable. For example, if the hot process stream is at a high enough temperature, the hot-side trim exchanger can be placed on the upstream side of the cross exchanger in order to make steam. If the signal lag is too long, a cascade control loop can be used to make the controls more responsive.

4.2 Direct Mixing Heat Transfer

Sometimes, heat transfer is performed by direct contact of the heating or cooling utility with the process material. The most common application of this unit operation is when the process material is a solid. Consider a process where a solid has been precipitated out of a solution. Even after filtration, some residual liquid will be left in the material. Often this liquid needs to be removed. A common solution is to pass a hot, inert gas over or through the solid. If the gas temperature is higher than the boiling point of the entrained liquid, then as the solid heats, the liquid trapped or absorbed in the solid material will vaporize out of the solid.

Normally, the process solid material flow rate is specified by the upstream process. The only available independent variables are on the utility stream—flow rate, temperature, and pressure. Most commonly, the utility stream temperature is specified in the original unit's design. Since the utility stream is an inert gas, which is compressible, both inlet and outlet flow streams must be controlled. If a compressor/blower/fan is integrated with this unit operation, then one of these two control valves can be replaced by the speed control of the compressor.

In our example, the primary objective is to vaporize the liquid out of the solid until a target "dryness" is reached. This parameter can be difficult to measure and the measurement will be nonresponsive. Therefore, the inlet utility gas control valve is usually paired to the inlet or outlet utility gas flow rate. If the solid's dryness can be measured online, then this analysis can be used to adjust the setpoint of the gas flow rate as a cascade control loop. This scenario is shown in Figure 4.16 where the density of the solid is used to infer "dryness." When an off-line measurement must be used, this adjustment is handled manually by the control operator. Another option is to use the drying vessel's bulk or outlet gas temperature instead of hot gas flow rate, as shown in Figure 4.17.

If the solids flow rate can be easily measured, it can be used instead of the gas flow rate. More commonly, the solids flow rate will be used as the feedforward (FF) measurement variable in a FF/FB controller along with the

Heat Transfer Unit Operations Control

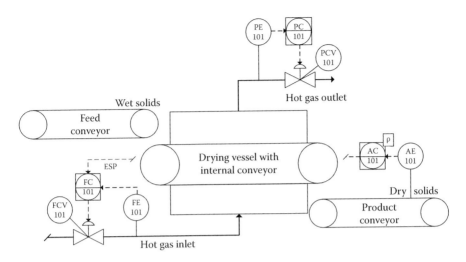

FIGURE 4.16
Using a hot inert gas to directly heat a solid.

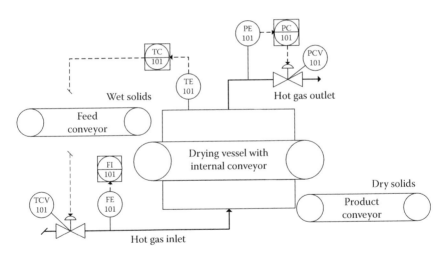

FIGURE 4.17
Using a hot inert gas to directly heat a solid with bulk dryer temperature control.

gas flow rate to adjust the gas flow rate control valve (Figure 4.18). Another option is to use a ratio controller that adjusts the gas flow rate to keep the ratio of gas-to-solids feed flow constant.

This leaves the outlet utility gas flow control valve. Since the utility fluid is compressible, the most common parameter is utility gas outlet pressure (see Figures 4.16 through 4.18). This insures that the utility gas actually flows through or over the material in the heat exchanger. Another option

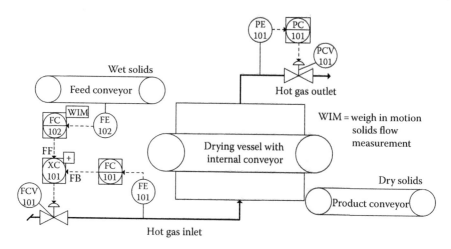

FIGURE 4.18
Scheme showing one way to control a hot inert gas used to directly heat a solid with weigh-in-motion solids flow rate used as a feed forward input to the hot gas flow rate control loop (WIM, weigh-in-motion solids flow measurement).

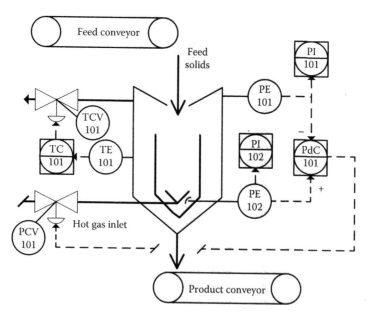

FIGURE 4.19
Using a hot inert gas to directly heat a solid via fluidization. The wet solids are fluidized by the hot dry gas. When the solids reach the required "Dryness," they rise out of the fluidization zone, fall down the sides of the vessel, and then exit from the bottom of the vessel.

is the temperature of the solid process material leaving the heat exchanger. This option is adopted when the material temperature is critical, such as in food processing or many pharmaceutical applications.

Consider as another example a solid material that needs to be cooled down. In this case, a cold inert gas may be passed over or through the material such that it will reach the desired temperature. The outlet process material temperature would be paired with the utility inlet control valve and the utility gas outlet pressure would be paired with the utility outlet control valve, which is very similar to the control schemes shown in Figures 4.16 through 4.18.

In some designs, heat transfer is improved by using the utility gas stream to fluidize a process solid that is in granular or particle form. In this case, the pressure drop across the vessel can be used as an indicator of fluidization efficiency. This pressure drop would be paired to the inlet or outlet utility gas control valve as shown in Figure 4.19. The other control valve would be paired to the heat exchanger's primary objective. This could be the temperature within the fluidization vessel (this option is shown in Figure 4.19) or a measure of the degree of "dryness" of the solid material.

4.3 Electrical Resistance Heat Transfer

Sometimes in applications where the process material must be heated, the heat generated by resistance to electrical current transfer through a wire is used as the utility heat source. Heat transfer occurs primarily by radiation, so these heaters are often known as radiant heaters. In some applications, the bare wires are directly exposed to the process material. This case is usually limited to the heating of solid materials.

More commonly, the wires are encased in a protective tube or sheath. The wires heat the sheath, which then heats the process material. Because heating is done in wires, many different shapes are possible. However, process control is performed similarly, regardless of the shape or installation of the heater.

As described in Section 2.1, there are two common ways to control the energy provided by the heater. The most common is to open and close a contact on the electrical supply so that the heater receives electric current for only a fraction of the time. By varying the fraction of time the heater is energized, the energy of the heater can be adjusted. The other common method uses an adjustable rheostat, a device that varies the resistance in the electrical circuit. By varying the resistance, the quantity of current that flows through the circuit, and thus through the heater, can be adjusted.

From a control perspective, the energy provided through the resistance heater is analogous to the energy provided using a utility fluid, such as steam. Instead of using a control valve on the utility stream to control its flow rate, the on/off time fraction or the rheostat setting is used as the independent

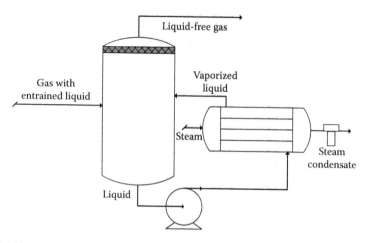

FIGURE 4.20
Using an external heat exchanger to vaporize liquid separated out of a gas stream.

variable. All of the control schemes described above for heating or vaporizing a fluid or a solid can be utilized with just this single replacement in the control scheme.

Consider the example of a liquid knockout drum. The basic control scheme is shown in Figure 2.4. It may be desirable to vaporize the liquid recovered in the drum. The liquid outlet of the drum could be pumped through a heat exchanger and controlled as shown in Figure 4.20. Alternately, if the bottoms fluid could be vaporized in the heat exchanger and then the process gas can be compressed in order to recycle it back into the drum. However, if the amount of liquid recovered in the drum is only a small fraction of the entire process stream entering the drum, another scheme can be used that eliminates the need for the outlet heat exchanger and compressor. We can install an electrical resistance heater in the bottom of the knockout drum. The liquid level in the drum can then be used to control the heater energy input (Figure 4.21).

4.4 Fired Heaters

If a process stream must be heated to temperatures that are too high to be reasonably reached using steam, a direct fired heater is often employed. The name comes from the fact that when energy is obtained in the form of heat from the combustion of the fuel, the hot flue gases (products of combustion) are directly used as the hot utility stream. In an indirect fired heater, the flue gas is used to heat or vaporize a separate utility stream (the most common being steam) that is then used for heat transfer applications.

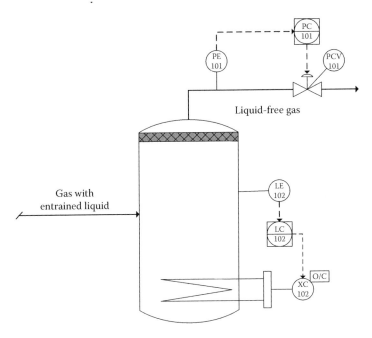

FIGURE 4.21
Using an electrical resistance heating coil to vaporize small quantities of liquid accumulating in a surge drum.

In the direct fired heater, the objective is to minimize the quantity of fuel that must be consumed in order to provide the energy required by the primary process fluid. The controls for a direct fired heater are the same as those described in Section 2.7 for a steam boiler. In this case, the fuel feed flow rate to the burner is controlled by the process fluid outlet temperature (if simply heating a fluid, see Figure 4.22).

Once the decision is made to utilize a direct fired heater, it is beneficial to recover as much of the energy out of the flue gas as possible. Once the flue gas has passed through the primary process fluid tube bank, it may still be at a fairly high temperature (limited by the approach temperature constraints of the process fluid inlet temperature). If there are other process streams that need to be heated, but at a lower temperature than the process fluid, these can be routed through a secondary tube bank located above the primary process fluid tube bank. In this case, there are no degrees of freedom available on the utility side of the heat exchanger. To add a degree of freedom, a bypass on the secondary process side can be installed. Figure 4.23 depicts the case where the secondary process stream does not change phase.

If there is too much duty available in the utility flue gas, the heater may be divided into two or more sections, with the secondary process tube bank

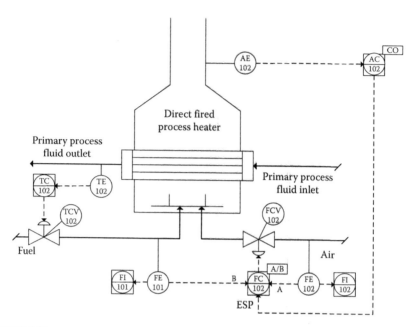

FIGURE 4.22
Control scheme when heating a process fluid in a direct fired heater.

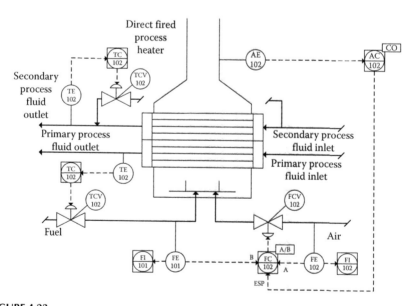

FIGURE 4.23
Direct fired heater with both primary and secondary process fluids; adding additional tube banks increases the amount of energy that can be recovered from the fired heater.

Heat Transfer Unit Operations Control

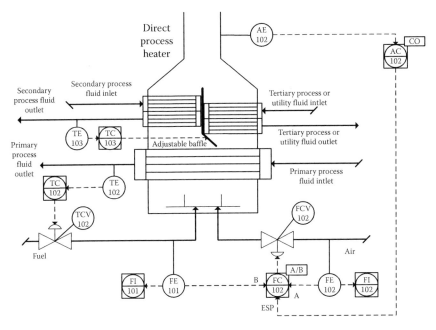

FIGURE 4.24
Heating a process fluid in a direct fired heater with excess heat routed through two secondary sections to heat two additional fluids. The quantity of flue gas going through the secondary sections is controlled using an adjustable baffle.

installed in only some of these sections. Baffles can be used to control the volume of flue gas entering the sections where the secondary heat transfer will take place. The secondary process fluid outlet temperature (no phase change) or pressure (vaporizing) can be paired with the baffle position control (Figure 4.24). In more complicated cases, multiple process streams may be heated in multiple sections.

If the flue gas entering the secondary heat transfer section of the heater is at a high enough temperature, tube banks containing BFW can be installed in the sections not used for secondary process fluid heat transfer to generate steam. Steam generation tube banks may also be directly used if no secondary process fluid heat transfer is included in the heater. To extract the maximum heat possible, a third layer of tube banks may be installed that is used to preheat the BFW at temperatures below the vaporization temperature. By using these extra tube banks, the flue gas temperature can be reduced to the range of 65°C–90°C (150°F–200°F), which provides just enough temperature to insure flue gas emission buoyancy while maximizing recovery of usable energy out of the stream prior to emission. This configuration with controls is depicted in Figure 4.25.

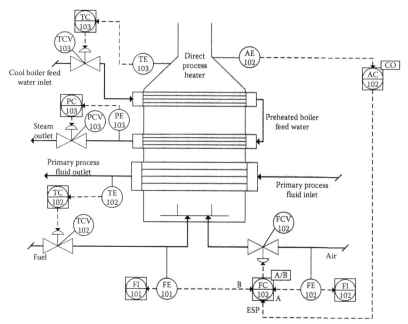

FIGURE 4.25
Heating a process fluid in a direct fired heater with excess heat used to generate utility grade steam.

4.5 Monitoring and Adapting for Heat Exchanger Fouling and Scaling

Indirect heat transfer requires that the process fluid(s) contact a solid surface so that energy can be transferred with another process stream or utility stream. In real processes, fluids are rarely completely free from impurities. As time passes, deposits or corrosion products can build up on the heat exchanger solid surface. For cooling water or processes that include a biological reaction step, bacteria can attach to the heat exchanger surface and then a colony of the bacteria can grow into a film. If the buildup is a mineral deposit, it is known as scaling. When the buildup is either an organic or biological deposit, it is known as fouling.

Fouling and scaling reduce the heat transfer efficiency of the heat exchanger. Since this reduction is time dependent, we need to make sure that we account for this phenomenon, where appropriate, in our heat exchanger control strategy. Fortunately, most of the control strategies we described above automatically adjust flow rates as efficiency decreases. For example, when cooling water is used to cool down a process stream, we pair the process stream outlet temperature to a control valve on the cooling water stream. Regardless of the exact heat transfer efficiency, the utility flow rate will be adjusted to

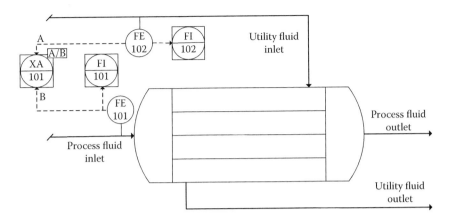

FIGURE 4.26
A simple heat exchanger fouling/scaling monitoring system using a XA, a calculated alarm block. If the ratio of utility to process flow rates exceeds a threshold value, the alarm is activated, indicating that the heat exchanger should be cleaned.

meet the cooling demand in the heat exchanger. This is true as long as we do not exceed the capacity of the cooling water supply system.

Our basic controls may adjust for fouling and scaling, but they do not mitigate the increased utility consumption (and therefore cost) that results. Thus, when we have a heat exchanger that is prone to fouling or scaling, it is usually good practice to monitor the heat transfer efficiency. When efficiency drops to a given tolerance level, the heat exchange function should be switched to a backup unit and the fouled exchanger taken out of service and cleaned. The simplest way to monitor this efficiency is to calculate the quantity of utility used per volume or mass of process fluid that moves through the heat exchanger. When the quantity gets too high, an alarm can be activated alerting the operators that it is time to switch exchangers. This simple system, shown in Figure 4.26, is adequate as long as the inlet temperature of both the process and utility streams do not vary substantially over time. If one or both vary substantially, a more complicated scheme is required. This scheme is described in Section 7.1.2 and shown in Figures 7.7 and 7.8.

4.6 Switching Controls for Parallel Heat Transfer Units

Sometimes, there are reasons to have banks of heat exchangers. For example, if they are used for an application where fouling or scaling occurs fairly rapidly. It can be useful to monitor the performance of the heat exchangers and to use that information to automatically switch from in-service to cleaning mode. Consider a bank of five parallel heat exchangers where four are in service and one is in cleaning/standby mode as shown in Figure 4.27.

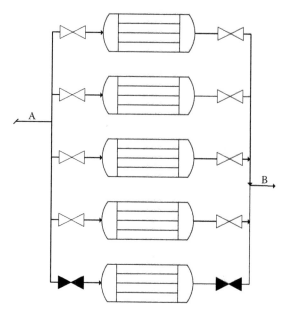

FIGURE 4.27
A bank of parallel heat exchangers with four in service and one serving as an installed spare.

A common mistake is to try to use a measure of the pressure drop across the individual heat exchangers. The logic here is that if the fouling or scaling occurs, it will lead to increased pressure drop through an exchanger. While this is true for any of the individual units, this strategy will not work when you have a bank of exchangers connected to common inlet and outlet manifolds, because, regardless of the condition within any specific heat exchanger, the pressure drop between the common manifold points on the inlet (labeled point A in Figure 4.27) and outlet lines (point B) will be identical. Instead, it is necessary to monitor the fraction of the mass or volumetric flow through each individual exchanger. Since the pressure will be equal through each pathway, the flow rate through each exchanger automatically adjusts to equalize the pressure drop by each pathway from point A to point B.

Thus, the key measurement parameter is flow rate rather than pressure. For this control, we measure the flow rate in each leg of the heat exchanger as well as the total flow in the common header (note: if there is substantial residence time in the heat exchangers, it may be more accurate to combine the individual flows together to give the total flow rate using calculation blocks; calculation blocks are described in Section 7.2). Then, we take the ratio of the individual flow to the total flow. When the fraction flowing through a given heat exchanger drops below the threshold value, a signal can be sent to a sequential logic controller to put the standby exchanger in service and take the fouled exchanger out of service. The scheme is shown in Figure 4.28.

Heat Transfer Unit Operations Control

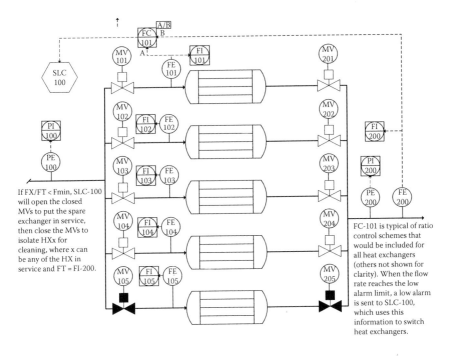

FIGURE 4.28
Scheme to monitor heat exchanger fouling or scaling and automatically swap in the spare for the fouled exchanger.

If fouling does not affect resistance to flow appreciably, then we can construct a scheme similar to the schemes discussed in Section 4.5. For example, we could measure the individual utility flow rate to each heat exchanger. As fouling increases, the quantity of utility needed to get the consolidated process outlet to the correct temperature increases. When the flow to one of the exchangers reaches a high limit, a signal can be sent to a sequential logic controller to switch the heat exchangers. The control through each exchanger can also be controlled using a special controller known as an allocation controller. This scheme is described in Section 7.6.

Problems

4.1 Describe the difference between heat transfer due to conduction, convection, and radiation.

4.2 A hot process fluid is cooled by vaporizing liquid nitrogen. Draw a schematic showing how to control such a system.

4.3 For the system in Problem 4.2, draw a schematic showing the safety automation system controls that should be provided.

4.4 A hot process fluid is cooled using a cold wastewater stream. There is more wastewater available than is required to perform the heat exchange. Draw a schematic showing how to control such a system.

4.5 For the unit operation in Problem 4.4, assume that the wastewater stream has insufficient pressure to route it through the heat exchanger and then on to its discharge location due to the pressure drop through the heat exchanger and downstream piping. Therefore, a variable speed pump is integrated into the system. Draw a schematic showing how to control the integrated pump/heat exchanger system.

4.6 For the unit operation in Problem 4.4, assume that the wastewater stream has sufficient pressure to route it through the heat exchanger but insufficient pressure after the heat exchanger to route the warm wastewater to its discharge location due to the pressure drop through the heat exchanger and downstream piping. Therefore, a variable speed pump is integrated into the system. Draw a schematic showing how to control the integrated pump/heat exchanger system.

4.7 Consider the process and control scheme depicted in Figure 4.3. Draw a schematic to show the analogous control scheme for the case where a condensate pot is used on the steam condensate return line instead of a steam trap.

4.8 Refrigerant loops use a thermodynamic cycle to provide cooling to temperatures below those that can be reached using cooling water. Consider the following refrigerant loop:

Cold ammonia refrigerant *vapor is heated* in a heat exchanger while *cooling* the process stream. The refrigerant then passes through a knockout drum to remove any possible entrained liquid from the warm refrigerant vapor. The vapor is then fed to a compressor to increase its pressure and temperature. The refrigerant is cooled and condensed in a heat exchanger using cooling water. The liquid ammonia is routed to a surge drum where any makeup refrigerant is added by pumping liquid ammonia into the drum. Liquid from the surge drum is adiabatically flashed across a valve that reduces its pressure and temperature. The flash pressure is correct if all of the liquid ammonia is vaporized during this activity. At this point, the refrigerant loop is complete and the cold ammonia vapor enters the heat exchanger to cool the process stream. Liquid collected *at a very slow rate* in the knockout drum is pumped to the surge drum.

Draw a schematic showing this system and its process controls.

Heat Transfer Unit Operations Control

4.9 For the system described in Problem 4.8, draw a schematic showing the safety automation system controls that should be provided.

4.10 For the system described in Problem 4.8, another way to recover the liquid collected in the knockout drum is to route a slipstream of hot compressor outlet vapor to the bottom of the drum and sparge the gas into the liquid. By doing this, some of the liquid will vaporize and the liquid level in the drum can be controlled by adjusting the rate of gas sparging into the drum. Draw a schematic showing how the controls would be modified for this option.

4.11 For the system described in Problem 4.8, another way to recover the liquid collected in the knockout drum is to install an electric heating coil in the bottom portion of the drum. By doing this, some of the liquid will vaporize and the liquid level in the drum can be controlled by adjusting the rate of energy input from the heater. Draw a schematic showing how the process and controls would be modified for this option.

4.12 For the system described in Problem 4.11, a steam heating coil might be more efficient for very large refrigerant loops instead of an electric heating coil. The steam heating coil operates the same way as steam utility in a shell and tube heat exchanger. The steam condenses in the coil as it heats and vaporizes the liquid in the drum. Draw a schematic showing how the process and controls would be modified for this option.

4.13 Consider the process and control scheme depicted in Figure 4.12. If the hot process stream is at a high enough temperature, some of the energy can be recovered in the form of useable steam. Draw a schematic to show this modified process and its control scheme. Include the safety automation system features as well as the routine process control features.

4.14 Consider the process and control scheme depicted in Figure 4.15. Draw a schematic to show how this scheme would be modified if the cold process fluid is partially vaporized in the cross exchanger and fully vaporized prior to exiting the trim exchanger.

4.15 Consider the process and control scheme depicted in Figure 4.13. Draw a schematic to show how this scheme would be modified if the hot process fluid were completely condensed in the cross exchanger.

4.16 A cold fluid is being vaporized by a system that consists of both a cross exchanger and a trim exchanger. The duty exchanged from the hot process fluid can vaporize most of the cold fluid with the rest accomplished using steam in the trim exchanger.

  4.16.1 Draw a schematic to show this scheme and its controls.

  4.16.2 Draw the safety automation system features.

4.17 For the system described in Problem 4.16, draw a schematic to show how you might modify this scheme if the hot process fluid is condensed in the cross exchanger.

4.18 Consider the process and control scheme depicted in Figure 4.15. Draw a schematic to show how this scheme would be modified if the hot fluid trim exchanger were used to generate usable steam upstream of the cross exchanger. (*Hint*: include a cascade loop(s) to minimize problems from time lag.)

4.19 A hot solid is to be cooled to a specified temperature using a cold nitrogen gas stream. The solid is conveyed through the cooler on a belt conveyor. Inside the cooler, the nitrogen passes over the solid countercurrent to the direction of solids flow. A recycle compressor is installed on the cooler nitrogen outlet line in order to provide the pressure required to move the nitrogen through the process side of the heat exchanger in a refrigerant loop (see Problem 4.8 for a description of such a refrigerant system) where it is cooled down to the target utility temperature. It is then recycled back through the solids. Draw a schematic showing the process and control scheme for this application.

4.20 Consider the process and control scheme depicted in Figure 4.22. Draw a schematic to show how this scheme would be modified if the primary process fluid is vaporized in the heater.

4.21 Consider the process and control scheme depicted in Figure 4.23. Draw a schematic to show how this scheme would be modified if the secondary process fluid is vaporized in the heater.

4.22 Consider the process and control scheme depicted in Figure 4.24. Draw a schematic to show how this scheme would be modified if a bypass is installed on the secondary process fluid line to allow control of the secondary process fluid temperature.

5

Separation Unit Operations Controls

One of the most common tasks in a process is to generate two or more streams out of one or more inlet streams. The whole point of a process is to transform one or more raw materials into one or more products, where the products have desirable properties compared to the raw materials. In some facilities, the primary objective of the process is to separate raw materials into various fractions that are more useful. For example, when crude petroleum oil is produced from a well, there is a fraction of the oil that is too volatile. If this fraction is not separated out of the oil, the oil will be unsafe for transport to the refinery. This fraction, known as the "associated gas," is typically removed from the oil by performing a single-stage separation, followed by a stripper (a distillation column with only the bottom section of the column). After the "associated gas" is removed from the crude, the gas can be fed to a process train where it is separated into natural gas (methane), ethane, propane/butane (liquefied petroleum gas, LPG), and heavier hydrocarbons. A combination of distillation and absorption is typically utilized to perform these separations.

In this chapter, we will look at how to control the most common unit operations used to perform separations.

5.1 Single-Stage Separation

5.1.1 The Flash Drum

In Chapter 2, we analyzed one type of single-stage separator in detail, the liquid knockout drum, which is typically performed in a pressure vessel having one inlet plus two outlets—one for the gas phase and one for the liquid phase. You might want to go back and review that material before continuing with this section.

A similar, slightly more complicated separation is the single-stage flash separator, known as a "flash drum." In this case, we will use the same equipment, a pressure vessel, to form a gas/liquid mixture that is at or near vapor-liquid equilibrium (VLE). This is known as a single VLE stage separator. In the most common case, the pressure vessel is well insulated so that it can be assumed that the unit operation is adiabatic. This is most commonly denoted as an adiabatic flash separator. The inlet stream is most commonly a liquid, although a vapor feed or two-phase feed can also be used.

Considering the case of a liquid feed, the pressure of the adiabatic flash separator is reduced substantially from the upstream process. At the lower pressure, a portion of the liquid will vaporize. The energy drawn from the fluid for vaporization will reduce the temperature of both the gas and liquid phases in the vessel. Because the flash is adiabatic, temperature and pressure are interrelated. If we specify one of these in our controls, the related condition of the other parameter is achieved. We cannot control them independently because they are related variables (see Section 2.6 for a discussion of related variables).

There are three independent variables:

1. The mass of material entering through the feed nozzle, M_I
2. The mass of material leaving through the top outlet nozzle, M_{O1}
3. The mass of material leaving through the bottom outlet nozzle, M_{O2}

Just as in the case of the knockout drum, the inlet mass flow rate is typically set by the upstream process. This leaves the same two independent variables for us to use—the top and bottom outlet streams. Thus, we specify control valves (CVs) on each stream. The primary objective of the unit is to separate the inlet fluid into two streams—one gaseous and one liquid. If the inlet stream is a single component (say water), the objective of the flash drum is to vaporize a portion of the stream. This is most commonly measured by taking the ratio of the mass in the overhead vapor to the mass of the inlet stream. This parameter is known as the flash ratio, f. For any given set of inlet conditions, there is only one pressure/temperature condition in the vessel for a given value of f.

If the inlet stream has multiple components, the objective is usually to generate two streams with different compositions, based on the VLE established by the flash operation. Since there is only one stage of separation, for any given pressure/temperature condition in the flash drum there can be only one composition (assuming a constant inlet composition). Therefore, if we measure the concentration of one component in either stream, this can be used to control the composition of both streams. Which stream we sample and what component we measure are case dependent. For example, we may want to measure component X in the liquid stream, or we may want to measure component Y in the gaseous stream. It may not be possible to perform an online measurement at all. We may need to take a sample to the laboratory. In this case, we would want to use an indirect measurement that can be correlated to composition. Two are obvious: temperature and pressure.

What other objectives are there for the flash drum? First, we want to insure good separation of the two fluids. That is, we want only liquid to leave from the bottoms outlet and only gas to leave from the overhead outlet. Second, we want to insure that there is adequate pressure to transfer the fluids to

Separation Unit Operations Controls

their next unit operations. Sound familiar? These are the same secondary objectives that we developed for the liquid knockout drum in Chapter 2. There is one important difference in this case—pressure is no longer an independent parameter. Instead, pressure is linked to temperature, f, and to the composition of the two phases.

One way to specify the controls is exactly the same as for the liquid knockout drum (see Figure 2.4). This scheme is commonly used for a single-component fluid flash, since pressure is much easier to measure than f and also more responsive. Since the parameters are related, either pressure or f can be used. We could also use temperature, but since pressure is the simplest and most responsive, it is normally chosen. If we did want to directly use f, we could use a ratio controller (see Section 2.7 for a description of ratio controllers). In this case, we would measure the overhead flow rate and the inlet flow rate. Both measurements would be routed into a ratio controller to calculate $f = m_{out,1}/m_{in}$. The ratio controller can then directly manipulate the overhead line CV. Usually, we use pressure for control but calculate f from the flow measurements as additional information that is useful in operational assessment and troubleshooting.

For multicomponent fluids where composition is important, a correlation between the key component composition and pressure is derived. In this case, pressure is an indirect measure of the true parameter, composition. However, since the variables are related, it can provide proper control. Even if we can measure composition, pressure is a more responsive parameter (see Section 1.5 on the dynamics of process parameters) and thus is preferred. In this case, we might want to use the analysis as the master loop to provide an external setpoint to a "slave" pressure control loop in a cascade control scheme (see Section 2.8 for a description of cascade control loops).

For the other independent variable, the most common scheme is to use the liquid level in the flash drum as the dependent variable for the control loop managing the CV on the liquid outlet line. If we use pressure or f to control the overhead flow, then holding the liquid level constant will directly adjust the bottoms flow rate to account for any changes in the fraction of the inlet fluid that is leaving as liquid after the flash. The flashing of a liquid from high to low pressure can introduce dynamics into the process. In fact, it can sometimes be difficult to get an accurate level measurement due to movements in the liquid surface in the flash drum. Fortunately, the controls will normally be effective as long as the level is kept approximately constant. So another advantage of using level control is its ability to dampen out dynamics from the liquid. If we need to have a more careful control of the liquid outlet stream, we can add a "slave" flow control cascade loop to the "master" level control loop. This will smooth out the flow to the next unit operation.

Sometimes it is not possible to get a reasonable level control measurement due to extreme dynamics in the flash drum. In this case, outlet flow control can be specified. Since the liquid level is not directly used in the control, it is important to find a spot to install a level measurement element or switch that will alarm before vapor enters the liquid outlet stream. This is often accomplished by adding a liquid level "pot" to the bottom of the flash drum. Figure 5.1 shows one control scheme for a flash drum with a liquid level "pot." In this scheme, analysis of one component in the overhead vapor is linked in a cascade control scheme with pressure for the overhead CV and flow control is linked to the bottoms CV. Low level and low-low level alarms are also included to avoid sending vapor into the liquid outlet line.

Now we have seen how the same basic unit operation can be used for two completely different purposes. However, their regulatory control scheme can be exactly the same. While this is often the case, there are many exceptions. It is recommended that you always go through a detailed analysis, similar to the one we performed in this section, to insure that you specify the best control scheme for your application.

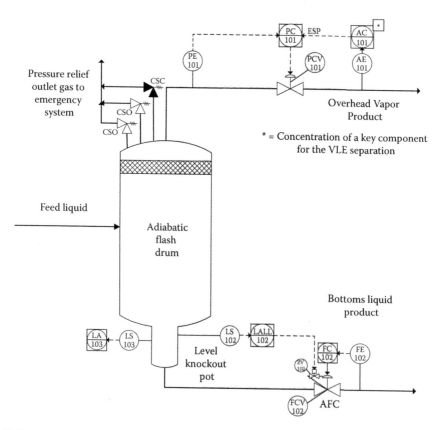

FIGURE 5.1

Control scheme for an adiabatic flash drum with a liquid level pot using the analysis of a key component and flow rate as the dependent variables. * denotes the concentration of a key component for the VLE separation.

5.1.2 The Phase (Gravity) Separator

The knockout drum we analyzed in Chapter 2 is a type of phase separator where the majority of the mixed phase inlet fluid is in the vapor phase, whereas the flash drum described in the last section is usually, but not always, a liquid feed. Another unit operation that uses a single-stage separation in a pressure vessel is the phase or gravity separator, which describes a single-stage separator where a significant fraction of the mixed phase inlet fluid is in the liquid phase. In fact, the inlet could be two liquid phases—organic and aqueous—with no vapor phase at all. Alternately, the inlet could have three phases—vapor, organic liquid, and aqueous liquid. Yet another variation could be an inlet stream that contains mostly liquid but also some solid materials.

For the configuration that has a mixed inlet phase with a significant fraction of a single liquid phase and the balance vapor, the controls are identical to those we outlined for the knockout drum. The relative amounts of vapor and liquid do not really affect the basic control strategy. So for these types of separators, the most common control configuration is pressure control on the vapor outlet stream and liquid level control on the liquid outlet stream (Figure 2.4).

Now consider a phase separator where the inlet is all liquid and where the liquid will partition into two immiscible phases, one organic (nonpolar) and one aqueous (polar), in the separator. A mass balance around the separator is:

$$M_I = M_{O1} + M_{O2} + M_A \tag{5.1}$$

where
  I is the inlet
  O1 is the organic outlet
  O2 is the aqueous outlet
  A is the accumulation or depletion

Equation 5.1 sets the relationship to control the accumulation/depletion term and we can assume that the inlet feed rate is set by the upstream unit operation. Thus, there are two independent variables for this unit operation, the two outlet streams. A common control configuration would install CVs on each outlet, unless there is a need to add a pump to move the outlet fluid to the next unit operation. In that case, pump speed control might replace a CV in the control scheme on either or both outlets.

The primary objective of the two-phase separator is to produce an organic liquid that contains the minimum concentration of polar species and to produce an aqueous liquid that contains the minimum concentration of nonpolar species. The ideal minimum concentrations are determined by the liquid-liquid equilibrium (LLE) compositions of the two phases. The time required to achieve near equilibrium conditions is a physical feature of the system and cannot be addressed by the control system.

Secondary objectives are to: (1) maintain the interface level between the phases such that no aqueous phase material enters the organic liquid outlet

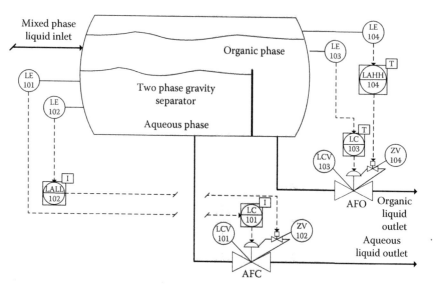

FIGURE 5.2
Control scheme for a two-phase separator where the organic phase has a lower density than the aqueous phase. I, interface level; T, total level.

and no organic phase material enters the aqueous liquid outlet and (2) maintain the total level in the separator such that the vapor space within the vessel is not pressurized beyond the limits of the vessel. Depending upon the composition of the organic phase, it will either have a greater or lesser density than the aqueous phase. The most common case is an organic phase with a lower density than the aqueous phase. In this case, we construct a level control loop that links the interface level measurement to the CV on the aqueous outlet line and a second level control loop that links the total level measurement to the CV on the organic outlet line. If a pump is included on either or both outlet lines, the appropriate level might be linked to the pump's speed or on/off control circuit instead of a CV.

The safety automation system would include an LALL, which would initiate the closure of the CV on the aqueous line, and an LAHH, which would initiate the opening of the CV on the organic line. A complete full control system for the case where the organic phase has a lower density than the aqueous phase, with no integrated pumping requirement, is shown in Figure 5.2.

5.2 Multistage Distillation Overall Concepts

The goal of a distillation system is to use the differences in the VLE compositions of a fluid containing two or more different components. Gas travels up the column through a series of stages, which may be trays or sections of packing. At each stage, the gas contacts a liquid that is traveling down the column. The gas and liquid are well mixed in the stage, allowing mass transfer between the gas and liquid phases to occur so that the overall mixture on that stage approaches VLE. As the vapor rises through the column, the concentrations of the more volatile compounds in the fluid are enriched and the concentrations of the less volatile compounds in the fluid are decreased. The opposite occurs to the liquid as it falls through the column.

VLE changes strongly as a function of temperature and weakly as a function of pressure. Therefore, the temperature of the stages increases from the top of the column to the bottom (sometimes substantially), while pressure increases slightly. The feed material enters the column at the point that is most effective in making the separation. A simple schematic is shown in Figure 5.3.

In a stripping column, the feed is a liquid and enters the column at the top. The feed provides the liquid that travels downward through the column. Some of the liquid leaving the bottom of the column is vaporized in a heat

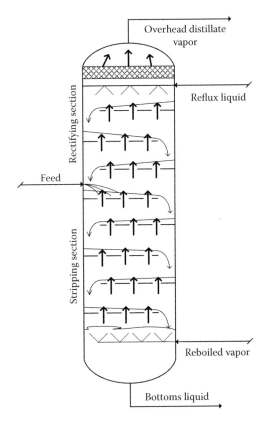

FIGURE 5.3
Vapor-liquid traffic in a trayed distillation column.

exchanger known as the reboiler. This vapor is then recycled back into the bottom of the column and provides the gas that travels upward through the column. In some configurations, a portion of the column bottoms liquid is routed directly to the next unit operation as the column bottoms product and a portion enters the reboiler, where all of the liquid entering the reboiler is vaporized. This is known as a total reboiler. In other configurations, all of the column bottoms liquid is routed to the reboiler, but only a portion of the liquid entering the reboiler is vaporized, creating a VLE system in the reboiler. This type of reboiler is known as a partial reboiler. Since the entire liquid enters the partial reboiler, this configuration adds one equilibrium stage to the separation process to supplement the stages in the column. These configurations are shown in Figure 5.4.

In a rectifying column, the feed is a vapor and enters the column at the bottom. The feed provides the vapor that travels upward through the column. Some or all of the gas leaving the top of the column is condensed in a heat exchanger known as the overhead condenser. The outlet of the condenser is sent to a reflux accumulator drum. If only a portion of the vapor is condensed in the overhead condenser, it is known as a partial condenser. The two-phase mixture is separated in the reflux accumulator drum with the vapor phase being the overhead product of the separation system. The liquid from the drum is routed to a pump that provides the energy to return the liquid to the top of the column. This liquid is known as the reflux and provides the liquid that travels downward through the column. The partial condenser adds one equilibrium stage that supplements the stages in the column.

When all of the vapor is condensed in the overhead condenser, it is known as a total condenser. The liquid from the overhead condenser is routed to

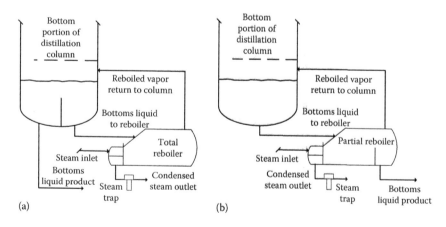

FIGURE 5.4
The bottoms system with adequate liquid level to provide pressure for the vapor to return to the distillation column: (a) total reboiler option and (b) partial reboiler option. *Note*: This type of bottoms system is known as a thermosyphon reboiler system.

the reflux accumulator drum where the liquid level in the drum is used to dampen out dynamics from the overhead portion of the distillation column. Part of the liquid from the drum is routed to the next unit operation as the overhead product, while the remainder is pumped back into the top of the distillation column as the reflux.

These two overhead column configurations, partial and total condensation of the column overhead vapor, are shown in Figure 5.5.

In a full distillation system, the feed enters somewhere between the top and bottom of the column and may be vapor, liquid, or a mixture. The stages above the feed point are known as the rectifying section of the column. An overhead condenser/reflux accumulator drum/reflux pump system is provided to process the overhead gas from the column and to provide the reflux liquid necessary for the stages above the feed point. If the reflux rate is sufficiently high, this liquid may also supplement the liquid from the feed to provide the liquid in the bottom portion of the column. The vapor traveling up the column comes from the feed and/or from vapor coming from the bottom portion of the column.

The portion of the distillation column below the feed point is known as the stripping section. The liquid traveling down the stripping section comes from the feed and/or from liquid coming from the rectifying section. A reboiler is provided at the bottom of the column to process the bottoms liquid and to provide the reboiled vapor return necessary for the stripping section of the column plus any supplemental vapor needed in the rectifying section of the column.

Ultimately, the objective of this separation unit operation is to split a single fluid stream containing at least two components into two streams (the overhead product stream and the bottoms product stream) that have different compositions. In the most rigorous case, a full distillation system is used to generate an overhead product stream with a very low concentration of "heavy" (in terms of VLE) components and a bottoms product stream with a very low concentration of "light" components. A simple example would be a feed stream containing pentane and hexane. A full distillation system might be used to generate an overhead stream with a maximum concentration of 2 wt% hexane and a bottoms stream with a maximum concentration of 2 wt% pentane.

Now consider this same example, but with a feed stream containing butane, pentane, hexane, and heptane. A full distillation system might be used to generate an overhead stream with a maximum concentration of 2 wt% hexane and heavier (hexane and heptane) and a bottoms stream with a maximum concentration of 2 wt% pentanes and lighter (pentane and butane). This is known as a multicomponent distillation system. For mixtures such as this (or more complex) that have more than two components, the goal of the separation is to essentially separate between a "light key," in this case pentane, and a "heavy key," in this case hexane.

In other cases, the goal might be to purify only one of the two product streams. While this may be true in a full distillation application, it is more

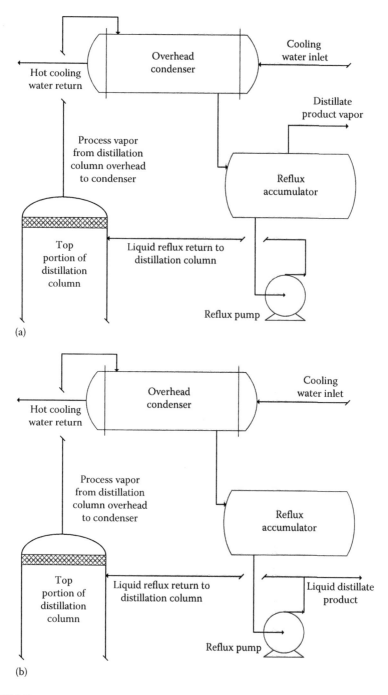

FIGURE 5.5
The overhead system for a distillation column: (a) partial condenser option and (b) total condenser option.

common for stripper systems where only the bottoms purity is tightly controlled and rectifying systems where only the top product purity is tightly controlled.

Because VLE is related to the temperature and pressure of the system, there are fewer degrees of freedom in a distillation system than you might think. First, using a mass balance around the entire system yields:

$$M_F = M_D + M_B + M_A \tag{5.2}$$

where
  F is the feed
  D is the overhead product
  B is the bottoms product
  A is the accumulation

As in previous analyses, we can assume that the feed rate is specified by an upstream unit operation. Equation 5.2 can then be used to calculate M_A from two independent variables, M_D and M_B.

In addition to these two overall independent variables, we need to analyze the intermediate systems to look for additional variables. A mass balance around the overhead condenser/reflux drum/reflux pump portion of the distillation system (Figure 5.5) yields:

$$M_O = M_D + M_R + M_{A,O} \tag{5.3}$$

where
  O is the column overhead vapor
  D is the overhead product
  R is the reflux recycled to column
  A,O is the accumulation of the overhead material

Equation 5.3 shows that the accumulation or depletion of overhead material is related to the amount of vapor traveling up the column and the amount of material leaving the reflux accumulator drum as either overhead product or reflux liquid. M_O is set by the column operation, which leaves M_D and M_R. We already identified M_D as an independent variable from the overall mass balance, so this equation adds one additional independent variable, M_R.

There is one more step needed for the overhead portion of this system; we need to perform an energy balance around the overhead condenser. An analysis for this type of indirect heat transfer is provided in Section 4.1. To summarize, there is either one or two additional independent variables depending upon the utility fluid. If there is no utility fluid phase change and the utility fluid is a liquid (the most common case, with cooling water as the most common utility fluid), then there is one independent variable and a CV is typically installed on the utility fluid outlet. If the utility fluid is a gas, then there are two independent variables and CVs are typically installed on both the utility fluid inlet and outlet lines. When a utility fluid is vaporized in the condenser (most commonly boiler feed water for high-temperature distillation systems or liquid nitrogen for cryogenic systems), there are two independent variables and CVs are typically installed on both the utility fluid inlet and outlet lines.

Similarly, if we perform a mass balance around the reboiler (Figure 5.4):

$$M_{CL} = M_B + M_V + M_{A,C} + M_{A,R} \quad (5.4)$$

where
 CL is the liquid leaving the bottom stage in the column
 B is the bottoms product
 V is the reboiler vapor return to the column
 A,C is the accumulation in the bottom of the column
 A,R is the accumulation in the reboiler

In this case, there are two places where accumulation occurs, one is the liquid hold-up in the bottom of the column and one is the reboiler. Some reboilers are configured to use the potential energy (the liquid head) in the column to provide the pressure necessary to drive the reboiler vapor back into the column, as shown in Figure 5.4. For these systems, the liquid level in the bottom of the column will automatically adjust to provide the pressure driving force necessary and thus is not independent.

Other distillation systems use a pump on the bottoms liquid to increase the pressure of the liquid entering the reboiler, which is then maintained at a high enough pressure so that the vapor can get back into the column. For these systems, the accumulation in the bottom of the column is an extra degree of freedom.

M_{CL} is set by the column operation, which leaves M_B and M_V. M_B has already been identified as an independent variable from the overall mass balance, so this equation adds one (or two if the reboiler is a forced type; one using a pump) additional independent variable, M_V.

We also need to perform an energy balance around the reboiler. Again, please refer to Section 4.1 for a detailed analysis of indirect heat transfer unit operations. To summarize, there will be either one or two additional independent variables based on the nature of the utility stream. When there is one additional variable, a CV is placed on the outlet of the utility stream. When there are two additional variables, CVs can be placed on both the inlet and outlet streams. The outlet CV may actually be a "steam" or "vapor" trap, which is an integrated mechanical level control loop.

Now let us identify common, reliable control schemes for distillation systems. Control schemes will be most stable if we keep the overhead and bottoms systems decoupled (see the end of Section 2.3 for a discussion of coupled vs. decoupled systems), so each of these will be considered separately.

5.2.1 Overhead System Controls for Distillation

The primary objectives of the overhead system may vary from application to application, but the most common are: (1) to generate an overhead product with a concentration of the "heavy key" that is at or below the purity specification and (2) to generate sufficient refluxing liquid to provide adequate liquid for the column. Another common set of objectives is to: (1) generate a sufficient quantity of overhead product to satisfy the downstream processes and (2) generate sufficient refluxing liquid to provide adequate liquid for the column. Secondary objectives of the overhead system are to provide adequate pressure in the overhead product for the downstream process and adequate pressure in the reflux liquid to return the liquid to the column. When the utility fluid is vaporizing or is a gas phase stream, an additional secondary objective is to provide adequate pressure in the utility outlet stream for the downstream utility requirements.

As described above, the overhead portion of the distillation system has either three or four independent variables: (1) the liquid reflux recycled to the column, (2) the distillate (overhead) product, (3) the utility fluid to/from the overhead condenser, and (4) the utility fluid from/to the overhead condenser if the utility fluid is compressible or undergoes a phase change.

The most important measurement variable in the overhead system is the composition of the heavy key in the overhead product (or flow rate of the overhead product if composition is not important). Since most analyses are nonresponsive variables, a cascade control system is used with online analysis when possible. When analysis is not possible, we try to find an indirect measure that correlates well in this application to composition. The decision of which independent variable to link with the overhead composition (or flow rate) depends upon whether the system has a total or partial condenser and if it has a total condenser, on the reflux ratio, as discussed below.

5.2.1.1 Overhead System with a Partial Condenser

When employing a partial condenser (Figure 5.5a), a good, stable control scheme is to cascade the master online analysis loop to a temperature control loop. The temperature measurement is the overhead vapor product from the reflux drum that is linked to the utility flow rate on the condenser. When online composition is not possible, overhead temperature is frequently used as the indirect correlating variable. Both of these work well because the VLE in the reflux accumulator varies with changes in the vapor and liquid temperature. If the utility stream in the condenser is a liquid that does not change phase, there is one CV (usually on the outlet line) and this is utilized in this loop. When the utility stream is a vapor or is a liquid that changes phase to a vapor (e.g., liquid methane in a very cold application), the CV on the inlet line is used in this control loop. The utility vapor outlet pressure is then linked to the utility stream outlet CV to insure that the utility vapor stream meets downstream utility requirements.

Continuing this control scheme, the liquid level in the reflux drum is used as the master control loop in a cascade with a reflux flow loop as the slave. The reflux flow rate is linked to the CV on the reflux return line (or to the speed control on the reflux pump). The pressure in the reflux drum vapor space is then linked to the CV on the overhead vapor product line to insure that the product meets the requirements of the downstream process.

One variation of this control scheme is shown in Figure 2.18.

5.2.1.2 Overhead Control Scheme with Total Condenser

With a total condenser (Figure 5.5b), the entire distillation overhead vapor is condensed and routed to the reflux drum. The outlet liquid from the reflux drum is divided into two streams: the reflux stream and the overhead product stream. For this scheme, the temperature of the condenser outlet stream is a poor control parameter since the vast majority of the energy removed from the overhead stream is the result of the phase change of the vapor to liquid. As long as there is liquid on the process side of the overhead condenser, the exchanger is working correctly. So a common scheme would include measuring the liquid level of the process liquid formed in the overhead condenser and using this level measurement to control the utility flow rate of the condenser. Since level is not a responsive variable, it is common to cascade the process fluid level control loop to a slave control loop using the cooling water flow rate to the utility flow rate CV. If the utility stream is a liquid and does not change phase, this CV is usually on the outlet utility line from the condenser. If the utility stream is a gas or changes from liquid to vapor in the condenser, then the flow CV is normally on the condenser utility inlet line. The utility vapor outlet pressure is then linked to a CV on the condenser utility outlet line. This scheme is shown in Figure 5.6.

The two most important parameters for the reflux drum are: (1) the composition of the heavy key in the overhead product and (2) the liquid level in the reflux drum (which is related to the total overhead flow rate from the column). To assist in dampening out any upsets in the overhead system (since it includes a recycle stream), the level control is purposely made "sluggish" by loop tuning. That is, the loop is nonresponsive. However, these types of level control loops tend to make large changes to the associated control variable (CV or pump speed control). These large changes can cause upsets in the system, the very thing we were trying to avoid by making the loop sluggish! We can do two things to help counter this. First, we can make adjustments from the level controller to the larger of the two outlet streams—either the reflux return or the overhead product stream. Second, we can add a flow cascade to help smooth out the adjustments. In other words, link the level control loop to the CV with the largest flow rate.

So to insure that this smoothing out of dynamics is effective, the level measurement and controller should be linked to the CV on the reflux return via a flow cascade if the ratio of reflux to overhead product is greater than 1 but

Separation Unit Operations Controls

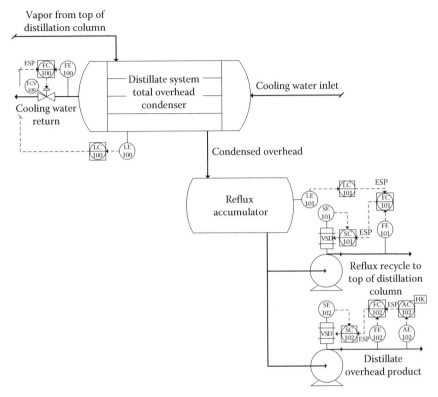

FIGURE 5.6
Control scheme for the overhead portion of a distillation system employing a total condenser, having a reflux ratio ≥1.0, and two independent pumps with variable speed drivers. HK is the heavy key.

to the overhead product CV if the ratio of the reflux to overhead product is less than 1.

If an online measurement can be made of the heavy key or an indirect parameter that correlates to the heavy key, this measurement would be linked through a cascade control scheme to a slave flow control loop associated with the reflux drum's other outlet stream. If the reflux rate is greater than one, this will be the flow control loop for the overhead product stream. If the reflux rate is less than one it will be the flow control loop for the reflux return stream. The slave flow control loop would link a flow measurement on the same stream as the CV associated with this cascade control scheme.

There will be at least one pump associated with the reflux outlet flows. Sometimes, one pump is used for both the reflux return and overhead product flows. In this case, speed control of this pump may replace the CV on the smaller of the two flow lines. This scenario is illustrated in Figure 2.20. At other times, the reflux drum pressure may be sufficient to deliver the overhead product to the next unit operation and the pump will only be located on the reflux return line. In this case, speed control of this pump may replace the CV on the reflux return line.

There are also situations where both the reflux and the overhead streams need a flow boost, but at very different conditions. In these cases, two pumps are installed. Usually, one pump is placed on each stream after they have divided. In rare cases, one of the pumps is placed on the common line and then a booster pump is put on the line that needs the additional flow/pressure boost. For all of these cases, pump speed controls can replace both CVs. The version employing two independent pumps with speed control is shown in Figure 5.6.

5.2.2 Bottoms System Controls for Distillation

The primary objectives of the bottoms system may vary from application to application, but the most common are: (1) to generate a bottoms product with a concentration of the "light key" that is at or below the purity specification and (2) to generate sufficient reboil vapor return to provide adequate vapor for the column. Another common set of objectives is to: (1) generate a sufficient quantity of bottoms product to satisfy the downstream processes and (2) generate sufficient reboil vapor return to provide adequate vapor for the column. Secondary objectives of the bottoms system are to provide adequate pressure in the bottoms product for the downstream process and adequate pressure in the reboiler to return the reboil vapor to the column. When the utility fluid is condensing, an additional secondary objective is to insure that utility vapor does not pass through the exchanger without condensing. If the utility fluid is a gas phase stream that does not change phase, an additional secondary objective is to provide adequate pressure in the utility outlet stream for the downstream utility requirements.

There are three common configurations: (1) separate bottoms product stream with a total reboiler (Figure 5.4a), (2) combined column bottoms stream feeding the reboiler that partially vaporizes the liquid, and (2a) the liquid head in the bottom of the distillation column is sufficient to provide the pressure needed to return the reboil vapor into column just below the lowest stage (Figure 5.4b) or (2b) there is a pump in the common bottoms line that provides the pressure needed to return the reboil vapor into the column (Figure 5.7). In any of these configurations there may be a pump on the bottoms product stream if required to get the fluid to the downstream process.

The most important dependent variable in the bottoms system is the composition of the light key in the bottoms product (or flow rate of the bottoms product if composition is not important). Since analyses are usually nonresponsive variables, a cascade control system is often used. When analysis is not possible, we try to find an indirect measure that correlates well in this application to composition.

When a total reboiler is utilized, the composition is linked to a slave control loop. The most common configuration uses the temperature of the bottoms

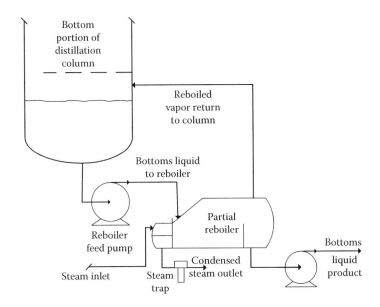

FIGURE 5.7
The bottoms system of a distillation column with a partial forced reboiler system (employing a pump to provide the pressure to get the vapor back into the column) and bottoms product transfer pump.

liquid linked to the CV located on the bottoms product line. The liquid level in the reboiler is then linked to the utility fluid flow CV. If the reboiler utility is a condensing fluid (like steam), this CV is located on the utility inlet line. A steam trap or a condensate pot with a level control loop is used on the utility outlet line. If the reboiler utility is a vapor stream that does not change phase, then the utility outlet pressure is linked to a CV on the utility outlet line to insure that the vapor leaving the reboiler meets the downstream utility requirements.

If a thermosyphon reboiler is used (the case where there is sufficient pressure in the bottom of the column to return the vapor to the column), then the liquid level in the bottom of the column is automatically adjusted in height to make the reboil recycle system work. When a forced reboiler is used, the column liquid level is linked to the reboiler pump speed control (or to a CV on the pump outlet line if a fixed speed pump is employed).

When a partial reboiler is utilized, there are two options for pairing the key measurement variables, bottoms product composition and reboiler process liquid level, to the two control variables. Just as was the case for the reflux drum, partial reboiler level control loops tend to make large changes to the associated control variable (CV or pump speed control). These large changes can cause upsets in the system. We can do two things to help counter this. First, we can make adjustments from the level controller to the larger of the two outlet streams (in terms of mass flow rate)—either the reboiled vapor return or the bottoms liquid product stream. Second, we can add a flow cascade to help smooth out the adjustments.

So, if the ratio of the mass flow of reboil vapor to the mass flow of the bottoms product liquid is greater than 1, the reboiler process liquid level is linked to the utility fluid CV, usually via a cascade with utility flow rate. For the CV on the bottoms product line (or speed controller if there is a variable speed transfer pump), composition is used in a cascade with the column bottoms temperature just like the total reboiler or with the bottoms product flow rate. This scheme is shown in Figure 5.8.

If the reboiled vapor-to-bottoms product ratio (on a mass basis) is less than 1, the reboiler process liquid level is linked to the bottoms product flow line CV (or speed controller if there is a variable speed transfer pump) and the composition is linked to the utility fluid flow CV. In both cases, cascades from these master controls to slave control loops are recommended. The level control would be linked to a slave control loop that would use the bottoms temperature linked to the bottoms product flow line CV.

The composition control would be linked to a slave control loop that would use the utility fluid flow rate linked to the utility fluid CV. If the reboiler utility is a condensing fluid (like steam), this CV is located on the utility inlet line.

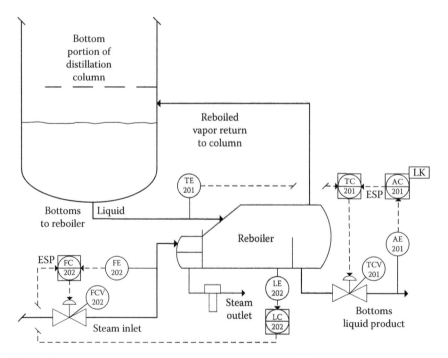

FIGURE 5.8
Control scheme for the distillation bottoms system with a partial thermosyphon reboiler and a mass ratio of reboil vapor to bottoms liquid product greater than 1. LK denotes the light key.

A steam trap or a condensate pot with a level control loop is used on the utility outlet line. If the reboiler utility is a vapor stream that does not change phase, then the utility outlet pressure is linked to a CV on the utility outlet line to insure that the vapor leaving the reboiler meets the downstream utility requirements.

If a thermosyphon reboiler is used, then the liquid level in the bottom of the column is automatically adjusted in height to make the reboil recycle system work. When a forced reboiler is used, the column liquid level is linked to the reboiler pump speed control (or to a CV on the pump outlet line if a fixed speed pump is employed).

A slight modification to the above designs can allow some elements of feed forward control to be embedded into the temperature slave control loop. Instead of using the bottoms liquid temperature, the temperature at one of the stripping section stages can be used. Usually process simulations are used to determine at which point in the column a change in temperature will have the greatest effect on the bottoms product composition. A thermocouple is then inserted as close to this point as possible in the column.

A more sophisticated variation of this strategy is to use the rate of change of the temperature rather than the absolute temperature as the slave control dependent variable. Most modern control systems allow the controller to be configured for rate of change control. This increases the response to dynamic changes in the column. In the past, this was accomplished by measuring the temperature at two adjacent stages in the column, then controlling to the change in the temperature difference between these two measurements.

5.3 Liquid-Based Absorption/Adsorption/Extraction/Leaching System Controls

Some substances, known as solutes, are highly attracted to certain liquids, known as solvents, due to the formation of weak chemical bonds. Separation processes can be designed to take advantage of this property. The substance containing the solute is contacted/mixed with the solvent in such a way that a portion of the solute (usually a significant portion) will transfer from the original substance into the solvent. When the solute is in a gaseous stream, this process is generally known as absorption. If the solute is in the liquid phase, the process is generally known as extraction and involves the extraction of the solute from either an organic liquid into an aqueous solvent or the extraction of the solute from an aqueous liquid into an organic solvent. When the solute is in a solid, the process is generally known as leaching. However, these labels are not rigorously followed and you can find extraction processes labeled as absorption and leaching processes labeled as extraction.

Similarly, solutes may be highly attracted to certain substances, either liquid or solid (known as sorbents), due to physical attractive forces. These processes are generally classified as adsorption processes. When the sorbent is a liquid, the controls are essentially the same as for absorption/extraction/leaching processes, so will be considered together in this section. In Section 5.4, we will consider those cases where the solvent or sorbent are solids. To simplify the discussion, we will use the terms "sorption" or "sorber" to describe general concepts that apply to all of these types of systems, and use the specific terms absorber, extractor, and leacher when describing concepts that are specific to those gas, liquid, and solid phase solutes, respectively.

In sorption processes, the substance containing the solute is mixed (gas or liquid) or immersed (solid) in the solvent/sorbent. In a single-stage sorber, the concentration of solute in the solvent/sorbent is the same in the entire sorption vessel. In a multistage sorber, the concentration of the solute in each phase changes within the sorption unit operation. The multiple stages may occur in the same vessel, similar to a distillation contactor, or may occur in a sequential series of vessels. The former is more common for absorption, while the latter is more common for leaching, with extraction being performed routinely in both configurations. In either configuration, the original substance and the solvent may travel through the stages concurrently or co-currently. As you can see, there are many variations!

Some sorption systems use the solvent only once, but the vast majority recover the solute back out of the solvent so that the solvent can be recycled back to the sorber. We will start with the simplest case and then build toward more complex multistage sorption systems that include recovery of the solute and recycle of the solvent/sorbent.

5.3.1 Single-Stage, Once-Through Absorption

Consider the simple single-stage gas absorber system shown in Figure 5.9. Gas containing the solute and liquid solvent are continuously fed and mixed together. Gas is also continuously removed from the top of the vessel and liquid continuously removed from the bottom of the vessel. A mass balance around the absorber is:

$$M_{I,a} + M_{I,b} = M_{O,a} + M_{O,b} + M_A \tag{5.5}$$

where
  a is the substance containing the solute
  b is the solvent
  I is the inlet to the absorber
  O is the outlet from the absorber
  A is the accumulation or depletion

Equation 5.5 is used to satisfy the accumulation term. $M_{I,a}$ is typically set by the upstream process, leaving three degrees of freedom for the system, $M_{I,b}$, $M_{O,a}$, and $M_{O,b}$. CVs are typically installed on three associated inlet and outlet lines to provide the required three independent variables.

Separation Unit Operations Controls

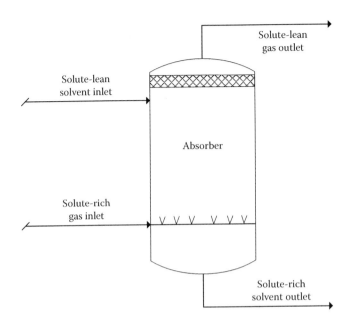

FIGURE 5.9
Single-stage gas absorption system.

The primary objective of the absorber is to reduce the concentration of solute in the original gas stream to meet a predetermined specification. If possible, a direct measurement of this concentration in the outlet gas stream should be made. When this is not possible, an indirect measurement can be used and correlated to the concentration using laboratory measurements.

In order to meet the concentration specification, sufficient solvent must be sent through the absorber to remove the desired quantity of solute. An efficient way to achieve this is with a ratio control strategy. When the concentration of solute in the original gas is not expected to vary rapidly or significantly, the inlet gas flow rate can be used as an indirect measure of the inlet solute rate. Many times, the concentration of the actual solvating compound in the solvent is not 100%. For example when mono-ethanol amine (MEA) is used as a solvent to absorb H_2S from a gas stream, it is blended with water and the concentration will typically be somewhere between 25% and 40% MEA. If this concentration does not change rapidly or significantly, the total solvent stream flow rate can be used as an indirect measure of the inlet solvent rate.

Combining these two together, we take the ratio of the solvent flow rate to the solute flow rate and use this to adjust the CV on the solvent inlet line. If the outlet gas solute concentration is available directly or indirectly, this measurement can be used as the master loop in a cascade with the ratio control (this is another example of a "ratio with feedback trim" control described in Section 2.8).

A more accurate control loop can be constructed by calculating the actual solvent molar flow rate and the actual solute molar flow rate for use in the

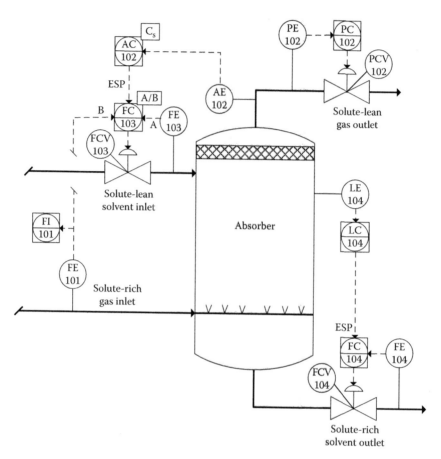

FIGURE 5.10
Control scheme for a single-stage gas absorption system. C_s denotes the concentration of the solute.

ratio control. However, as long as there is reasonable feedback control from the gas outlet, this additional complication is not normally required.

Secondary objectives in the absorber are: (1) maintaining the pressure in the vessel below its bursting pressure; (2) generating sufficient pressure in the gas outlet stream to allow flow to the next unit operation, if one exists; and (3) maintaining an acceptable liquid level in the absorber. To accomplish these objectives, we use the vessel's overhead pressure to control the gas outlet CV and the vessel's liquid level to control the bottoms liquid outlet CV. The complete control scheme is shown in Figure 5.10.

5.3.2 Single-Stage, Once-Through Extraction

In this unit operation, two immiscible liquid streams are mixed together to allow the solute to transfer from the originating liquid into the solvent.

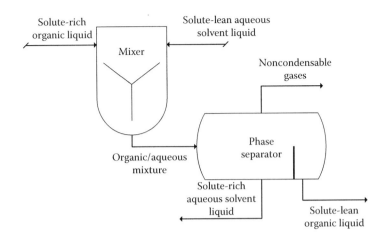

FIGURE 5.11
Single-stage once-through liquid-liquid extraction system.

The streams are then allowed to phase separate. This can be performed in a single vessel or by using separate extractor and settler vessels. We will consider the simpler of the two cases and use separate vessels as shown in Figure 5.11.

A mass balance around the entire system (ignoring the noncondensable gas stream, which would usually be on simple pressure control) is:

$$M_{I,a} + M_{I,b} = M_{O,a} + M_{O,b} + M_A \tag{5.6}$$

where
 a is the substance containing the solute
 b is the solvent
 I is the inlet to the extractor
 O is the outlet from the separation vessel
 A is the accumulation or depletion

Equation 5.6 is used to satisfy M_A and $M_{I,a}$ is usually defined upstream, leaving three degrees of freedom. CVs can be installed on the solvent inlet and the two outlet streams as the three independent variables.

A mass balance just around the mixer yields:

$$M_{I,a} + M_{I,b} = M_{O,c} + M_A \tag{5.7}$$

where
 a is the substance containing the solute
 b is the solvent
 c is the mixed stream
 I is the inlet to the extractor
 O is the outlet from the extractor
 A is the accumulation or depletion.

This equation introduces one additional degree of freedom, which is accommodated by installing a CV on the mixer outlet line.

A mass balance just around the settler yields:

$$M_{I,c} = M_{O,a} + M_{O,b} + M_A \tag{5.8}$$

No new degrees of freedom are introduced with Equation 5.8.

As with the absorber, the primary objective is to reduce the concentration of solute in the original substance to meet a predetermined concentration. If possible, the direct measurement of this concentration in outlet stream "a" should be made. When this is not possible, an indirect measurement should be used. Like the absorber, the measurements for the solvent-to-solute ratio should be obtained directly or indirectly and used to calculate the ratio, which can then be used to control the solvent inlet flow rate. The outlet stream concentration can be used as a feedback trim (either from an online measurement or from a lab measurement).

Secondary objectives are to: (1) maintain a reasonable liquid level in the mixer, (2) perform an acceptable phase separation in the settler, (3) avoid any of the lighter (less dense) phase liquid leaving in the settler bottoms stream, and (4) maintain an acceptable total liquid level in the settler. The first of these can be accomplished by matching the mixer total level measurement with the mixer outlet CV; the second, by matching the settler interface level measurement with the settler heavy (more dense) phase outlet stream CV; and the third, by matching the settler total level measurement with the settler light phase outlet stream CV. It may also be desirable to include a gas outlet on either the extractor or the settler or both to avoid pressure buildup due to noncondensable material. The associated vessel pressure should be matched with a CV on the gas outlet stream for control of these variables.

A low-low interface level safety control is included for the heavy phase outlet stream CV to insure that no light phase material leaves in that stream. The complete control scheme for the case where an aqueous solvent is the heavy phase is shown in Figure 5.12.

5.3.3 Multistage Absorption with Solvent Recovery

Now let us consider a more complicated example, the process scheme shown in Figure 5.13. A gas stream containing the solute is routed upward through a trayed or packed column where it contacts the liquid solvent moving downward through the column. By the time the gas leaves the top of the contactor, the solute concentration has been reduced. The inlet solvent is often termed the "lean" solvent, because it has the ability to absorb solute out of the gas phase.

The solvent leaving the absorber is known as the "rich" solvent, because it has absorbed a measurable concentration of the solute out of the gas.

Separation Unit Operations Controls

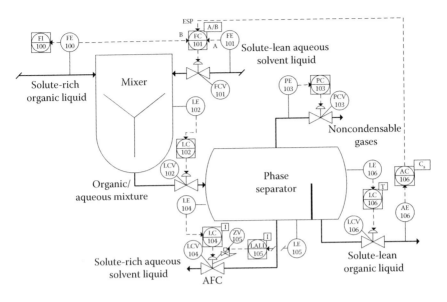

FIGURE 5.12
Control scheme for a single-stage once-through liquid-liquid extraction system. C_s denotes the concentration of the solute; I denotes the interface level; T denotes the total level.

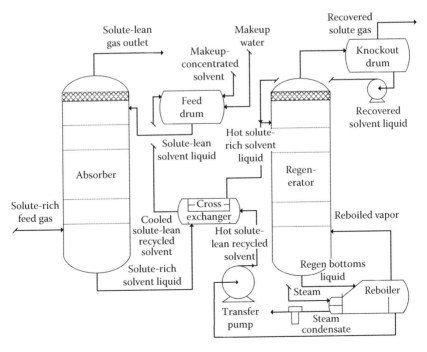

FIGURE 5.13
Gas extraction system with solute recovery and solvent recycle.

The rich solvent is then routed to a regeneration column. For some systems, the solubility of the solute in the solvent is temperature dependent, and a change in temperature will release most of the solute out of the solvent. The phase containing the solute is then routed to a separator to remove any entrained/residual solvent and is then available as a purified stream. For the example shown in Figure 5.13, solubility is inversely proportional to temperature.

The solvent from the regeneration column is returned to its optimum absorption temperature and then recycled to the absorber. Energy can sometimes be saved by using a rich/lean solvent cross exchanger. In some systems, this is adequate to return the lean solvent to the desired temperature. If not, a trim exchanger using a utility is employed. Similarly, the lean solvent temperature may be further adjusted before the regeneration column as required. A pump must be included somewhere in the solvent recycle loop. Its location depends upon the relative operating pressures of the absorber and regenerator. In the example shown in Figure 5.13, the regeneration is operating at a lower pressure than the absorber, so a pump is required on the solvent recycle line.

Another way to regenerate the solvent is to pass another gas stream through the solvent so that solute can desorb out of the solvent by concentration gradient.

As an example of how to design the controls for these types of systems, let us consider the specific scheme shown in Figure 5.13. An overall system mass balance is:

$$M_{g,I} + M_{ns,I} + M_{mw,I} = M_{g,O} + M_{rs,O} + M_A \tag{5.9}$$

where
  g is the original gas
  rs is the recovered solute gas
  ns is the new concentrated solvent
  mw is the makeup water
  I is the inlet
  O is the outlet
  A is the accumulation or depletion

Equation 5.9 is used to satisfy M_A, while $M_{g,I}$ is typically set by the upstream process, leaving four degrees of freedom. CVs are typically installed on the new concentrated solvent inlet, the makeup water inlet, the solute-lean gas outlet line from the top of the absorber, and the recovered solute outlet line from the top of the knockout drum after the regenerator overhead.

The new solvent solution is typically more concentrated than the target concentration in the absorption system. To understand why we would want to adjust the concentration of the solvent solution within the absorption system rather than in a separate "premix" unit operation, we need to examine a component mass balance around the system for water:

$$M_{gw,I} + M_{nsw,I} + M_{mw,I} = M_{gw,O} + M_{rsw,O} + M_A \tag{5.10}$$

where
  gw is the water in the solute-rich process gas
  rsw is the water in the recovered solute gas produced
  nsw is the water in new solvent solution
  mw is the water added separately from the solvent stream
  I is the inlet
  O is the outlet
  A is the accumulation or depletion

Usually, the solute-rich process inlet gas is not saturated with water. As a result, water from the solvent solution is absorbed into the gas in the absorber. The exiting solute-lean process gas is saturated or nearly saturated with water. If this gas is valuable, such as a light hydrocarbon gas, this water is undesirable and much of it may be recovered out of the gas stream in a downstream unit operation such as by cooling the gas down so that water will condense out of the stream. When this is the case, the recovered water may be recycled back to the absorption system. However, it is unlikely the amount of water recovered and recycled exactly matches the water absorbed by the gas. Therefore, it is necessary to include a means to adjust the water content in the solvent. If this gas is not valuable, then the water just leaves the system with this gas. In this case, there will be a continuous makeup requirement for new water into the solvent.

Further, the recovered solute gas stream will also be saturated with water vapor. This stream may also be cooled after recovery in the regenerator and, if so, then some of the water may precipitate out of the gas. This water can be recovered and recycled back into the solvent. It should be noted that the recovered solute gas will also be saturated with solvent and some of this will also precipitate back out with the water. However, the concentration of solvent in this recovered water is unlikely to exactly match the solvent concentration used in the absorption system. Therefore, provision must be made to make solvent concentration adjustments in the system. Using a concentrated solvent makeup stream and a separate water makeup stream allows this adjustment to be easily made.

Before we describe the details of the control scheme, there are a couple of other important features to highlight. This absorber system includes a recycle loop for the solvent. There must be at least one vessel in the loop where dynamics can be dampened. Also, there must be at least one CV in the loop that must be controlled using flow control rather than level control. This sets the flow rate for the entire recycle loop. All of these features can be accommodated by including a makeup or surge drum in the solvent recycle loop. The drum is usually placed just before the absorber. It is at this point that makeup solvent and/or water is added to the system. This is also a good point to set the recycle rate for the entire system.

So let us look at a typical, stable control scheme for this system. Please refer to Figure 5.14 to see each element of the scheme as it is described below. The primary objective of the system is to reduce the concentration of solute in the original gas to match a given specification. An online measurement of the solute concentration in the solute-lean process gas leaving the absorber is the primary control variable. If this concentration cannot be directly measured, an indirect measure can be used and calibrated using laboratory analyses. If this is not possible, then laboratory data should be used to manually adjust the external setpoint to the ratio controller (FC-201 in Figure 5.14) described below.

As a secondary objective, we want to use the smallest quantity of solvent we can to meet the primary objective. The more solvent we use, the more energy that must be expended in the regenerator and the more losses we will incur. So, the solvent-to-solute ratio is also important. If the solute concentration in the solute-rich process inlet gas is fairly stable or does not change too rapidly, the ratio of the flow rates of the solvent entering the absorber (FE-201) to the solute-rich process gas entering the absorber (FE-100) can be used. This ratio is then paired to a CV on the solvent outlet from the feed drum (FCV-201). The solute concentration in the solute-lean process gas outlet (AE-101) is used to provide feedback trim (AC-101) to the ratio controller (FC-201).

The solute-lean process gas outlet from the absorber is typically controlled by the absorber pressure (PE/PC/PCV-101). The value of pressure used should be a value that allows efficient operation of the absorber and also provides sufficient pressure to accommodate downstream operations if possible. The rich solvent collects in the bottom of the absorber below the gas inlet. The level of this solvent can be measured (LE-202) and used to control (LC-202) the flow rate of solute-rich solvent leaving the absorber either directly or in a cascade with the flow rate (FE/FC/FCV-202).

The solute-rich solvent is routed from the absorber to the cross exchanger (sometimes known as the rich/lean heat exchanger). For this example, let us assume the case where the solute solubility in the solvent solution decreases with increasing temperature. Therefore, the goal is to heat the rich solvent stream as high as possible prior to the regenerator in order to minimize the energy required in the regenerator reboiler. No control is needed in the heat exchanger.

While not shown in Figure 5.14, the gas leaving the top of the absorber column is often sent through a cooler to condense water/solvent out of the gas and then to a separator where the condensed liquid is recovered from the gas.

From the cross exchanger, the rich solvent is routed into the top of the regenerator, which is a stripping-type distillation column. The "cold solvent" travels down the column where it is heated by the stripping gas stream

Separation Unit Operations Controls

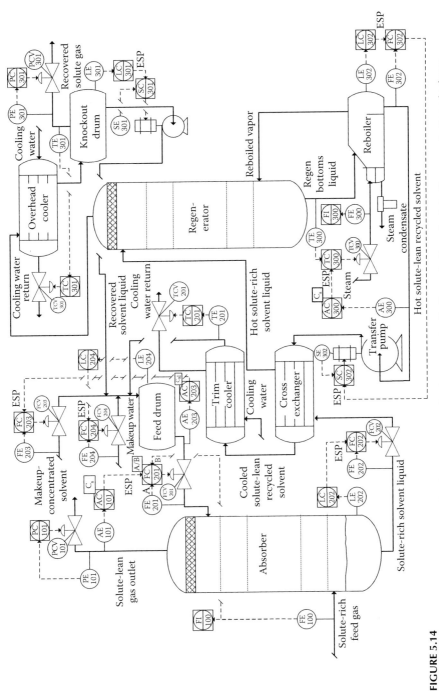

FIGURE 5.14
Controls for a gas absorption system with solute recovery and solvent recycle. C_s, Concentration of solute; C_{sl}, concentration of solvent.

traveling up the column. As it heats up, the solvent releases solute into the gas stream. Hot solvent liquid from the bottom of the column is routed to a partial reboiler where a fraction is vaporized and recycled back into the stripper. The primary objective of the regenerator is to reduce the concentration of solute in the solvent solution to a predetermined specified level. This is measured (AE-300) by analyzing the concentration of the solute in the lean solvent leaving the regenerator when possible, or by using an indirect measurement calibrated with laboratory samples. The secondary objective in the regenerator is to minimize the amount of steam consumed. Therefore, the outlet solute concentration is typically matched (AC-300) to the steam inlet CV on the reboiler. Since this analysis is usually an unresponsive variable, it is typically configured in a cascade loop with the stripper bottoms temperature (TE/TC/TCV-300). I have also included a flow indicator (FE/FI-300) on the steam inlet line so that steam consumption can be monitored.

The lean solvent exiting the stripper is routed via a pump through the cross exchanger where it is cooled. For this example, I have included a variable speed pump (SE/SC-302). Since concentration is matched to the reboiler steam CV, the level of solvent in the reboiler (LE/LC-302) can be paired to the pump motor speed via a cascade with the lean solvent flow rate (FE/FC-302).

No controls are required for the cross exchanger. For this example, I have added a trim exchanger after the cross exchanger to cool the lean solvent to the desired absorption temperature. The lean solvent is cooled to the optimum absorbing temperature using cooling water as the utility. The cooling water flow rate CV (TCV-201) is controlled using the lean solvent outlet temperature (TE/TC-201). The solvent is then routed into the feed drum. Another option is to use the feed drum outlet solvent temperature as the measurement variable in this control loop.

The primary objective of the feed drum is to adjust the concentration and amount of solvent available in the solvent recycle loop. The solvent concentration (typically using an indirect measurement such as pH) is matched (AE/AC-203) to the makeup solvent inlet line CV via a flow cascade (FE/FC/FCV-203), and the feed drum level (LE/LC-204) is matched to the makeup water inlet line CV via a flow cascade (FE/FC/FCV-204). The feed drum outlet CV is controlled by the solvent-to-solute ratio controller (FC/FCV-201) described earlier.

The recovered solute gas leaving the regenerator is typically routed through a heat exchanger where cooling water is used to reduce the gas to a specified temperature, which will allow the solvent and water to condense but minimize the quantity of solute that will be in the liquid phase. A two-phase mixture of recovered gas and aqueous liquid leaves the heat exchanger and is routed into a separator. The temperature of the recovered gas leaving the separator (TE/TC-301) is matched to the heat exchanger cooling water CV (TCV-301), while the pressure of the recovered gas leaving the separator is matched to the separator gas outlet CV (PE/PC/PCV-301). The pressure of the liquid leaving the separator is increased so that it can be recycled to the solvent feed drum. The liquid level in the separator (LE/LC-301) is matched to the speed controller (SE/SC-301) on this variable speed pump.

5.4 Solid-Based Absorption/Adsorption/Extraction/Leaching System Controls

Another common class of separation systems uses a solid material, often known as the "sorbent," that has a chemical or physical attraction to the solute. The gas or liquid containing the solute is passed through a bed containing the sorbent, and the mass transfer driving force of the system results in solute leaving the original stream and bonding to the sorbent. Three common types of systems are: (1) once-through systems, (2) fluidized bed systems, and (3) fixed bed regenerated systems.

Efficient sorbents have a very high surface area-to-volume ratio. The most common way of achieving this is to use porous solids. A disadvantage of porous solids is that they are more fragile than nonporous solids. Shape can also increase surface area. For example, spherical particles have a higher surface area-to-volume ratio than cylinders, which have a higher ratio than rectangles. An emerging field is the use of nanoparticles. Because of their small size, nanoparticles have a much higher surface area-to-volume ratio than macroscale particles. Yet because the nanoparticle is nonporous, it is more rugged than a larger, porous solid.

5.4.1 Once-Through Solid Sorbent Systems

There are a number of ways that a solid sorbent can be placed in contact with a gas or liquid containing a solute. The most common way is to pass the fluid through a fixed bed of the sorbent material. Controls for regenerated fixed bed systems are described in Section 5.4.3. Once-through fixed bed systems are just a simpler subset of that more complicated arrangement. The solute-rich gas or liquid passes through the solid sorbent bed where solute is bound to the solid by chemical or physical attraction. If the concentration of solute is very dilute in the process fluid, a sorbent bed may last many years before it needs to be replaced. An example of this is the removal of trace quantities of organics from an aqueous stream using a sand filter (a packed bed containing screened sand particles). For these systems, controls are often not required. The process water flows upward through the sand filter and leaves the vessel with a reduced organics content. The flow rate is set by the upstream process and since water is incompressible, there are no further degrees of freedom that require control.

Another common technique is to contact the solid and a liquid together in a mixing vessel and then allow the solids to settle out from the liquid in a gravity separator. This stream may then require filtering, pressing, or drying to remove entrained liquid from the used sorbent before it is disposed, depending upon the application.

Consider the once-through adsorption system shown in Figure 5.15. The primary objective is to achieve a target outlet solute concentration in the liquid. A secondary objective is to use as little sorbent as possible to achieve this concentration. A mass balance around the system is:

$$M_{P,I} + M_{Sr,I} = M_{P,O} + M_{Sr,O} + M_A \tag{5.11}$$

where
P is the process fluid
Sr is the solid sorbent
I is the inlet
O is the outlet
A is the accumulation

The process fluid inlet is typically set by the upstream process and M_A is satisfied by Equation 5.11. This leaves three degrees of freedom. If the organics content of the inlet process fluid is relatively constant, then the sorbent inlet rate can be based on the ratio of the sorbent to process inlet flow rate. The process fluid outlet solute concentration can be used as a

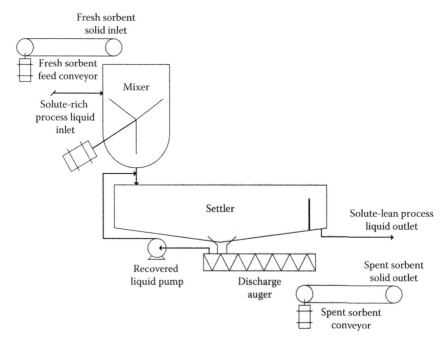

FIGURE 5.15
A once-through mixer/settler type adsorption system.

feedback trim, if it can be measured or inferred from an indirect measurement. Otherwise, laboratory data should be used to adjust the ratio controller setpoint.

The rate at which the used solid sorbent leaves the settler is usually determined by the physical characteristics of the sorbent particles as well as the configuration of the settling vessel (e.g., rakes may be used to guide the solids into a discharge line). For the example shown in Figure 5.15, I have employed a screw-type discharge augur emptying onto a conveyor. Liquids pressed out of the solids in the augur are pumped back into the settler. For this type of system, the mass of used sorbent leaving the system can be controlled by the speed of the augur motor. Liquid recycle from the augur can be controlled using the pump discharge pressure matched to an outlet CV for a fixed speed pump (as in Figure 5.15) or to the pump's speed control for a variable speed pump. The process outlet liquid CV is usually matched to the settling vessel's liquid level. One additional degree of freedom is added by a mass balance around the mixer. The level in the mixer can be used to control the rate of transfer of the liquid/solid mixture from the mixer into the settler.

The entire control scheme is shown in Figure 5.16.

5.4.2 Fluidized Bed Solid Sorbent Systems

If the sorbent particles have sufficient strength, they can be used in a continuous unit sorption operation. Fluidized bed systems are the most common. In a fluidized bed, the gas or liquid stream containing the solute flows upward through a bed of solid sorbent particles. At high enough velocities, the friction drag of the fluid on the particles will cause the particles to move in the upward direction of the flow. This is known as fluidizing the particles.

Gravity acts against the drag force and enables the fluid to "slip" past the particles. If the velocity is too high, the particles become entrained in the fluid and will leave the contactor. However, there is a range of velocities where the drag induced is sufficient to raise the particles but is insufficient to completely overcome the opposing gravitational force. Under these conditions, the particles "float" in the fluid.

Fluidizing the particles has two advantages: (1) it increases the mass transfer between the fluid and the particle, and (2) it allows particles to be diverted out of the contactor for regeneration.

Consider the system shown in Figure 5.17. The fluid containing the solute enters the bottom of the contactor and flows upward, fluidizing a bed of solid sorbent particles. Solute is removed from the fluid by sorption onto the particle surface, which increases the density of the particles. If the fluid velocity is controlled properly, particles that are nearing their saturation with solute will become too heavy to be fluidized and will drop to the bottom of the contactor. A continuous stream of rich sorbent particles is removed from the contactor and routed to a regenerator vessel.

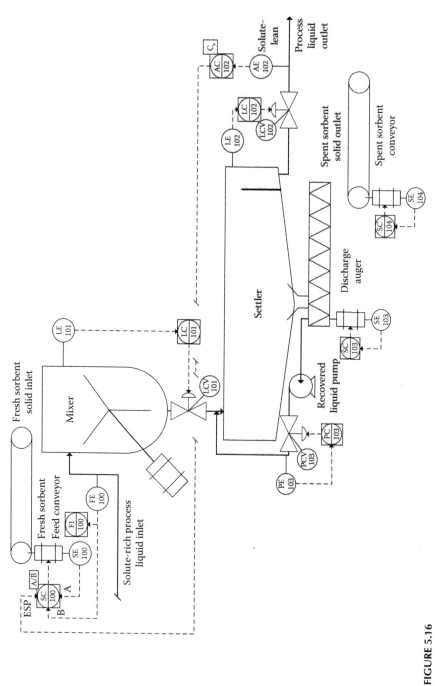

FIGURE 5.16
Control scheme for a once-through mixer/settler type adsorption system. C_s denotes the concentration of solute.

There are a number of ways that the particles can be transferred from the contactor to the regenerator. For the system shown, the particles fall into a cyclone to separate them from the fluid. The separated particles fall from the cyclone into an inert gas stream where they are entrained. The particle-laden inert gas is then routed to the regenerator. The process gas from the cyclone flows back up into the adsorber.

Once the solute-rich sorbent particles enter the regenerator, the solute will desorb off of the particles by mass transfer. Some systems can perform this by changing the temperature of the particles compared to the contactor, reducing the sorption capacity of the sorbent and releasing solute. In other systems, a regeneration fluid provides a mass transfer gradient that leads to desorption.

In the example shown in Figure 5.17, the particles in the regenerator are fluidized by a hot, inert regeneration gas. As the particles lose solute, their density decreases and they rise. In some systems, particles are drawn off of the top of the fluidization zone and transferred back to the contactor. In others, the lower density particles become entrained in the regeneration fluid and leave the regenerator with the fluid. For the example shown in Figure 5.17, as the lighter particles rise, they encounter a wider spot in the vessel (in this example the fluidization zone is shown as occurring inside of a liner inside the vessel) where the velocity of the gas decreases. At this point, fluidization is no longer maintained and the particles fall by gravity. Many of these will fall outside of the fluidization zone where they can be collected in the bottom of the vessel. A cyclone (shown) or filter is then used to remove the particles from the regeneration fluid so that they can be returned to the contactor. In Figure 5.17, this is accomplished using an inert transport gas, which may be the solute-lean process gas.

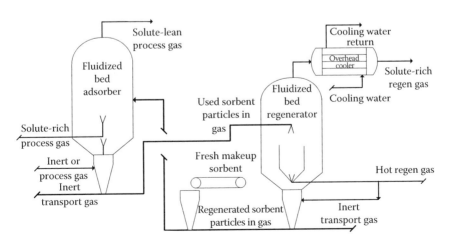

FIGURE 5.17
A continuous fluidized bed adsorption/regeneration system.

Now let us turn to the controls for the system shown in Figure 5.17. An overall mass balance is:

$$M_{p,I} + M_{r,I} = M_{p,O} + M_{r,O} + M_A \tag{5.12}$$

where
p is the process fluid
r is the regeneration fluid
I is in
O is out
A is the accumulation or depletion

The inlet process conditions are typically set by the upstream operations, and the accumulation term is accommodated by Equation 5.12, leaving three independent variables. To cover these, CVs can be placed on the process outlet stream, the regeneration gas inlet stream, and the regeneration gas outlet stream. If there is a pump or compressor associated with the regeneration fluid, motor speed control may be substituted for one of the regeneration system CVs (inlet if the motive force is applied before the regenerator as shown in Figure 5.17 or outlet if the motive force is applied after the regenerator).

Now consider just the contactor:

$$M_{p,I} + M_{sr,I} = M_{p,O} + M_{sr,O} + M_{A,C} \tag{5.13}$$

where sr is the solid sorbent.

Two new independent variables have been introduced by Equation 5.13: the solid sorbent inlet and outlet rates. For this system, the sorbent outlet rate is not truly independent. Recall that sorbent remains fluidized until the density of the sorbent exceeds the fluidization point. At this point, the sorbent falls to the bottom of the contactor and exits. This leaves only the sorbent inlet rate, which is typically controlled by the rate of transport gas.

Similarly, a mass balance of the regenerator yields:

$$M_{r,I} + M_{sr,I} = M_{r,O} + M_{sr,O} + M_{A,r} \tag{5.14}$$

No new independent variables are introduced by Equation 5.14.

The primary objective of the sorbent system is to reduce the concentration of solute in the process fluid to meet a given target level. In order to accomplish this, the velocity of the process fluid must be maintained at a value that allows sorbent particles to be fluidized until they reach a certain density that is related to solute loading in the particles, but not beyond this point. Similarly, another associated objective is to flow sufficient regeneration fluid through the regenerator particle bed to fluidize the particles while their density is above a certain point that is related to a lean solute loading in the particles. The flow rate should be maintained high enough that when the density of the lean sorbent particles drops below a specified threshold, the particles are entrained out of the fluidization zone of the regenerator.

Another objective is to produce a fluid out of the regenerator that has a specified concentration of solute. This concentration is typically determined based on other process conditions.

To meet the process fluid concentration specification, the most powerful dependent variable is the sorbent particle-to-solute feed rate ratio. For the system shown in Figure 5.17, the system is essentially self-controlled by the rate of adsorption onto the sorbent and the rate of desorption of solute in the regenerator. As long as sufficient sorbent particles are present in the adsorber, the solute will be adsorbed out of the process gas. The rate at which particles are regenerated does not have to equal the rate at which particles become full of solute.

If the process gas inlet rate cannot be controlled, an inert gas can be injected either into the bottom of the adsorber or into the bottoms cyclone to provide a degree of freedom that can be used to adjust the fluidization level in the adsorber, insuring good fluidization even when inlet process gas rates change. Similarly, the hot regeneration gas flow rate can be adjusted based on the fluidization level in the regenerator or using the overhead CV.

Another way to maintain particle fluidization in the contactor within the correct range of velocities is via the process fluid outlet CV. If the process fluid is incompressible, a surge drum has to be added to either the inlet or outlet of the process stream (unless inherent in the upstream or downstream process) to accommodate variations in the fluid flow rate that are introduced in the contactor. A third common method is to measure the pressure drop across the fluidization zone and to use this value as the dependent variable instead of fluidization zone height.

Since sorbent transfer between the two vessels is accomplished by inert gas fluidization, testing should be performed to optimize the inert gas flow rates for the two sorbent transfer streams. The complexity of trying to determine the optimum flow rate continuously and to use this determination is beyond the scope of the basic controls we are trying to describe in this chapter. Therefore, a simple flow control scheme is specified.

The complete control loop is shown in Figure 5.18.

5.4.3 Fixed Bed Regenerated Solid Sorbent Systems

The most common way to use a solid sorbent to remove a solute from a process fluid is to pass the fluid through a fixed bed of the sorbent material. For continuous processes, multiple parallel contactors are required, configured as a semi-batch operation such as the system shown in Figure 5.19. In a well-designed system, all of the solute is removed from the process fluid in the first 30–100 cm of the bed when it is first put into service. As that first parcel of sorbent becomes saturated, solute begins to enter the sorbent located

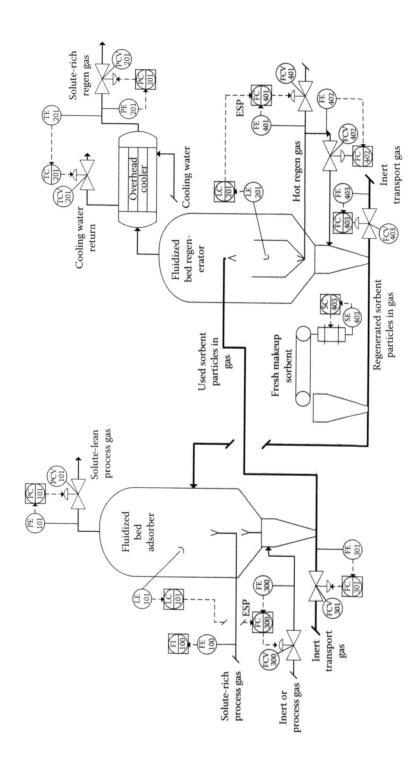

FIGURE 5.18
Control scheme for a typical continuous fluidized bed adsorption/regeneration system.

Separation Unit Operations Controls

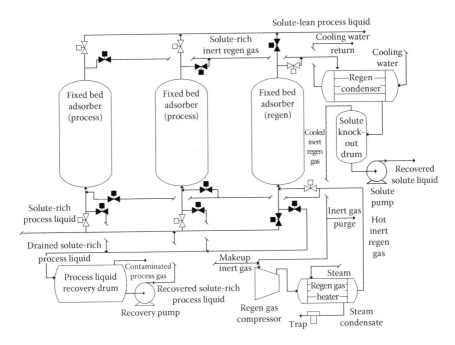

FIGURE 5.19
A typical semi-batch fixed bed adsorption/regeneration system.

further into the bed. If left long enough, all of the sorbent would become saturated and no additional solute would be removed from the process fluid. However, at some point prior to this point, the concentration of solute in the outlet process fluid will begin to rise and eventually reach the allowable maximum concentration limit. This point is known as the "breakthrough" point. The process gas must be rerouted to another fresh sorbent bed at or before the breakthrough point to stay within the desired limit.

Once a bed has reached the breakthrough point, and the process fluid has been redirected, the bed can be regenerated. The most common regeneration strategies are to (1) reduce the bed pressure, which will cause most of the solute to desorb out of the sorbent (this is known as pressure swing adsorption or PSA), (2) change the bed temperature (for many sorbents the sorption capacity of the material is temperature dependent, so changing the temperature up or down can cause much of the solute to desorb out of the sorbent; this is known as temperature swing adsorption, TSA), and/or (3) pass a fluid through the bed, causing solute to desorb out of the sorbent due to the concentration gradient imposed on the bed.

Now let us consider a control scheme for one variation of this type of process. As shown in Figure 5.19, three parallel beds are used to remove the solute from a liquid phase stream. A logic flow diagram (see Section 1.8 for a description of these diagrams) for one of the beds in this semi-batch

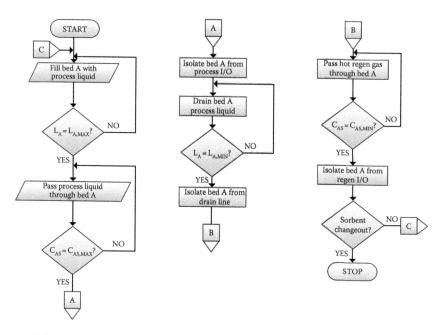

FIGURE 5.20
The logic flow diagram for one of the absorber beds in the semi-batch fixed bed adsorption unit operation shown in Figure 5.19.

unit operation is shown in Figure 5.20. At any time, two of the beds are in sorption mode and one bed is in regeneration mode. When one of the beds reaches breakthrough, it is taken out of service and the regenerated bed is placed in service.

Once out of service, the spent bed is drained of liquid to a liquid recovery drum. From this drum, the liquid is pumped back into the inlet of the in-service sorption beds. The gas from the liquid recovery drum contains volatile material contained in the process liquid. This gas would typically be processed further in downstream operations, depending upon its composition.

Figure 5.19 is an example of a system where temperature is used to regenerate the bed, specifically the case where solubility is inversely related to temperature. An inert gas, such as nitrogen, is routed through a compressor to increase its pressure and then through a heat exchanger to heat up the gas. It is then routed to the out-of-service sorbent bed containing the drained, spent sorbent. The hot inert gas is routed through the bed and solute desorbs out of the sorbent into the gas stream. In this example, the solute-rich hot inert gas is routed through a condenser to cool down the gas and condense out the solute. The two-phase mixture is routed to a knockout drum, where the liquid solute is separated from the inert gas. The solute is then routed to downstream processing. The cool inert gas from the knockout drum is recycled back to the compressor.

If there are trace contaminants sorbed into the sorbent and then desorbed into the inert gas, a gas purge must be used to avoid contaminant building in the inert gas recycle loop.

Now let us design a control strategy for the scheme shown in Figure 5.19. Because this is a semi-batch unit operation, we must consider the control scheme for each step of the batch sequence. The sequential events table corresponding to the process scheme in Figure 5.19 and the logic flow diagram in Figure 5.20 is shown in Table 5.1.

An overall mass balance of the system is:

$$M_{p,I} + M_{ig,I} = M_{p,O} + M_{ig,O} + M_{s,O} + M_A \tag{5.15}$$

where
p is the process fluid
s is the solute fluid
ig is the inert gas (makeup in and purge out)
I is in
O is out
A is the accumulation or depletion

As usual, we assume that the inlet process fluid conditions are set by the upstream process, leaving five independent variables. The accumulation term is satisfied by Equation 5.15, leaving the makeup inert regeneration gas inlet, process liquid outlet, solute liquid outlet, and inert gas purge outlet streams. However, since the process fluid in this example is an incompressible liquid, $M_{p,O}$ is not independent, but is the flow of process fluid entering the unit operation less the amount that transfers into the sorbent bed. This leaves three variables: makeup inert regeneration gas, inert gas purge, and solute outlet. In this example, we have placed CVs on the inert makeup and purge lines and have used speed control of the recovered solute liquid pump for the other control variable.

Just before each adsorber bed is placed into process service, it must be filled with liquid. Level elements are included for each bed to determine when the bed has been filled to the correct level. These are used as the termination criteria for steps 1, 2, 3, 6, 7, 11, and 12 in the sequence (Table 5.1).

The primary objective of the system is to reduce the concentration of solute in the process fluid to a specified level. If this can be directly or indirectly measured, then an online analyzer can be placed on the common sorbent bed outlet manifold. When a fresh bed is placed in service, most of the solute is removed in the first part of the bed. Once this part gets saturated, sorption occurs farther along the bed. At a certain point, there will be insufficient unsaturated sorbent left to remove the solute down to the desired concentration. When that point is reached, the analyzer measurement is used as the termination criteria for steps 5, 10, and 15 in the sequence (Table 5.1).

TABLE 5.1

The Sequential Events Table for the Semi-Batch Fixed Bed Adsorption Control Scheme Shown in Figure 5.20

Step	Description	MV 101	MV 102	MV 103	MV 104	MV 105	MV 201	MV 202	MV 203	MV 204	MV 205	MV 301	MV 302	MV 303	MV 304	MV 305	Termination Criteria
1	Fill Bed A with liquid; Bed B in process, drain Bed C	O	C	C	C	C	O	O	C	C	C	C	C	O	C	C	If LI-303 = Lmin AND LI-103 < Lmax, Go to 2 ELSE If LI-103 = Lmax AND LI-303 > Lmin, Go to 3
2	Fill Bed A; Bed B in process; Bed C isolated	O	C	C	C	C	O	O	C	C	C	C	C	C	C	C	If LI-103 = L_{MAX}, Go to 4
3	Bed A isolated; Bed B in process; drain Bed C	C	C	C	C	C	O	O	C	C	C	C	C	O	C	C	If LI-303 = L_{MIN}, Go to 4
4	Bed A in process, Bed B in process, regen Bed C with hot gas	O	O	C	C	C	O	O	C	C	C	C	C	C	O	O	If AI-401 = $C_{S,MIN}$, Go to 5
5	Bed A in process, Bed B in process, Bed C isolated	O	O	C	C	C	O	O	C	C	C	C	C	C	C	C	If AI-400 = $C_{S,MAX}$, Go to 6
6	Bed A in process, drain Bed B, fill Bed C	O	O	C	C	C	C	C	O	C	C	O	C	C	C	C	If LI-203 = Lmin AND LI-303 < Lmax, Go to 7 ELSE If LI-303 = Lmax AND LI-203 > Lmin, Go to 8
7	Bed A in process, Bed B isolated, fill Bed C	O	O	C	C	C	C	C	C	C	C	O	C	C	C	C	If LI-303 = L_{MAX}, Go to 9

(Continued)

TABLE 5.1 (*Continued*)
The Sequential Events Table for the Semi-Batch Fixed Bed Adsorption Control Scheme Shown in Figure 5.20

Step	Description	MV 101	MV 102	MV 103	MV 104	MV 105	MV 201	MV 202	MV 203	MV 204	MV 205	MV 301	MV 302	MV 303	MV 304	MV 305	Termination Criteria
8	Bed A in process, drain Bed B, Bed C isolated	O	O	C	C	C	C	C	O	C	C	C	C	C	C	C	If LI-203 = L_{MIN}, Go to 9
9	Bed A in process, regen Bed B with hot gas, Bed C in process	O	O	C	C	C	C	C	C	O	O	O	O	C	C	C	If AI-401 = $C_{S,MIN}$, Go to 10
10	Bed A in process; Bed B isolated, Bed C in process	O	O	C	C	C	C	C	C	C	C	O	O	C	C	C	If AI-400 = $C_{S,MAX}$, Go to 11
11	Drain Bed A; fill Bed B; Bed C in process	C	C	O	C	C	O	C	C	C	C	O	O	C	C	C	If LI-103 = Lmin AND LI-203 < Lmax, Go to 12 ELSE If LI-203 = Lmax AND LI-103 > Lmin, Go to 13
12	Bed A isolated, fill Bed B, Bed C in process	C	C	C	C	C	O	C	C	C	C	O	O	C	C	C	If LI-203 = L_{MAX}, Go to 14
13	Drain Bed A, Bed B isolated, Bed C in process	C	C	O	C	C	C	C	C	C	C	O	O	C	C	C	If LI-103 = L_{MIN}, Go to 14
14	Regen Bed A with hot gas; Bed B in process; Bed C in process	C	C	C	O	O	O	O	C	C	C	O	O	C	C	C	If AI-401 = $C_{S,MIN}$, Go to 15
15	Bed A isolated, Bed B in process, Bed C in process	C	C	C	C	C	O	O	C	C	C	O	O	C	C	C	If AI-400 = $C_{S,MAX}$, Go to 1

Once a used bed is taken out of service, hot inert gas is routed through the bed. The flow rate of hot gas flowing to the bed is controlled by maintaining the outlet pressure from the regeneration gas heated at the specified value. This insures that adequate gas passes through the bed for regeneration. Because the inert gas is compressible, the inlet pressure to the compressor must also be controlled. The pressure of the recycled gas from the solute knockout drum is kept at a specified level by controlling the speed to the compressor. The makeup gas rate is controlled to match the quantity of purge gas leaving the system. The purge gas rate is controlled by measuring the concentration of a contaminant in the inert gas and using this value to determine the correct purge rate. Because measuring contaminant concentration requires an analysis which is likely to be a nonresponsive variable, it is cascade to a flow control loop. The temperature of the inert gas leaving the regeneration gas heater is used to control the flow of steam entering the heater.

The hot inert gas flows through the used adsorber bed until the concentration in the gas leaving the bed reaches a specified minimum value. An analyzer is installed on the common outlet manifold and this signal is used as the termination criteria for steps 4, 9, and 14 in the sequence.

The solute-rich inert gas is routed through a condenser to reduce the solubility of the solute in the gas. Most of the solute will condense out of the gas in the condenser and is collected in the solute knockout drum. The temperature of the inert gas leaving the solute knockout drum is used to set the flow of cooling water passing through the condenser. The liquid level in the solute knockout drum is controlled using the speed control on the recovered solute liquid pump.

The complete control scheme is shown in Figure 5.21 with the associated sequential events table (Table 5.1). Notice that Table 5.1 allows for the possibility that it might take longer to fill one of the vessels than it would take to empty one of the vessels. It also allows the reverse possibility; it could take longer to empty a vessel than to fill a vessel. This is done using ".AND.," ".OR.," and ".ELSE." statements in the sequential logic script along with the "TRUE/FALSE" evaluation that compares the measurement variable to the termination criterion for that measurement. These are similar to how these statements work in Basic or other programming languages. When ".AND." connects two evaluation statements, both evaluations must read "TRUE" for the "GO TO" action to be invoked. Otherwise, the sequence remains at the current step. When ".OR." connects two evaluation statements, the "GO TO" action will be invoked if either evaluation statement reads "TRUE." The ".ELSE." connector will link a separate set of evaluations to a first set. If all of the evaluation statements in the first set read "FALSE," then evaluation of the set that follows the ".ELSE." connector is addressed. For further information on the control script used in this textbook, please see Appendix C.

So, for example, the termination criteria for step 1 in Table 5.1 states that if Bed C is drained before Bed A is filled, go to step 2 in the sequence. However, if Bed A is filled before Bed C is drained, go to step 3. If neither of these two

Separation Unit Operations Controls

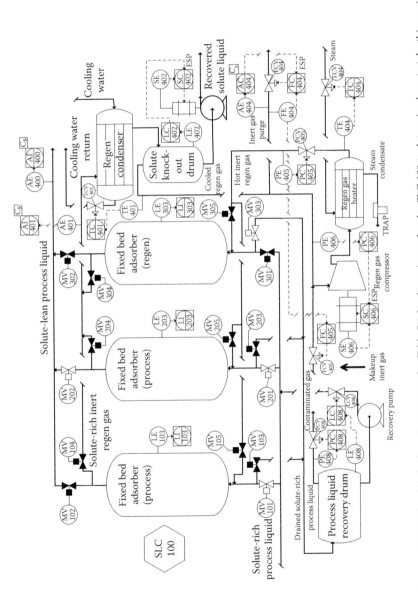

FIGURE 5.21
Control scheme for a semi-batch fixed bed adsorption system. See Table 5.1 for the actions taken by SLC-100 to determine how to take this semi-batch unit operation through its sequence. AE-400 measures the concentration of the solute in the outlet process liquid, while AE-401 measures the solute concentration in the regeneration gas outlet. AE-404 measures the concentration of a key contaminant in the regeneration air and is used to set the purge and makeup gas rates.

criteria is met, the sequence remains at the current step until either Bed C is drained or Bed A is filled.

Another objective for this system is to minimize the amount of energy needed for regeneration. The energy inputs are:

1. Utility usage in the inert gas heat exchanger
2. Power usage in the inert gas compressor

The correct way to do this requires a more sophisticated control strategy than we have learned to this point in the text, but we will come back to this example in Chapter 9.

Problems

5.1 Define the following terms: stripper, rectifier, light key, heavy key, absorption, extraction, leaching, adsorption, solute, solvent, sorbent, VLE, flash, adiabatic, flash ratio, and thermosyphon.

5.2 A gaseous process stream is fed into a flash drum in order to reduce its temperature to the exact inlet condition necessary for a downstream reactor. In unusual cases, some of the gas may condense in the flash drum. Since this is a rare condition, the level in the flash drum is allowed to build up to a specific level, at which point a pump on the bottoms outlet line is started to transfer the liquid to a recovery system. Liquid is removed from the drum until a specific low level is reached, at which point the pump is stopped. Because the pump rarely operates, an installed spare is not required. Draw a schematic showing the process and controls for this system.

5.3 Consider the scheme shown in Figure 5.1. Draw a schematic to show how this scheme would be modified if "f" is used as the measurement variable instead of pressure for the overhead control loop.

5.4 Consider the scheme shown in Figure 5.2. Draw a schematic to show how this scheme would be modified if the liquid feed contains three separate liquid phases: a light organic phase, an aqueous phase, and a heavy organic phase.

5.5 Consider the scheme shown in Figure 5.2. Draw a schematic to show how this scheme would be modified if variable speed pumps are integrated with this unit operation to transfer the two outlet fluids to the next steps in the process.

5.6 Consider the scheme shown in Figure 5.2. Draw a schematic to show how this scheme would be modified if the inlet fluid separates into a vapor phase, a light organic liquid phase, and an aqueous liquid phase. This unit operation is known as a three-phase separator.

5.7 A common application of the three-phase separator (Problem 5.6) is to separate an organic liquid from a gas stream in a system where the inlet fluid has trace quantities of polar compounds (e.g., water). When this is the application, a liquid level "pot" similar to that shown in Figure 5.1 (but installed at the low point on a horizontal pressure vessel) is typically used. In this application, the liquid level pot is known as a "water boot." If the quantity of water collected in the water boot is small and a pump is required to transfer the boot's contents to its next location, an on/off control scheme is usually employed. Draw a schematic showing the process and controls for a three-phase separator that employs a water boot with on/off pump on the water discharge line.

5.8 A demulsifier is a two-phase separator designed to separate polar and nonpolar liquids that have formed an emulsion. An emulsion is a fine dispersion of minute droplets of one of the two liquid phases in the other liquid phase. While the formation of emulsions may be purposeful, such as in many paints, sometimes detrimental emulsions occur during processing. While emulsions contain two liquid phases, they act as a single liquid phase and often have very different properties than either phase would have in isolation. A common example is the formation of emulsions in many crude oils produced from offshore oil wells.

To "break" the emulsion, special chemicals are injected into the crude oil and it is then routed to a two-phase separator where an electrostatic grid is used to ionize the polar phase, which helps the phases to separate. The emulsion chemicals are expensive, so their use should be optimized. Laboratory tests are typically used to find the minimum chemicals-to-crude feed ratio that will allow efficient separation at the separator's design temperature. The strength of the electrostatic field generated by the grid can also be optimized by controlling the amperage flowing through the grid. Draw a schematic showing the process and controls for this special two-phase separator, including the chemical injection (with fixed speed pump) and electrostatic field generating systems (hint: use the symbol "I" for current in your control loop for the electrostatic field generating system).

5.9 In complex heterogeneous organic fluids, the smaller organic compounds "dissolve" into the large ones. Essentially, these smaller compounds fit in the gaps between the larger organic compounds in the fluid, adding mass but little or no volume to the mixture. Stripper columns are often used to remove these compounds from the bulk, heavy organic liquid. For some fluids, using a reboiler in the bottom of the stripper column results in decomposition of some of the heavy organic compounds into lighter (smaller) ones, which reduces the quantity of larger organics produced. If this is an undesirable effect, an alternative to the traditional reboiled vapor system needs to be used.

A common alternative is to inject steam in the bottom of the column. The steam provides the energy to heat the liquid and also contributes to the vapor phase traveling up the column, increasing separation efficiency. The vapor leaving the top of the stripper is then cooled so that the steam will condense. The condensed steam plus any organics that condensed in the cooler are then removed from the overhead product gas in a three-phase separator. The organics can be pumped back into the top of the stripper with the feed stream, while the water is pumped to treatment for reuse or disposal. Draw a schematic showing the process and controls for this system.

5.10 Consider the scheme shown in Figure 2.17. Show how this scheme would be modified to such that the master level control loop cascaded to a slave flow loop on the reflux accumulator with a variable speed pump.

5.11 Consider the scheme shown in Figure 2.17. Show how this scheme would be modified if liquid methane were vaporized in the overhead condenser as the cooling utility stream.

5.12 The scheme shown in Figure 5.6 is configured for the case where the ratio of the reflux to overhead product is greater than 1. Show how this scheme would be modified if the ratio of the reflux to overhead product is less than 1.

5.13 Show how the scheme shown in Figure 5.6 would be modified if fixed speed pumps with minimum flow recycle lines are used instead of variable speed pumps.

5.14 Show how the scheme shown in Figure 5.8 would be modified if a forced reboiler is used instead of a thermosyphon reboiler (hint: add a pump in the bottoms liquid line between the column and the reboiler).

5.15 In a distillation column with an older control system, the temperatures on trays 4 and 5 (from the bottom) are used to calculate the rate of change of temperature in the column. This rate of change is used as the slave measurement parameter for the reboiler steam inlet flow rate. Show the process and control scheme for the bottom of this distillation column if it employs a thermosyphon partial reboiler and if the ratio of reboil vapor to distillate bottoms is less than 1.

5.16 Acetic acid is sometimes produced in a fermentation process. Liquid-liquid extraction is used to remove the acetic acid from the dilute aqueous fermentation broth. Because the acetic acid concentration in the broth is very dilute, a multistage countercurrent system is often employed.

The acetic-acid-rich broth is pumped into a continuously stirred vessel where it contacts a partially enriched organic solvent stream pumped from the second-stage phase separator. In the extractor, some of the acetic acid transfers from the broth into the solvent. From the first-stage extractor, the organic/aqueous mixture enters a phase separator (Figure 5.11 shows one extraction/separation stage).

The aqueous phase from the first-stage phase separator is pumped to a second continuously stirred vessel where it contacts a partially enriched organic solvent stream pumped from the third-stage phase separator. Some of the acetic acid transfers from the aqueous phase into the solvent. The mixture then enters the second-stage phase separator.

From the second-stage phase separator, the aqueous phase is pumped to a third continuously stirred vessel where it contacts regenerated organic solvent. The mixture is then sent to the third-stage phase separator.

The aqueous phase from the third-stage phase separator is pumped to treatment.

The acetic-acid-rich solvent from the first-stage phase separator is pumped to a multistage, single-vessel regenerator where hot water is used to reextract the acetic acid out of the organic solvent. The acid-lean solvent is then cooled and pumped to a surge drum where any fresh solvent can be added. The outlet of the surge drum is routed to the third-stage extractor. Pressurized regeneration water is heated by steam in a shell and tube heat exchanger. The temperature of the water is controlled based on the concentration of acetic acid remaining in the solvent after regeneration (as measured by an automatic online acid titrator). The quantity of hot water used is controlled so that the water phase leaving the regenerator is at a specific acetic acid concentration.

Draw an acceptable process and control scheme for the entire system acetic acid recovery system based on this description and assuming that all pumps have fixed speed motors.

5.17 Consider the process and control scheme depicted in Figure 5.14. A common addition described in the text is to add a cooler (using cooling water) to the solute-lean gas outlet stream followed by a knockout drum in order to reduce the water content of the gas. If the outlet temperature from of the gas leaving the knockout drum is the same as the temperature of the solute-rich feed gas entering the absorber, and if the water recovered in the knockout drum is pumped back into the feed drum, controlling the solvent concentration is much easier. Draw a schematic showing how Figure 5.14 would be modified to add these features.

5.18 Consider the process and control scheme shown in Figure 5.16. If this were one of three stages in a countercurrent adsorption system, draw a schematic showing an entire three-stage system.

5.19 A common way to regenerate many fixed bed adsorption systems is to suddenly and drastically reduce the pressure in the bed that is full of the solute. This is known as a pressure swing adsorption (PSA) system. Since the solute is only weakly bound to the sorbent, in many cases almost all of the solute will desorb out of the sorbent into the vapor phase as the pressure drops. The vaporized solute is removed from the adsorber vessel using a downstream compressor. Before being placed back into service, the regenerated adsorber vessel must be repressurized back to the process pressure conditions. In some systems, this is accomplished by simply opening the inlet block valve and allowing process gas to fill the vessel until the correct pressure is reached. Draw a schematic, logic diagram, and sequential events table showing the process and controls for the PSA system described, including the downstream compressor.

5.20 Refer to the description in Problem 5.19. In some applications, repressurization of a regenerated bed in a PSA system cannot be accomplished by simply opening the inlet block valve as this introduced too large of an upset in the downstream process. One way to handle this situation is to compress a small slipstream of the solute-lean downstream process gas and gradually pressurize the bed back to process conditions. When this bed is placed in service, it will not change the net flow of solute-lean process gas to the downstream process, avoiding the upset. Draw a schematic, logic diagram, and sequential events table showing the process and controls for the PSA system described in Problem 5.19 as modified by the description provided in this question.

6

Reaction Unit Operations Control

When present, chemical or biological reactions are the most important unit operations in a process control scheme. It is through reactions that raw materials are transformed into products. Efficient processes maximize the fraction of the inlet raw material that is converted into valuable products and minimize the fraction that ends up as a waste stream.

The majority of chemical reactions are accomplished using a catalyst to facilitate the desired reactions and inhibit unwanted side reactions. Whenever possible, continuous reactors are used. The most common are variations of the basic plug flow reactor (Figure 6.1). The reactants flow concurrently through a fixed (Figure 6.1a) or fluidized catalyst bed (Figure 6.1b), which also acts to keep the fluid well mixed. At the end of the reactor, the fluid contains products plus unreacted reactants and unwanted side products.

For easily reversible reactions, it may be better to perform the reaction in discrete stages (Figure 6.2). After each stage, the fluid is processed to separate out the products and unwanted side products from the unused reactants. Then the unused reactants are fed into a second stage, often with supplemental fresh feed of one or more of the reactants.

There are many specialty reaction configurations that we cannot address in this textbook. However, the principles described here should be widely applicable.

6.1 Continuous Flow Reactors

In this type of reactor, the raw materials are continuously input. As they flow through the reactor they react to form the products, which are then continuously removed from the reactor. Usually, one of the two reactants is the limiting reactant. When possible, conditions (stoichiometry, reaction temperature, etc.) are specified so that the limiting reactant is the most valuable reactant in the system, but there are many exceptions to this guideline. The residence time in the reactor is normally set by the upstream flow rate of the limiting reactant. A key parameter is the ratio of reactants (i.e., the stoichiometric ratio) fed to the reactor.

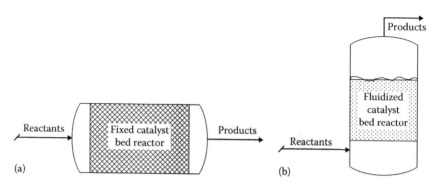

FIGURE 6.1
Plug-flow-type reactors: (a) through a fixed catalyst bed and (b) through a fluidized catalyst bed.

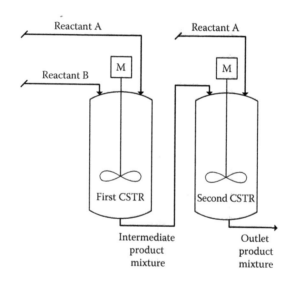

FIGURE 6.2
A multistage reactor system.

Any reactants that are not the limiting reactant are typically present in excess of the stoichiometric amount necessary for the reaction to maximize conversion of the limiting reactant. This leads to another key parameter(s), the ratio of the nonlimiting reactant(s) to the limiting reactant.

There is usually an optimum temperature or temperature range for each reaction and most reactors are designed to operate isothermally. Thus, in many reactors the temperature inside the reactor is carefully controlled. Reactors containing exothermic reactions may have cooling jackets (Figure 6.3a) or tubes (Figure 6.3b) where a cold utility fluid can extract the excess heat caused by the reactions. When the reaction is endothermic, a hot

Reaction Unit Operations Control

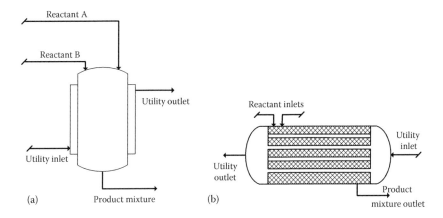

FIGURE 6.3
Maintaining isothermal reaction conditions using (a) a jacket or (b) heating/cooling tubes (in this example the shell is filled with catalyst).

utility fluid may be used in the jacket or tubes to provide the makeup energy necessary to maintain the reactor's temperature.

While less important, most reactions are also performed at near-optimum pressures. Pressure control is critical when there is a volume or phase change in the reactor. In some catalytic reactions, unwanted by-products can plug up the catalyst. This reduces operating performance and can lead to unsafe operating performance. Pressure drop can often be used to monitor this condition and initiate action before unsafe conditions are reached.

6.1.1 A Heterogeneous Binary Reaction of Two Liquid Reactants

Consider the reaction system shown in Figure 6.4. A limiting reactant, "A," is fed to a continuous flow reactor filled with catalyst. The second reactant, "B," is fed co-currently with "A" into one end of the reactor. In this configuration, tubes are installed in the reactor where cold water is used to remove the exothermic heat of reaction. The configuration in Figure 6.4 assumes that the reaction temperature is high enough to generate useful steam in the cooling tubes. The steam generated is used in an associated heat exchanger to preheat reactant "B" to near reaction temperatures.

The outlet stream from the reactor shown in Figure 6.4 is a single liquid phase. This stream is routed through a cross exchanger with the reactant "A" feed to preheat this stream to near-reaction conditions. It is then routed to downstream unit operations to separate the product stream into products, useful by-products, unwanted by-products (waste), and unreacted reactants. We will discuss how to construct an advanced control strategy for the entire system in Chapter 9. For now, let us focus just on the reactor unit operation.

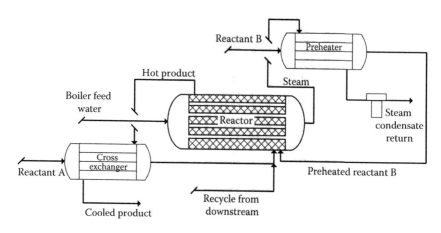

FIGURE 6.4
A fixed bed reactor for an exothermic reaction with two reactants, both of which are preheated. Reaction occurs on the shell side of the reaction vessel.

A mass balance of the reactor is as follows:

$$M_{A,I} + M_{B,I} + M_{R,I} = M_{P,O} + M_{Ac} \qquad (6.1)$$

where
  A, B are the reactants
  R is the recycle stream that will contain some quantity of all reactants and products (depending upon the separations performed)
  P is the product stream
  Ac is the accumulation or depletion of mass in the system
  I is inlet
  o is outlet

As per previous examples, we assume that the inlet flow of "A" is set by the upstream conditions and that the flow rate of the recycle stream is set by conditions in the downstream separation process. If the reaction is a liquid phase reaction, then the fluid is incompressible, so the flow rate of the product is no longer independent of the inlet flow rates. Equation 6.1 is used to account for the accumulation term. This is common for this type of reactor, where the reaction is taking place on the "shell" side of a "shell and tube" design. We will cover other cases below and in the "Problems" section at the end of the chapter. This leaves one degree of freedom, which is accommodated by a control valve (CV) on the "B" inlet line (or using speed control of an associated pump).

The key parameters are the ratio of "B" to "A" and the reactor temperature. Therefore, the CV on the "B" inlet is usually controlled using the ratio of the flow of "B" to the flow of "A." If the recycle stream ("R") is mostly "B," an intermediate calculation (see Section 7.1) may be used to sum the flow of "B" and the flow of "R" prior to taking the ratio.

The actual desired objective is to maximize the primary product ("P") while minimizing the wastes ("W"). A number of different options exist. A simple scheme, based on the process shown in Figure 6.4, would be to take the ratio of the measurement of the flow rate of the final "P" product to the flow of "A" into the system and use this in a cascade with the slave control loop. This will work well if the time lag in the separation system is not too long. If the generation of "W" is important, the ratio of the flow rate of "W" produced to either the flow rate of "P" produced or the flow rate of "A" consumed can be used. Another option is to take the difference between the quantity of "P" produced and the quantity of "W" produced and use this as the numerator in a ratio with the quantity of "A" consumed.

If time lag is a problem, another option may exist if the concentration of the most important product (or most undesirable waste by-product) can be measured in the product outlet stream. This concentration can be used in a master control loop that adjusts the ratio controller setpoint in order to meet a specified outlet concentration. Another variation is to measure the concentration of "A" in the outlet and adjust the ratio controller setpoint to keep the concentration of unreacted "A" at a specified maximum outlet concentration. This last option is shown in Figure 6.5.

The other important parameter to optimize is the reactor temperature. In Figure 6.4, boiler feed water is fed to the cooling tubes and steam is extracted from the tubes. Since a compressible fluid is produced, CVs are required on both the inlet and outlet streams. The outlet steam pressure is typically used with the outlet CV to insure that the steam produced is at the correct conditions for use.

The inlet water flow rate is typically coupled with the reactor temperature. Because this parameter is so important, it is common to measure the temperature at multiple points along the reactor and to use the average temperature as the primary dependent variable. A special controller, known as a calculation block (CALC block), accepts multiple input temperature readings and calculates the average, which is then output to the temperature controller for use in the feedback control loop. We will learn more about how to use CALC blocks in Section 7.1. This scheme is shown in Figure 6.5. If temperature is not quite this critical, the temperature of the outlet product stream can be used.

Also included in Figure 6.4 are two heat exchangers used for preheating the reactant streams. For the cross exchanger, no control is possible unless a bypass line or a trim exchanger is installed. A bypass would be used if the duty available from the hot product stream is too high for the inlet reactant A, which would then be at too high of a temperature for the reactor. The bypass of hot product around the cross exchanger would be controlled and coupled to the reactant A inlet temperature. A trim exchanger would be used if the duty available from the hot product stream is too low to get the inlet

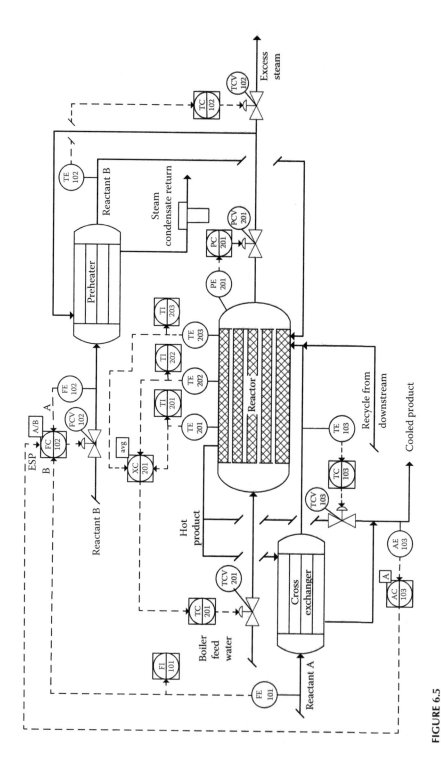

FIGURE 6.5
Control scheme for an isothermal (using cooling tubes) fixed bed reactor with an exothermic reaction and two reactants, both of which are preheated.

reactant A stream to the optimum temperature for the reactor. In this case, the heated inlet reactant A stream would be heated further either before or after the cross exchanger, most commonly with steam. Figure 6.5 shows the case where a bypass is employed.

The controls for the reactant B preheater are similar to those described for the cross exchanger. Figure 6.5 shows the case where a bypass is employed because the quantity of steam generated in the reactor is higher than the demand of the reactant B preheater.

6.1.2 Safety Control Systems

There are two primary process safety concerns that the control system can mitigate: (1) temperature excursions and (2) overpressure of the reactor.

Temperature excursions can occur either because of a loss of cooling water in the cooling tubes or because of a runaway reaction. The first step is to insure that there are adequate temperature readings in the reactor. In addition to the normal process temperature elements, an independent set of triple redundant temperature elements should be placed at the point where the maximum reaction temperature is anticipated or at multiple locations along the catalyst bed. A two out of three voting scheme is then used to determine the correct temperature reading at the measurement point.

This is another example of a CALC block and is similar to the one specified in Section 6.1.1 for the reactor temperature control system. For the safety controls, the temperature readings from three temperature elements all reading essentially the same temperature are compared to each other. The CALC block selects the two temperatures that are the closest to each other and averages their values. The third, and least accurate temperature measurement, is ignored. If this average reading exceeds the high, high temperature limit (TAHH) at any measurement point, safety action needs to be initiated. One of these actions could be to open the inlet cooling water flow CV completely. If the heat of reaction is low enough, this action may be sufficient.

If the heat of reaction is high, then additional action should be taken simultaneously. A common action is to introduce an inert fluid into the reactor. For gas phase reactors, this can work very quickly to quench the reaction, but in a liquid reactor, this action may be too slow. For a liquid reaction, the reactor can be dumped by opening a valve on a drain line or multiple drain lines located on the bottom of the reactor. This can be further enabled and the reaction can be quenched by using a pressurized inert fluid (liquid or gas) to increase the speed of flow of the reaction fluid out of the reactor while filling the reactor with an unreactive fluid. If these fluids are hazardous or flammable, they must be routed to a containment system. If not hazardous, a chemical sewer may be adequate.

The second reason for a temperature excursion is loss of cooling water. In this case, a TAHH limit would air-fail-open the boiler feed water inlet CV to the reactor cooling tubes.

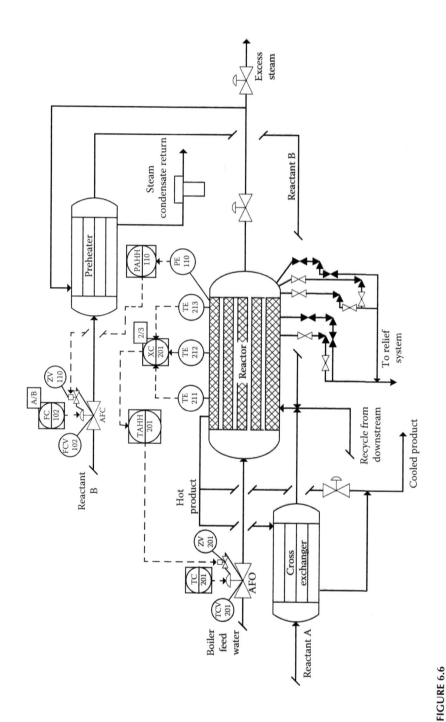

FIGURE 6.6
Typical safety system for a fixed bed reactor with an exothermic reaction and two reactants, both of which are preheated.

The second primary safety concern is overpressure of the reactor. This can occur because of fouling/plugging of the catalyst bed or because of a drastic increase/decrease in the volume of fluid in the reactor due to a higher-than-anticipated reaction rate. For the first case, we would measure the pressure in the inlet section of the reactor, say right before the start of the catalyst. If the high, high limit (PAHH) is reached, CVs on the inlet feed lines would be closed.

For the second case, we measure the pressure in the back section of the reactor. If the PAHH is activated, we would open the drain line or lines on the bottom of the reactor so that the excess liquid can leave the reactor before it reaches its upper pressure safety limit.

Typical safety controls are shown in Figure 6.6. In this case, triple redundant temperature readings from the reactor bed are configured in a two out of three (2/3) voting logic block. The block selects the two temperature readings that are the closest to each other and sends the average value to the TAHH. If the temperature signal exceeds the setpoint, the TAHH "air fails open" the CV on the boiler feed water inlet line. PAHH-110 monitors plugging in the reactor by measuring the pressure in the inlet section of the reactor. In addition to these safety control elements, relief valves must be installed on both the shell side (two since no phase change; one in service, one an installed spare) and tube side (three since the water is vaporizing; two in service, one an installed spare) of the reactor.

6.2 CSTR-Type Reactors

Another common type of reactor is a continuous stirred tank reactor (also known as a continuous-stirred-type reactor) or CSTR. In this reactor, the incoming liquid reactants are mixed together and a product stream matching the composition of the mixture is continuously withdrawn. If a catalyst is used, it is often present in a slurry form. These solids must be filtered out of the product stream, either in the reactor or in the product outlet line. The advantage of post reactor filtration is that the recovered catalyst can then be regenerated, if necessary, prior to recycling it back to the reactor.

Sometimes, the reaction nears equilibrium due to the concentration of products in the reactor before reaching the desired conversion. In these cases, an initial CSTR feeds a separation unit operation where product is removed from the mixture. This mixture can then be fed to a second CSTR. It is also common to add additional quantities of one or more of the reactants (usually the limiting reactant) to the second CSTR. This sequence can be repeated as many times as necessary to achieve the desired conversion.

Consider the CSTR system shown in Figure 6.7. Reactants "A" and "B" are fed into the first CSTR, with "B" being fed in excess to the stoichiometric amount required for the reaction:

$$A + 2B \rightarrow C \tag{6.2}$$

The temperature of the CSTR is controlled using a cooling water jacket surrounding the reactor. The outlet stream, containing unreacted A, unreacted B, and the product C, is sent to a distillation unit operation where a high purity stream of C is recovered as the bottoms product, while most of the A and B, plus a small quantity of C, is recovered as the overhead product. The overhead product is routed to the second CSTR along with additional "A." The product from the second reactor is mixed with the product from the first CSTR and fed to the distillation unit operation. The second CSTR also has a cooling water jacket.

The overall system mass balance is:

$$M_{A,1} + M_{B,2} + M_{A,3} = M_{O,C} + M_{Ac} \tag{6.3}$$

where
  1 is the A feed to CSTR 1
  2 is the B feed to CSTR 1
  3 is the A feed to CSTR 2
  O,C is the outlet product of C from the distillation system
  Ac is the accumulation or depletion in the system

Let us assume that the total feed rate of "A" to the system is fixed, but can be split between the two reactors. The quantity of C produced is set by the overall extent of reaction and the accumulation term is satisfied by Equation 6.3. This leaves two independent variables, one related to the feed rate of "B" and the other related to the split between the two "A" feed streams. CVs can be placed on the "B" inlet and either of the "A" inlet streams to account for these. If pumps are used to feed either "A" or "B" into the CSTRs, then the associated speed control can be used in lieu of the applicable CV (not shown).

Distillation system controls are covered in detail in Chapter 5 and so will not be repeated here. The distillation system is used to obtain the target purity of "C" in the distillation bottoms stream and to minimize the quantity of "C" in the overhead stream. To set the feed rate of "B" to the first CSTR, a ratio controller can be used that takes the ratio of the flow rate of "B" to the total flow rate of "A" (prior to the split). This control is shown in Figure 6.8.

Reaction Unit Operations Control

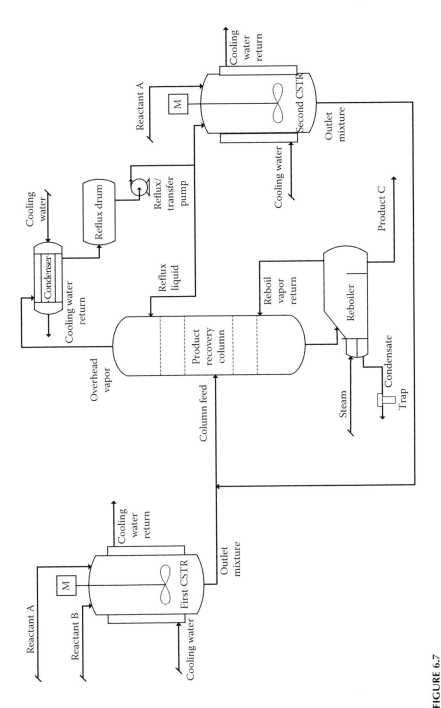

FIGURE 6.7
A two-stage CSTR reactor system with an intermediate distillation product purification step.

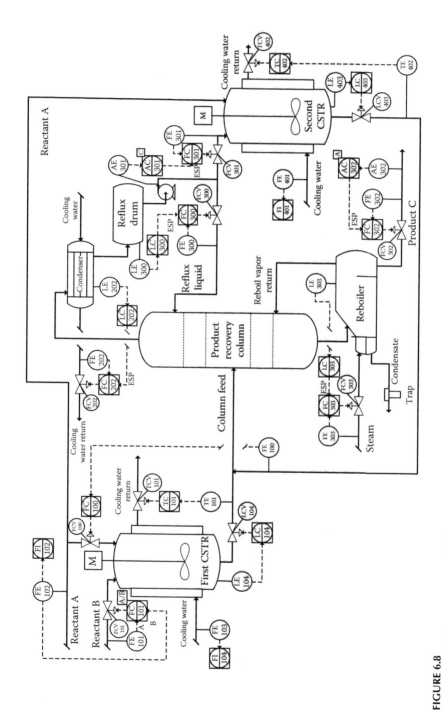

FIGURE 6.8
Typical controls for a two-stage CSTR reactor system with an intermediate distillation product purification step.

However, unless the feed rates are always perfect (an unlikely scenario), there will be a buildup of "A" or "B" in the system. A convenient place to measure this is in the distillation overhead stream, either with an online analyzer (preferred if possible), an indirect measure using a property that can be correlated to the relative concentrations of A and B in the stream, or with a periodic laboratory analysis. The concentration of "B" in this stream should be kept constant and should correspond to the target excess concentration of "B" that insures that the maximum quantity of "A" is converted into "C" in the system.

The system will be most economical if the load on the distillation column is minimized, that is, if the quantity of steam and cooling water used in the recovery system is the lowest necessary to meet the purity specifications. We will learn how to use more advanced optimization techniques to accomplish this in Chapter 9. For now, let us assume that through experimentation we have determined the recycle flow rate that approximates the optimum condition. We can then use this flow rate to adjust the split of "A" between the two reactors (FC-100 in Figure 6.8).

Another option involves the case where there is an unwanted side reaction that can form a waste product "D." In this case, a good control strategy would be to measure the quantity of "D" in the product stream and adjust the split of "A" to the two reactors to meet a target concentration of "D" (this setpoint could be optimized with an advanced control strategy to minimize "D" production).

Both CSTRs include cooling water jackets to remove the heat of reaction. In both cases, the temperature within the reactor or in the product outlet stream is measured and used to control the cooling water flow rate. In addition, the residence time in each CSTR is controlled using the liquid level in each vessel. These are shown in Figure 6.8.

6.3 Batch Reaction Systems

Batch reactions and batch reaction steps embedded in continuous processes are very common. Many reactions cannot be effectively carried out in a continuous system, including most biological reactions. In a batch reaction, the reactants are blended together at a specific temperature and pressure and then held at these conditions for a specified period of time so that the reaction can near (or reach) completion. Sometimes one of the reactants is added at intervals or continuously to the reactor in order to drive a reversible reaction toward higher yield of the product or to inhibit unwanted by-product formation (an example of such a system was described in Sections 1.2 and 2.11).

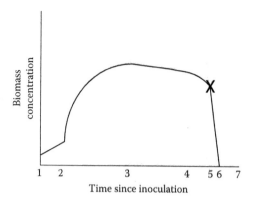

FIGURE 6.9
Typical batch cycle for biological reaction systems. Numbers are located near the end of each step: (1) inoculation, (2) accommodation, (3) growth, (4) production, (5) death, (6) removal of biomass, and (7) sterilization.

A unique subset of batch reactors is those used for biological systems. In these systems, an organism is used to produce the desired product from the raw materials. Biological transformations often involve a seven-step cycle (Figure 6.9): (1) inoculation (seeding the reactor with the organisms), (2) accommodation (allowing time for the organisms to adjust to their new environment), (3) growth, (4) production (when the organisms synthesize/expel the target product material), (5) death, (6) removal of biomass from the reactor, and (7) sterilization of the reactor (to prevent mutated or diseased strains from developing). When the biological step dominates the entire process, it often makes sense to operate the entire process in batches.

There will be significant differences in the inputs into the bioreactor for many of the steps in the process. For example, during the exponential growth phase it is important to feed the correct quantities of nutrients and carbon sources that will encourage the bacteriological culture to reproduce into healthy target quantities of biomass. Sometimes, biomass is the ultimate product; for example, it might be cultured for use as an animal feed. In this case, the reaction sequence may be terminated at the end of this stage.

For most bioreactors, the end product is synthesized by the microorganism. In this case, the rate of growth is monitored and when the growth phase reaches the target concentration or begins to slow down, the inputs are changed in order to encourage the microorganisms to synthesize the target product, which may be an antibiotic, an alcohol, a fatty acid, or other chemicals. The target product can sometimes be removed continuously during this stage, but for many bioreactors, the product remains in the aqueous phase of the bioreactor or in the organisms themselves.

At some point, the ability of the microorganisms to produce the target product declines. When the productivity drops below some target value, the reaction is terminated. At this point, the contents of the reactor are removed in one or more steps, depending on the system. A common option is to flush the reactor's liquid and solid contents out of the reactor and through a filter where the solids are separated from the liquid. A press or other secondary dewatering technique may be used to recover additional liquid, if the product is present in the liquid phase.

If the product is located within the cells of the microorganism, then the wet biomass may be dried further prior to processing. The cells may be ruptured by high-pressure pressing or using a solvent so that the product can be recovered out of the cells. A final purification step(s) is usually required to generate a commercial grade product.

From a control perspective, many of the sequential control techniques described for batch processes in the previous chapters can be used for batch reactors.

6.4 Batch Reactors in Continuous Processes

For large-scale production, it is more efficient to maximize the ability to use continuous process steps. In these cases the bioreaction step is embedded into the continuous process in a semi-batch manner, similar to the fixed bed sorption systems described in Chapter 5. If the residence time in the bioreactors is very long, many reactors in parallel may be utilized. Let us look at a fairly complicated example, the production of lactic acid. Lactic acid is a building block chemical that has become more important in recent years due to the development of polylactic acid (PLA) as a replacement for polyethylene or polycarbonate in certain applications such as food storage bags. PLA is preferred because it is not produced from fossil-fuel-derived raw materials, it is biodegradable, and the degradation products are not considered to be harmful to human health or the environment.

Lactic acid production can be divided into three main process sections: (1) seed yeast production, (2) lactic acid production, and (3) product recovery and purification. Seed fermentation (growth of biomass) of the yeast biocatalyst requires a supply of sterile deionized (DI) water, fermentation media, nutrients, and air. Production fermentation has similar requirements, but under anaerobic conditions, to promote lactic acid production. Crystalline dextrose has a dual use as the carbohydrate source (media) for both seed fermentation and production fermentation as well as the raw material for lactic acid production.

The seed production section is shown in Figure 6.10. A sugar source such as corn steep liquor is mixed with deionized water and dextrose syrup to produce a culture media. The culture media is pumped through a continuous sterilization system and routed to the seed reactors. Air is blown through

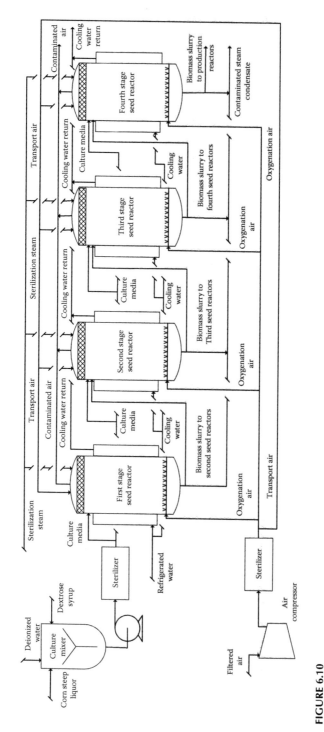

FIGURE 6.10
Typical seed reactor configuration for the production of lactic acid.

a sterilizer system and then fed to the seed reactors, as the yeast utilized need a well-aerated environment. The yeast consume oxygen and dextrose during their growth phase in order to increase the mass of yeast in the reactor. A 10-fold increase in yeast biomass is expected during a 24 h period. Generation of biomass for the production reactors is accomplished by transferring the growing biomass through a series of seed reactors of increasing volume. For example, if the growth time in each reactor is 24 h, then four reaction steps, each accommodating a 10-fold volume increase compared to the previous reactor, can be used. The growth phase is the terminal phase in the seed reactors. Sterile utility air is used to transfer the effluent from each seed reactor to the next larger-sized reactor in the process.

During yeast growth production, a constant temperature in the seed fermenters is maintained by removing the heat of fermentation via a cooling jacket. In this example, cooling for the seed fermenter cooling jackets is provided by refrigerated water utility.

For each seed reactor there are four batch steps in the sequence: (1) filling, (2) yeast growth, (3) draining, and (4) sterilization. Steps 1, 3, and 4 can all easily be performed in less time than the approximately 24 h time period of step 2; therefore, only two parallel reactors are required for each reaction step in the sequence for a total of eight seed reactors. Since the sequence for each reactor is identical, we will only describe the sequence for the initial, smallest volume reactors in the series. The sequence for the other three series reactor steps would be similar.

The control scheme for the culture media blending and sterilization, the air sterilization, and the initial seed reactors is shown in Figure 6.11 with the corresponding sequential events table shown in Table 6.1.

Culture media is continuously prepared and supplied to one or both of the reactors. There is a significant increase in media demand when one of the reactors is filling, so the culture mixer must have sufficient volume to accommodate this sudden change. Most likely, the culture media mixing operation would be common for all of the reactors and so would be of significant size. The flow rate of corn steep liquor into the mixer is controlled by the level in the mixer, as measured by LE-100 via a cascade loop using LC-100 to adjust the setpoint for the slave controller FC-100. Ratio controllers FC-101 and FC-102 are used to blend in the correct amounts of deionized water and dextrose syrup, respectively.

Eight steps are required to complete the entire cycle using two reactors. Conceptually, a bed is filled with an inoculation dose of yeast from the laboratory plus culture media (step 1) and then immediately is placed in biomass growth mode (steps 2–5). When the reactor reaches its capacity of biomass, the biomass/media slurry is pressurized to the second-stage seed reactors using sterilized compressed air (step 6). This step ends when the slurry level in the reactor reaches its minimum level. The bed is now under a slight pressure and full of air. The next step is to sterilize the reactor to avoid propagation of any diseases or viruses that may have developed in the bed (step 7).

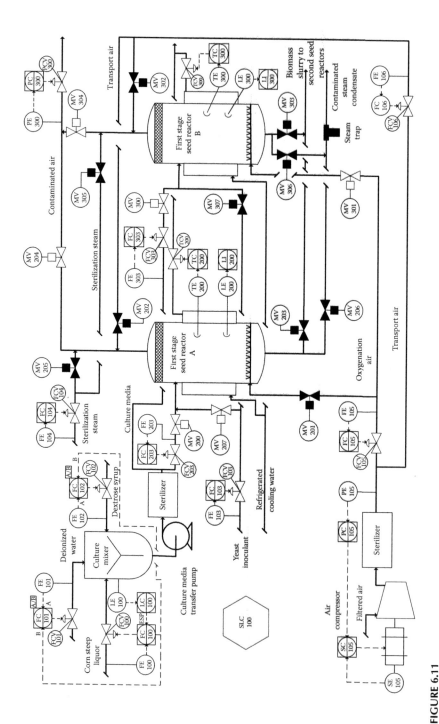

FIGURE 6.11
Control scheme for the first stage of a semi-batch seed reactor configuration for the production of lactic acid.

TABLE 6.1
Sequence of Events Table for SLC-100 for the First Stage of a Semi-Batch Seed Reactor Configuration for the Production of Lactic Acid

Step	Description	MV 200	MV 201	MV 202	MV 203	MV 204	MV 205	MV 206	MV 207	MV 300	MV 301	MV 302	MV 303	MV 304	MV 305	MV 306	MV 307	Termination Criteria
1	Fill Bed A with yeast and sterilized culture media; Bed B in growth mode	O	C	C	C	O	C	C	O	O	O	C	C	C	C	C	C	LI-200 = L_{target}
2	Bed A in growth mode; drain Bed B to second-stage reactor	O	O	C	C	O	C	C	C	C	C	O	O	C	C	C	C	LI-300 = L_{MIN}
3	Bed A in growth mode; sterilize Bed B	O	O	C	C	O	C	C	C	C	C	C	C	C	O	O	C	Sterilization time for B is complete as verified by lab samples
4	Bed A in growth mode; Bed B idle	O	O	C	C	O	C	C	C	C	C	C	C	C	C	C	C	LI-200 = $L_{NearMax}$
5	Bed A in growth mode; fill Bed B with yeast and sterilized culture media	O	O	C	C	O	C	C	C	O	C	C	C	O	C	C	O	LI-300 = L_{target}
6	Drain Bed A to second-stage reactor; Bed B in growth mode	C	C	O	O	C	C	C	C	O	O	C	C	O	C	C	C	LI-200 = L_{MIN}
7	Sterilize Bed A; Bed B in growth mode	C	C	C	C	C	O	O	C	O	O	C	C	O	C	C	C	Sterilization time for A is complete as verified by lab samples
8	Bed A idle; Bed B in growth mode	C	C	C	C	C	C	C	C	O	O	C	C	O	C	C	C	LI-300 = $L_{NearMax}$

For some processes, this sterilization step may only be performed periodically rather than every cycle. The bed is now ready for service and is held in this state (step 8) until the other seed reactor nears its maximum capacity of biomass/culture media.

Figure 6.11 shows the positions of the remotely operated block valves (MV-xxx) for the first step in Table 6.1. Reactor A is filled at a controlled rate with yeast as specified by the setpoint of FC-103 via MV-207. Culture media is also added to the reactor at the same time at a controlled rate as specified by FC-203 via MV-200 until a target level, as measured by LI-200, is reached. Because the flow rate into the reactors may change substantially during the filling step, the sequential logic controller, SLC-100, can change the setpoint for FC-203/303 automatically to adjust the rate of culture media addition during the filling step. It may be necessary to include a second media transfer line of a different diameter to accommodate the desired flow rate during the fill step. In that case, a second feedback control loop would be included on the second line. This is more likely to be required for the larger reactors used in stages 3 and 4 than in the small first-stage reactors and so is not included here.

The contaminated air outlet line on the top of Bed A is open (MV-204) so that air displaced from the reactor during filling can leave the reactor. A pressure controller, PC-100, is used to control the pressures in both reactors. The temperature in each reactor is controlled using TC-200/300, which adjust the cooling water rate through the reactor jackets depending upon the temperature in the respective reactor.

Once one cycle has been completed, both reactors will be in service and Bed B will be nearing the end of its growth cycle, as measured by LI-300, during step 1 in Table 6.1 sequence. During this step, culture media is metered into the bed using FC-303 via open MV-300. Air is routed into reactor B to provide oxygen to the growing biomass culture via MV-301. The delivery pressure of the air is controlled by PC-105 cascaded to the compressor speed controller, SC-105. The quantity of air sparged into either reactor during a biomass growth step is controlled by FC-105. Contaminated air exits reactor B via MV-304. PC-300 is used to maintain the back-pressure in the reactor.

Step 1 is terminated when reactor A has been filled to the appropriate level with yeast and culture media. At step termination, MV's open and close to the status shown for step 2 in Table 6.1.

In step 2, reactor A enters its biomass growth mode. During this time, culture media (via MV-200) and air (via MV-201) are metered into the reactor and contaminated air is allowed to leave. The biomass/culture media slurry in reactor B is pushed out of the bottom of the reactor via MV-303 using a controlled flow (FC-106) of transport air via MV-302. This continues until the level in reactor B, as measured by LE-300, reaches the minimum level target. Once this target is reached, step 2 terminates.

In step 3, reactor A remains in biomass growth mode. A controlled flow rate (FC-104) of steam is introduced into reactor B via MV-305. Contaminated steam condensate drains from the bottom of reactor B via MV-306. There is no good way to automatically determine when the bed is sterilized, so step 3 will go for a specified length of time, as determined by laboratory testing, to insure complete sterilization.

In step 4, reactor A remains in biomass growth mode until the biomass/culture media slurry level in reactor A is near its maximum as measured by LE-200, which terminates this step. Reactor B is isolated during this step and remains idle until reactor A has completed its growth cycle.

In step 5, reactor A remains in biomass growth mode while reactor B is filled at a controlled rate with yeast as specified by the setpoint of FC-103 via MV-307. Culture media is also added to the reactor at the same time at a controlled rate as specified by FC-303 via MV-300 until a target level, as measured by LI-300, is reached. When the correct level is reached, the step terminates.

In step 6, the biomass/culture media slurry in reactor A is pushed out of the bottom of the reactor via MV-203 using a controlled flow (FC-106) of transport air via MV-202. This continues until the level in reactor A, as measured by LE-200, reaches the minimum level target. Once this target is reached, step 6 terminates. Reactor B is in growth mode, with a controlled flow (FC-303) of culture media and a controlled flow (FC-105) of oxygenation air introduced into reactor B via MV-300 and MV-301, respectively. Contaminated air leaves reactor B via MV-304.

In step 7, a controlled flow rate (FC-104) of steam is introduced into reactor A via MV-205. Contaminated steam condensate drains from the bottom of reactor A via MV-206. There is no good way to automatically determine when the bed is sterilized, so step 7 will go for a specified length of time, as determined by laboratory testing, to insure complete sterilization. Reactor B remains in growth mode.

In step 8, reactor A is isolated and remains idle until reactor B reaches its near maximum level as measured by LE-300. The sequence then returns to step 1.

Problems

6.1 Consider the reactor control scheme shown in Figure 6.5. Draw a modified scheme if the inlet feed line for reactant B includes a variable speed pump.

6.2 Consider the reactor control scheme shown in Figure 6.5. Draw a modified scheme if the outlet product is in the vapor phase instead of in the liquid phase.

6.3 Consider the reactor control scheme shown in Figure 6.5. Draw a modified scheme if the composition of "A", used in the ratio-with-feedback trim control is replaced by the difference between the quantity of "C" and the quantity of "W," an unwanted waste, produced in the product stream.

6.4 Consider the reactor control scheme shown in Figure 6.5. Draw a modified scheme if the energy required to heat reactant "A" to its desired feed temperature is larger than the maximum quantity of heat that can be released by cooling the product stream in the cross exchanger.

6.5 Consider the reactor control scheme shown in Figure 6.5. Draw a modified scheme if the energy required to preheat reactant B is greater than the energy that can be provided by the steam generated in the reactor.

6.6 Two gas phase reactants are added to the bottom of a fluidized bed reactor, containing a catalyst. There are both minimum and maximum flow rates within the reactor in order to maintain the fluidization at the correct level in the reactor. Reactant "A" is the limiting reactant in the reaction:

$$A + 2B \rightarrow 2C \tag{6.4}$$

The reaction conversion rate is 40% per reactor pass, with no unwanted or side reactions. The reaction is endothermic and energy must be added to the reactor in order to maintain the reaction temperature within the optimum operating range. There is an increase in volume due to the reaction that must be controlled. Both reactants should be preheated to a specified temperature.

    6.6.1 Develop a process scheme for this reaction system.

    6.6.2 Develop a control scheme for your reaction system.

6.7 Consider the reactor scheme described in Problem 6.6. Modify the process and control schemes if the reaction is exothermic and energy must be removed from the reactor to maintain the reaction temperature at its optimum point.

6.8 Consider the reactor scheme described in Problem 6.6. Draw a safety control system for this reaction system.

6.9 Consider the reactor scheme described in Problem 6.7. Draw a safety control system for this reaction system that includes an inert quench gas capability.

6.10 Consider the reaction control scheme shown in Figure 6.8. Draw a modified scheme to use the difference between the quantity of "C" and the quantity of "W," an unwanted waste stream, produced in the product stream as feedback trim to the appropriate ratio controller. Justify your design.

6.11 Consider the reaction control scheme shown in Figure 6.8.

    6.11.1 Add a second distillation column to remove the unwanted by-product "W" from the product mixture. Justify the position of the column relative to the other unit operations in this system.

    6.11.2 Draw a control scheme for the new distillation system. Highlight any changes to the original control scheme shown in Figure 6.8 that you suggest for the modified system.

6.12 Two nonpolar liquid reactants, A and B, are reacted together via the slightly exothermic reaction:

$$A + 2B \rightarrow C + D \quad (6.5)$$

where C is a valuable gas phase product and D is a valuable nonpolar liquid phase product with a higher volatility than either A or B.

The single-stage conversion is 15% based on the depletion of "A," which is the more expensive reactant. Therefore, a five-stage CSTR reaction system is employed. Products C and D are removed between stages and additional B is added to each stage. Reactants should be preheated prior to each reaction stage.

6.12.1 Develop a process scheme for this reaction system. Highlight any assumptions you make in order to optimize the energy consumption by the system.

6.12.2 Develop a control scheme for your reaction system.

6.13 Consider the reaction system from Problem 6.12. Modify the process and control scheme if the reaction is highly exothermic.

6.14 Consider the reaction system from Problem 6.12. Modify the process and control scheme if the reaction is highly endothermic.

6.15 Consider the reactor control scheme shown in Figure 6.11. Modify the sequential events table (Table 6.1) for the case where the sterilization step is only performed every third time the reactor is emptied instead of every time as specified in Table 6.1.

6.16 A number of companies are trying to develop a commercially viable process to generate biofuel oils from algae. After the completion of the batch biological cycle shown in Figure 6.9, the algae are filtered out of the aqueous growth media and then conveyed to a press where part of the oil is removed from algae along with entrained water (a two-phase liquid mixture). The algae are then introduced into a leaching vessel using a screw conveyor where they settle through a mixture containing a hexane and ethanol solvent mixture. The remaining oil from the algae is leached out into the solvent. The bioreactor step is configured as a semi-batch step within the overall continuous process.

6.16.1 Draw a logic flow chart for this semi-batch system.

6.16.2 Develop a process scheme for this reaction system. Highlight any assumptions you make in order to optimize the energy consumption by the system.

6.16.3 Develop a control scheme for your reaction system as well as a sequential events table. Show the position of all on/close or on/off devices for the first step of the process on your diagram.

7
Other Control Paradigms

When possible, it is best to use simple, straightforward feedback control loops along with feed forward and cascade controls to mitigate the impact of disturbance variables. However, there are circumstances where other configurations of basic regulatory control will improve the stability of the process. The most common of these will be introduced in this chapter. In Chapter 9, we will cover more sophisticated control strategies, where complex controls are layered onto an underlying basic regulatory control scheme in order to increase the efficiency of the process.

7.1 Using Intermediate Calculations in Control Loops

Up to this point in the book, we have usually assumed that single-parameter measurement would be sufficient to allow adequate control of the process. This assumption is adequate for the vast majority of control scenarios. However, there are times when multiple measurements need to be combined in some fashion for adequate control. For these cases, a feature in all modern distributed control systems (DCS) can be used: the *calculation block*.

The *calculation block*, often referred to as the CALC block or CALC controller, was included in the first DCS software to accommodate orifice- and venturi-type flowmeters. These types of instruments are described in Chapter 10. In both cases, these meters take advantage of the following relationship, known as the Bernoulli equation:

$$\frac{P}{\rho} + \frac{v^2}{2} + gH + f = \text{Constant} \tag{7.1}$$

where
 P is the fluid pressure
 v is the fluid velocity
 H is the height compared to a reference point
 f is a term related to the friction of the system and any deviations from ideal flow

By measuring the pressure before and after a restriction in a horizontal flow field, the pressure drop between the two points is related to the velocity of the fluid:

$$\Delta P \propto v^2 \quad \text{or} \quad \sqrt{\Delta P} \propto v \tag{7.2}$$

and since the cross-sectional area of the pipe containing the flowmeter is constant,

$$\sqrt{\Delta P} \propto (v)(A_x) = G \tag{7.3}$$

where
A_x is the cross-sectional area of the pipe
G is the volumetric flow rate

In those first DCS, the software engineer setting up the control scheme had to include a CALC block to take the square root of the flowmeter signal (which was really the pressure drop signal). The output from the CALC block was then used in the flow controller block. In most modern DCS, this CALC block is added by simply telling the flow controller block what type of flowmeter is being used as the inlet stream.

However, the CALC block feature is still available and can be used for many other applications. Each CALC block performs a single mathematical operation—add, subtract, multiply, divide, square root, square, and so on—although there are some special functions that are a bit more sophisticated, like the "average" and the "2/3 voting" functions. The outlet from one CALC block can feed into a second CALC block. For example, it might be desirable to adjust the reading from a venturi-type flowmeter for deviations in the temperature of the fluid compared to some reference condition. In this case, the ratio of the reference temperature to the actual fluid temperature can be determined in one CALC block. The flow signal would be multiplied by this ratio in a second CALC block. The outlet from this second CALC block would be input into the flow controller block (which would actually use another CALC block to take the square root of the reading prior to using it in the controller algorithm). This example is shown in Figure 7.1.

Notice in Figure 7.1 that the CALC blocks are not visible to the control operator. This is common for simple, routine operations or where display of the CALC block operation might add confusion. However, the option is available to allow the operator to actually see the CALC block operation, if this is desired.

Let's look at a few more examples.

Other Control Paradigms

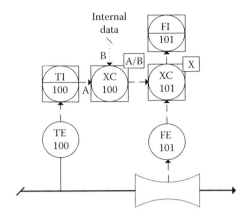

FIGURE 7.1
Using CALC blocks to correct a flow rate reading for temperature.

7.1.1 Using CALC Blocks in Blending Applications

Consider a blending unit operation where two streams are blended together to produce a single outlet stream with a specified concentration of a key component in the combined stream, as shown in Figure 2.6. We analyzed a simplified version of this system in Section 2.4 when we learned about disturbance variables. In that example we were only concerned with the total mass flow rate of the inlet and outlet streams. But now, we are trying to get the outlet concentration of a component (let's call it A) to match a specified value.

In Section 2.5, we learned how to use the flow rate of one of the inlet streams (let's call it stream 1) as a disturbance variable in a feed forward/feedback (FF/FB) control scheme to adjust the flow rate of the other inlet stream (let's call it stream 2). In that example we assumed that changes in the inlet stream 1 concentration were small enough that they could be ignored.

Now, let's consider the case where both the flow rate and the composition of A in the inlet stream can change sufficiently to cause a disturbance in the outlet stream's concentration. One way to handle this is to measure both the flow rate and composition of the inlet stream, and compare each to its own setpoint to calculate two controller outputs. These outputs can then be added together in a CALC block, and the combined output can be used as the feedforward input for a FF/FB control scheme. This scheme is shown in Figure 7.2.

If the order of magnitude of the deviations are sufficiently different, the approach just described becomes unstable and a different approach should be used. For this case, the mass flow of component A in the inlet stream can be calculated by multiplying the inlet concentration by the inlet flow rate in a CALC block. The outlet of this block can then be used in the feed forward

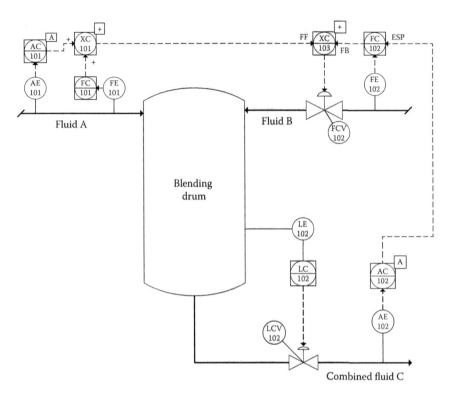

FIGURE 7.2
Using CALC blocks to determine the correct rate of blending of two variable streams. Both the composition and flow rate in fluid A are compared to their setpoints. Then the controller outputs are summed to provide a feed forward adjustment to the fluid B inlet flow rate. The composition of A in the product outlet stream is used as the feedback trim to adjust the setpoint for the feedback flow controller, FC-102.

controller to calculate the output due to the disturbance. This configuration is shown in Figure 7.3.

Let's add one more complication to our example: variation in the concentration of B in stream 2 as another disturbance variable. So now we have three disturbance variables, stream 1 flow rate, stream 1 composition, and the composition of "B" in stream 2. One way to handle this is to measure both the flow rate and composition in inlet stream 1 and the composition in inlet stream 2, and compare each to its own setpoint to calculate two outputs. Since analysis is a nonresponsive measurement, we still use flow rate as the feedback loop and use the combined outputs from the feedforward controllers for fluids A and B for the feed forward contribution. This control scheme is shown in Figure 7.4.

Another approach is to calculate the mass flow of component A in its inlet stream and the mass flow of component B in its inlet stream, and then calculate

Other Control Paradigms 223

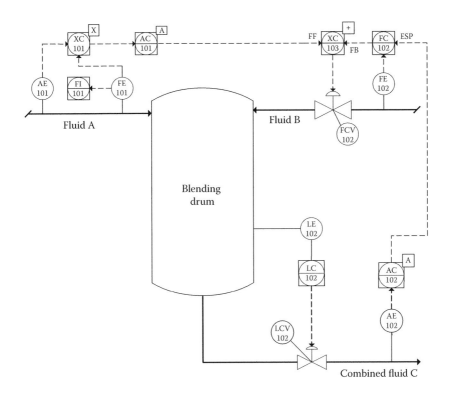

FIGURE 7.3
An alternate CALC block configuration. The flow rate of A in fluid A is calculated and compared to a setpoint and used in the FF/FB loop with the flow of fluid B.

two controller outputs. The two outputs would be added together and used in the FF/FB controller (Figure 7.5). The disadvantage of this scheme is that it requires four measurements just to calculate the disturbance variable. As a consequence, there is a greater chance of having a bad measurement. It is also more likely that the controls will become unstable.

A better solution is to control the concentration of B in stream 2 first by adding a second blending operation, if required, upstream of the current blender. This converts a disturbance variable to a control variable, which will improve the ability to tightly control the current blender (Figure 7.6).

7.1.2 Using CALC Blocks to Monitor Heat Exchanger Performance

Let's look at one more example of how CALC blocks can be used. In Section 4.5, we looked at how to monitor and adapt the operation of a heat exchanger due to fouling or scaling. A more complicated case occurs when either the process or utility stream temperatures vary substantially over

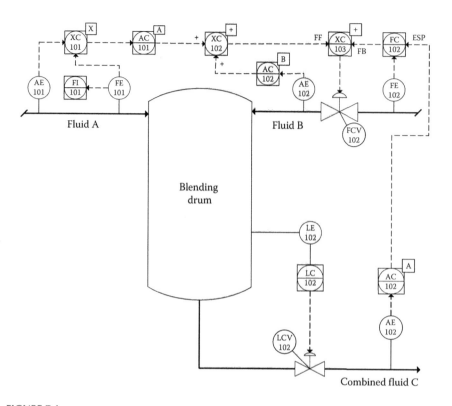

FIGURE 7.4
Another alternate CALC block configuration. The flow rate of A in fluid A and the composition of B in fluid B are calculated, compared to setpoints, and used in the FF/FB loop.

time, requiring a more sophisticated monitoring system than the one shown in Figure 4.26. The heat flux in the exchanger can be calculated in two separate ways:

$$q = \dot{m}_f \overline{C}_{p,f} \Delta T_f = UA\overline{\Delta T_x} \tag{7.4}$$

where
 q is the heat flux
 $\dot{m}_f$ is the mass flow rate of one fluid stream
 $\overline{C}_{p,f}$ is the average heat capacity of that fluid stream
 ΔT_f is the difference in the temperature of that fluid between the inlet and outlet of the exchanger
 U is the overall heat transfer coefficient for the exchanger
 A is the cross-sectional heat transfer surface area of the exchanger
 $\overline{\Delta T_x}$ is the average temperature driving force between the two fluids in the exchanger

Other Control Paradigms

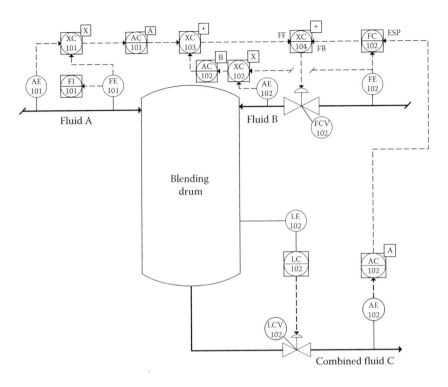

FIGURE 7.5
Another alternate CALC block configuration. The flow rate of A in fluid A and the flow rate of B in fluid B are calculated and compared to setpoints and then used in the FF/FB loop.

Rearranging this equation yields:

$$U = \frac{\dot{m}_f \bar{C}_{p,f} \Delta T_f}{A \Delta T_x} \tag{7.5}$$

$$U \propto \frac{\dot{m}_f \Delta T_f}{\Delta T_x} \tag{7.6}$$

We can use CALC blocks to determine $U(t)$, which can then be compared to a minimum acceptable value. When U reaches this minimum, an alarm can be activated to inform the operators that the heat exchanger should be taken out of service and cleaned. This control scheme is shown in Figure 7.7. Note how many measurements are required for this system. The temperature of both inlet and both outlet streams are measured and CALC blocks are used to determine the temperature change of the process fluid (XC-100) as well as the temperature driving force at each end of the exchanger (XC-101 and XC-102). The average temperature driving force of the heat exchanger is then calculated using XC-103. Finally, XA-104 calculates the overall heat transfer

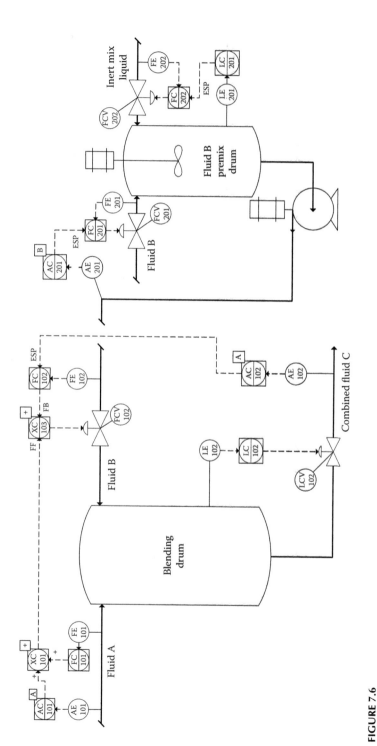

FIGURE 7.6
A blend system where B is premixed with an inert fluid prior to mixing with fluid A to increase the stability of the overall control system to changes in composition or flow of either fluid.

Other Control Paradigms

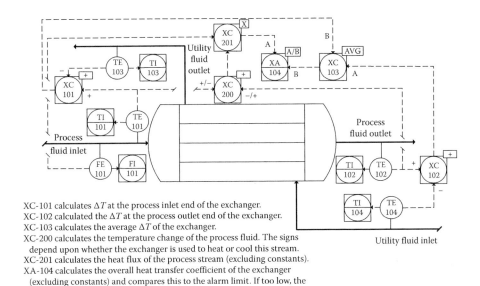

XC-101 calculates ΔT at the process inlet end of the exchanger.
XC-102 calculated the ΔT at the process outlet end of the exchanger.
XC-103 calculates the average ΔT of the exchanger.
XC-200 calculates the temperature change of the process fluid. The signs depend upon whether the exchanger is used to heat or cool this stream.
XC-201 calculates the heat flux of the process stream (excluding constants).
XA-104 calculates the overall heat transfer coefficient of the exchanger (excluding constants) and compares this to the alarm limit. If too low, the alarm is activated.

FIGURE 7.7
System to monitor the heat transfer efficiency in a heat exchanger prone to fouling or scaling where the temperature of one or more of the inlet streams varies widely. This scheme calculates "UA" from Equation 7.4. Since "A" is a constant, "U" can be tracked to monitor overall heat exchanger performance.

coefficient and compares it to a lower alarm limit. If "U" is at or below this setpoint, the heat exchanger can no longer operate efficiently and the alarm notifies the operator that the exchanger should be taken out of service and cleaned.

Another way to monitor heat exchanger performance is to determine the amount of heat added to or removed from the process stream (heat flux) per unit quantity of utility consumed using the first equality in Equation 7.4. This scheme, shown in Figure 7.8, requires fewer measurements than the scheme shown in Figure 7.7 and so is more reliable.

7.2 Inferential Control

Inferential control is a technique that can be used if we cannot directly measure the desired measurement parameter. This method is also known as *model-based control*. In inferential control, one or more process measurements are used to infer an unmeasured measurement variable.

Usually, an analysis of the physical and/or chemical properties of the process fluid (or solid) will reveal a related, measureable parameter that we

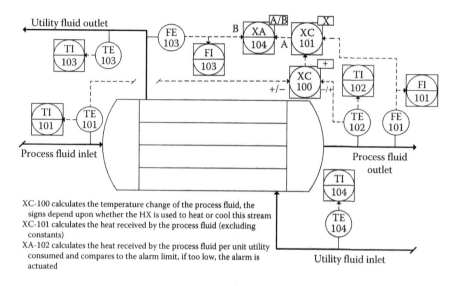

FIGURE 7.8
An alternative system to monitor the heat transfer efficiency in a heat exchanger prone to fouling or scaling where the temperature of one or more of the inlet streams varies widely. This scheme calculates the heat transferred per unit quantity of utility consumed.

can use to infer the effect that the independent (control) variable has on the dependent (measured) variable. We've already looked at a couple of examples of this in previous sections. For example, in the acid waste neutralization example in Section 2.4, we measured the pH of an aqueous fluid to ascertain if the amount of neutralizing base was correct. This is an indirect measure. The true parameter is the concentration of acidic species in the fluid. But that measurement is typically slower, if possible at all, more expensive, and less reliable than measuring the fluid pH.

Let's look at another example. Consider a system where three gas phase streams are combined and fed to a compressor. If the composition of one or more of the three gases varies significantly, this could significantly change the mass of material in the combined stream due to changes in the gas density. The power needed in the compressor is related to the mass being processed. So if the density changes significantly, this sudden change could cause a surge in the outlet pressure (up or down), cause either or both of the independent control variables to become unstable, and/or cause physical damage to the compressor itself. Measuring the composition of the three streams is too slow to react to this situation. A quick inferential variable is to measure the density or specific gravity of the combined stream. This measurement can then be used as a feed forward signal in an FF/FB arrangement with one of the control loops. Figure 7.9 shows a typical control scheme when gas density is used as an inferential measure of gas composition.

Other Control Paradigms

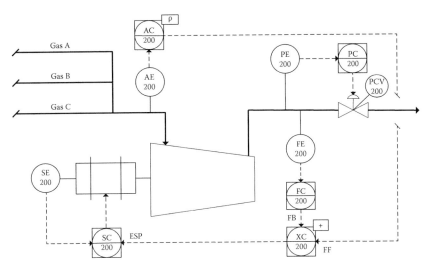

FIGURE 7.9
Inferential control scheme to protect a compressor from changes in inlet gas density.

When gas and liquid phases are both present, temperature can often be used as a reliable indirect measure of composition. Consider the partial condenser on the overhead of a distillation system as shown in Figure 5.5a. This control system was used in Section 2.8 to introduce cascade control loops to regulate the flow of reflux back to the column. Here we focus on the controls in this same scheme for the heat exchanger (see Figure 2.17). The partial condensation can be considered to be at equilibrium and is equivalent to one 100% efficient separation stage within the distillation column. Because the system is at or very near equilibrium, there is only one possible composition for a given temperature and pressure of the two-phase fluid. So, we can measure the outlet fluid temperature or pressure as an indirect measure of composition. Unfortunately, measurement in a two-phase fluid is often unreliable, so instead of measuring the temperature or pressure in the condenser outlet stream, the measurement is taken in the gas outlet stream (the distillate product) from the reflux drum located immediately downstream of the condenser.

As another example, consider the situation where we are entraining a powder in a gaseous stream in order to move the powder from one unit operation to another. The most logical dependent variable is the mass flow rate of powder in the piping. However, this would be very difficult to measure directly. Instead, we might measure the flow rate of the entire stream using a venturi meter (to avoid disturbing the entrained solids) and measure the rate of buildup of powder on a filter using a small slipstream sampled from the piping. We could use the average weight of powder trapped on the filter (say over the period of a minute) along with the overall flow rate to infer the mass flow rate of the powder in the piping.

Another version of this technique uses a *soft sensor*. A soft sensor is a virtual measurement calculated from measured process variables, laboratory values, online simulations, and/or model equations. For example, we might want to use the degree of conversion of a raw material into a product in a reactor as the dependent variable to control the feed rate of one reactant to the reactor. Laboratory samples might be used to determine the composition of the reactor outlet, which could be combined with flow rates to calculate a conversion. The soft sensor would use the last available composition data from the laboratory combined with the real-time flow rates to infer changes in the conversion rate. Figure 7.10 illustrates how to depict a control scheme using a laboratory input.

Perhaps it is difficult to analyze the outlet composition, even in the laboratory. If the reaction is highly exothermic, it might be possible to use the heat of reaction, coupled with changes in the reactor temperature (or changes in the cooling water flow rate necessary to maintain a constant temperature) to estimate the conversion. This is depicted in Figure 7.11.

In more sophisticated versions, an online model of the unit operation is built and used to predict changes in operation based on changes in the available real-time measurements. The output of the simulation is the inferred dependent variable used in the controller.

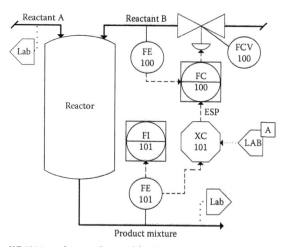

XC-101 is a soft sensor that uses laboratory composition analyses and product flow rate to calculate the conversion of reactant A in the reactor and uses the result to adjust the quantity of reactant B fed to the reactor.

FIGURE 7.10
Example of how to depict a *soft sensor* that uses multiple laboratory-generated input data.

Other Control Paradigms

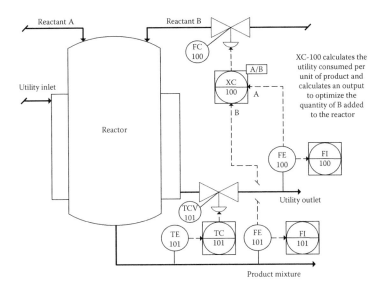

FIGURE 7.11
Using the quantity of cooling water consumed per unit quantity of reactant A to infer the optimum quantity of reactant B to feed to a reactor.

7.3 High and Low Select Controls

In certain circumstances, we want to take multiple measurements and select one of them for use as the measurement variable in a control scheme. Modern control systems include CALC blocks that can select the lowest value amongst multiple inputs known as a *low selector or low select controller*, or select the highest value amongst multiple inputs known as a *high selector or high select controller*. This type of CALC block is also known as an *auction controller*. The symbols for these two control blocks are shown in Figure 7.12.

A common use of the high selector is to control the temperature of an exothermic reaction in a fixed bed reactor. Consider a system where a reactor contains a fixed bed of catalyst and where the heat of reaction is removed

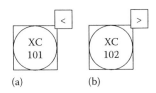

FIGURE 7.12
Symbols for (a) low select and (b) high select CALC blocks.

using a cooling jacket surrounding the reactor bed. As the reactants flow through the fixed bed of catalyst, their reaction will usually be greatest at the point in the bed where the catalyst activity is highest. As the catalyst bed ages due to fouling, scaling, and/or other deactivating mechanisms, the point of greatest reaction activity will move toward the end of the reactor. For an exothermic reaction, the temperature will be highest at this point in the reactor.

If the reaction is part of a food production process, this reaction (or cooking) temperature may be critical to avoid ruining (burning) the food product. Another application is when there is a danger that the reaction will become uncontrollable above a certain temperature (i.e., a runaway reaction or spontaneous combustion). Yet another example is when the normal reaction temperature is within a couple of hundred degrees of the safe operating temperature of the reactor. For many carbon steels, the metal will undergo a crystal structure phase change if the temperature rises above a certain point, which substantially changes the size and strength of the metal.

For the systems described above, a series of temperature measurements are taken along the length of the reactor. All of the measurements are inputs to a high selector CALC block. The highest temperature is then used in a feedback temperature control loop, which adjusts the cooling water rate through the jacket. This scheme is shown in Figure 7.13.

For an endothermic reaction, steam or a heat transfer fluid may be used to add energy to the reactor in order to maintain the reactor at the desired reaction temperature. For this case, a low selector would be utilized.

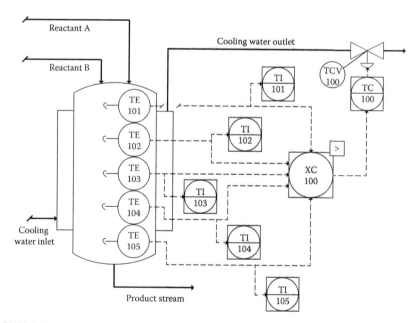

FIGURE 7.13
Reactor temperature control scheme employing a high-temperature select or CALC block.

Other Control Paradigms 233

7.4 Override Control

For most applications, we use the measurement of a dependent variable to control the position of a single independent variable. However, there are times when the correct matching of dependent and independent variable changes depending upon process conditions. The most obvious case is for batch processes, and we have already seen how sequential logic controls can be used to address this case. But there are other cases where this circumstance also occurs in continuous processes. The next three sections introduce special control configurations that can be used for many of these special cases.

The first of these types of controllers is the *override controller*. In this scheme some type of decision block is used to determine which of two controller outputs to apply to an independent variable. Consider, for example, a process where a slurry is routed from a mixing drum through a slurry pump to a downstream unit operation. The system was designed such that at normal design conditions, the liquid level in the mixing drum is used to control the speed of the outlet pump. However, regardless of the actual throughput of the unit, the flow rate through the pump must be above a certain minimum level to avoid having solids settle out of the fluid in the piping. So in this case, the *override controller* will compare the flow rate in the piping to a minimum required flow rate. If the rate exceeds the minimum, the level controller output is routed to the speed controller. However, if the rate is less than or equal to the minimum flow, then the flow controller output is routed to the speed controller. The control scheme is shown in Figure 7.14.

7.5 Split Range Control

A *split range controller* directs the output from a controller to one of two parallel control variables based on a decision parameter. In the most common application, the output is sent to one control variable when the value of the measurement variable is below a certain criterion and sent to a different control variable when the value of the measurement variable is above a certain criterion. However, the decision doesn't have to be based on the measurement variable. For example, the criterion might be the "valve position" (or fraction of available speed) of the active control variable.

Sometimes, the flow rate through a certain section of piping changes drastically with changes in the process conditions. This could be because the same piping is used for two different steps in a batch or semi-batch process or could be due to other circumstances. If the flow rate change is too great, a control valve (CV), incorporated into the piping will not operate well under both conditions.

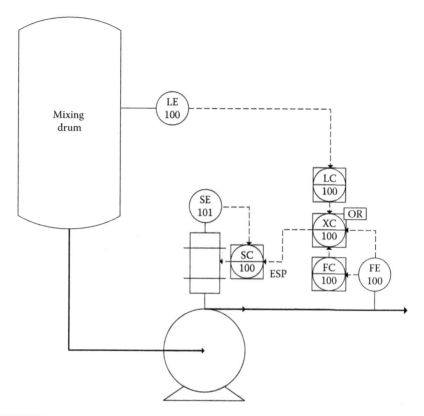

FIGURE 7.14
Override control scheme to insure that the flow of slurry through the pump meets or exceeds the settling velocity of the solids in the slurry. OR denotes an override controller. XC-100 selects FC-100 if the measurement, FE-100, falls below the specified minimum.

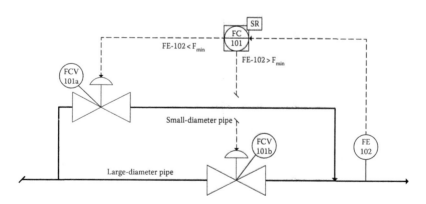

FIGURE 7.15
Use of a split range controller to improve control under two drastically different flow conditions.

Other Control Paradigms 235

To handle this condition, a second, smaller diameter pipe can be installed in parallel with the primary piping at the site of the CV. A second CV is also installed on the smaller diameter pipe loop. A split range controller can then be configured with a flow measurement variable (MV). If the flow rate is in the range of the high flow condition, the output from the controller is sent to the CV on the larger pipe. If the flow rate is in the range of the low flow condition, the output from the controller is sent to the CV on the smaller pipe. This scheme is shown in Figure 7.15. By convention, the CV that handles the lower criterion value is labeled "a" and the CV handling the higher criterion value is labeled "b."

7.6 Allocation Control

Allocation control is a slightly different way of using a single measurement and a single controller with multiple CVs. In split range control, the output of the controller is sent to one single independent variable at a time based on the threshold criterion. In *allocation control*, a fraction of the controller output is sent to two or more independent variables simultaneously. Consider a system where three pumps are used to transfer a slurry from an upstream unit operation to the same downstream unit operation. In order to maintain even wear and tear on the pumps, it may be desirable to have all three variable speed pumps operate under the same conditions. In Section 3.1, we examined one way to accomplish this: using the speed measurement of one pump as input to the speed controller of another pump. An allocation controller can accomplish the same thing by having 1/3 of the controller output sent as an external setpoint to each pump's speed controller. Another use of this controller is to even out the flow of a process fluid through a bank of parallel heat exchangers. Figure 7.16 shows a control scheme for five parallel exchangers (four in service and one an installed out-of-service spare) using an allocation controller. The total flow is measured and used to calculate an output in the controller. Twenty-five percent of the output is sent to each control valve to adjust the flow rate through the associated heat exchanger. The operator can change the denominator in the allocation control calculation if the number of exchangers in service changes.

7.7 Constraint Control

Constraint control is a more sophisticated application of the principles used in the controllers described above. When the operation of an independent variable nears its high or low limit (e.g., when a control valve approaches

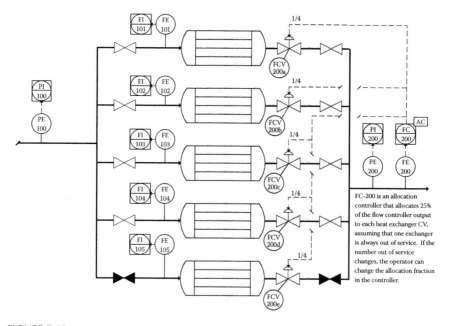

FIGURE 7.16
Allocation controller used to control the flow rate through four parallel heat exchangers with one installed spare. *Note*: the allocation does not need to be set to the same values, as shown here; any allocation that sums to 100% can be used.

100% open or 100% closed), that independent variable has reached a constraint in its ability to control the process. In split range control, when such a constraint is reached, the controller output is sent to a separate independent variable. However, sometimes a more sophisticated reconfiguration of the control scheme is required.

Consider a distillation system where meeting specified overhead and bottoms product compositions are the primary objectives of this separation unit operation. In most cases, the reboiler is oversized (has a higher design duty) compared to the overhead condenser, because it is more expensive to provide cooling utility capacity than heating utility capacity (although there are many exceptions to this guideline).

Here's an example. A distillation system that is being operated above its original design capacity is shown in Figure 7.17. This system has a partial condenser and a reflux rate that is greater than one. Under this scenario, the CV on the cooling water flow leaving the overhead condenser may approach 100% open. At this point, the rate of cooling water flow through the overhead condenser is no longer an independent variable. One of the objectives of the distillation system can no longer be achieved. If nothing is done, this will impact the overhead product purity. But what if this purity is more important than the bottoms product purity? In general, it is better to keep the overhead

Other Control Paradigms

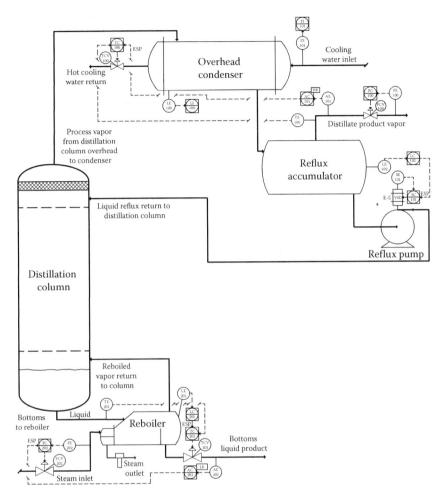

FIGURE 7.17
Typical control scheme for a distillation column with both overhead and bottoms composition control specifications. LK denotes the light key. HK denotes the heavy key.

and bottoms systems separate (decoupled), as explained in Chapter 5, but for this circumstance this isn't possible. In this case, we would want to reconfigure the overall control scheme to match the overhead product composition with the steam flow rate to the reboiler (Figure 7.18).

A *constraint controller* is a decision block that reconfigures the control scheme if the constraint is reached. Because multiple changes are made simultaneously, this is considered an advanced control strategy and the octagon symbol is used to denote this controller. For documentation on a P&ID, the configuration after the constraint is reached would typically be shown on a supplemental P&ID and referenced from the P&ID that depicts the normal configuration.

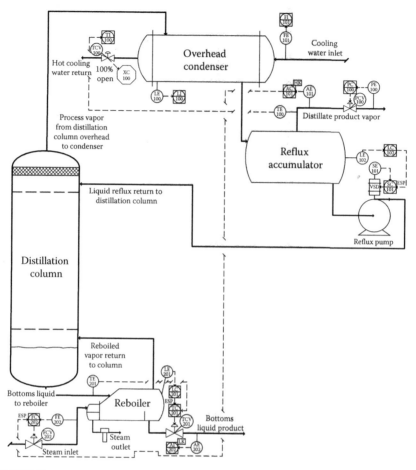

FIGURE 7.18
Revised control scheme for the Figure 7.17 distillation column when the column overhead condenser is constrained at maximum cooling water flow and the overhead product purity is more important than the bottoms product purity. XC-100 is a constraint controller that reconfigures the distillation control scheme to use overhead heavy key, HK composition as an external setpoint for the bottoms steam input rate instead of using bottoms light key, LK composition.

Problems

7.1 For a relatively stable gas, the ideal gas law can be used to calculate a "control density" for the gas. This density can then be used to adjust control loops to account for changes in the density of the gas as it moves through a process. Use a series of CALC blocks to modify the control scheme shown in Figure 3.10 so that changes to the steam flow rate to the turbine can be adjusted to account for changes in the inlet process fluid gas density.

Other Control Paradigms 239

7.2 Consider the control scheme shown in Figure 7.9. If an online density analyzer is not available or cannot be used for some reason, then the density can be estimated using the ideal gas law. A "compressibility factor" can be added to the calculation to allow the calculated density to be calibrated to measured densities at the same condition. Modify the control scheme shown in Figure 7.9 to use this calculated density to replace AE/AC-200 in this scheme.

7.3 Consider the control scheme shown in Figure 7.9. If the composition of the three gases differs significantly from each other, a change in the relative ratio of the flow rates of the three streams may cause problems in the operation of the compressor. The power needed in the compressor is related to the mass being processed in the compressor. So if the density of three streams is significantly different, then a change in flow rate in one of the three streams will change the mass of the blended composition fed to the compressor even if the overall volumetric flow rate has changed little. This sudden change could cause a surge in the outlet pressure (up or down), send either or both of the independent control variables unstable, and/or cause physical damage to the compressor itself. One way to handle this would be to calculate the ratios of the two smallest flow rate streams to the largest one in feed forward ratio controllers, then summing these two ratios in an FF/FB scheme with one of the control variables. This requires three flow rate measurements. Draw this control scheme.

7.4 Please modify the process and control scheme shown in Figure 7.13 to reflect the equivalent controls for an endothermic reaction.

7.5 Please modify the process and control scheme shown in Figure 7.13 if the cooling water flows in tubes embedded within the catalyst bed.

7.6 For the control scheme shown in Figure 7.14, add LALL and LAHH level safety system controls.

7.7 On an offshore oil platform, hot oil systems are often used as the heating utility fluid instead of steam. The cool oil flows through tubes that pass through the heat recovery section of the outlet ducting from a fired turbine where the oil is heated to the required utility service temperature. The rate of flow of oil through the heat recovery zone is controlled by the total required utility demand of the platform. This can be most easily accomplished by measuring the pressure of the oil after the recirculating pump and using this to adjust the rate of flow through the pump. However, if the flow rate is too low, the oil will get too hot while in the heat recovery tubes, which will lead to coking in the tubes. Therefore, a minimum flow rate of oil must always flow through the tubes. Any excess hot oil is routed to an air cooler to remove the excess heat. Draw the process and control scheme for this system.

7.8 Consider the system described in Problem 7.7. If there are multiple parallel processing systems on the platform, there may be drastically different hot oil circulation rate requirements depending upon how many of the parallel systems are in service. In this case, a piping manifold with parallel CVs of different sizes may be used. Draw the process and control scheme for this system if a split range controller is used with two parallel CVs.

7.9 Consider the system described in Problem 7.8. Draw the process and control scheme for this system if a constraint controller is used, such that when the smaller of two parallel CVs reached 100%, then the controller output is routed to the larger CV.

7.10 In Chapter 5, we learned that constructing the overhead distillate systems controls typically involves two separate cascade control loops where the two master measured variables are reflux drum level and distillate produce overhead composition. For a total condenser, the level control is matched to the reflux flow rate if the reflux-to-distillate ratio is greater than 1 and is matched to the distillate flow rate if the reflux-to-distillate ratio is less than 1. But what do we do if the reflux-to-distillate ratio is right around 1? For this situation, most control designers simply match the level controller to the reflux flow controller. However, if very careful overhead control is required, an allocation controller can be used instead. In this scheme, the output from the drum level controller is routed to the allocation controller. The distillate product composition controller output is used as an external setpoint for the allocation controller. Based on this setpoint, the allocation controller routes a portion of the level control output as an external setpoint to the reflux flow controller and the remainder of the level control output as an external setpoint to the distillate flow controller. This special type of allocation controller is known as a Ryskamp controller. Please draw this control scheme.

7.11 Please refer to the process and control schemes shown in Figures 7.17 and 7.18. Sometimes, plants change their operation and unit operations are reused in ways not originally envisioned. If this happens for a distillation system, it may be possible for the system to become reboiler duty constrained (i.e., limited by the rate of steam flow through the reboiler) instead of overhead condenser constrained. Please draw a process and control diagram comparable to Figure 7.18 for the case of a distillation system that becomes reboiler constrained when the bottoms product purity is more important than the overhead product purity.

7.12 Please refer to Problem 7.11. Modify your scheme for the case where the overhead product purity is more important than the bottoms product purity.

8
Controller Theory

In the previous sections, we have seen how basic regulatory control can be applied to process applications. This chapter provides an introductory understanding of how real-time controllers work. This can be helpful in understanding the capabilities of regulatory control and also in troubleshooting problems that might be encountered with existing installations.

8.1 On/Off Control

The simplest controllers are *on/off controllers*, which may also be known as *run/stop*, *start/stop*, and *open/close* controllers in certain specific applications. The output from this type of controller is either 0% or 100%. The measurement input is converted into a decision variable known as the *error*, "*e*." There are a number of options for how to generate *e*. For example, the measurement value can be divided by a value that defines the maximum range of the controller, essentially giving a ratio of the value to the maximum. Or, the difference in the value of the input measurement compared to a setpoint can be calculated to define *e*.

When the value of *e* exceeds the maximum allowable value, the controller output is 100%. If the value of *e* is less than the minimum allowable value, the controller output is 0%. When the value is between the minimum and maximum values, the controller output stays at its existing value (0% or 100%). The difference between the maximum and minimum allowable values is known as the *band gap*.

On/off controllers are everywhere: the thermostat on your home furnace and/or air conditioner, the float and valve mechanism on your home toilet, the temperature control of your oven. The most common application of the on/off controller in the process industries is for control loops that use electrical current as the energizing force for the unit operation.

For example, if a heat transfer unit operation uses a resistance-type heater, where the heat is generated from electrical power, an on/off type of controller, coupled with an electrical relay, is commonly used. A temperature measurement from the process substance enters the controller. If the value of *e* calculated from this temperature (e.g. the difference between the temperature measurement and the setpoint) is above the maximum allowable

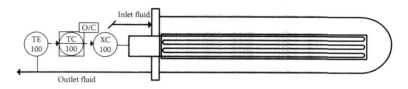

FIGURE 8.1
A typical on/off electrical resistance heater controller. *Note*: "O/C," which stands for "open/close," is commonly used to designate this controller since "O/O" ("on/off") is vague.

value, the controller output is 100%. This output value causes the relay to open, stopping the electricity flow to the heater, thus decreasing the energy transfer to the process substance. When the value of e drops below the minimum allowable value, the controller output changes to 0%. This output value causes the relay to close, allowing electricity to flow to the heater, and thus increasing the energy transfer to the process substance. This control scheme is shown in Figure 8.1.

Another common application is for level control loops where the absolute value of the level is not important, only that the level be maintained between maximum and minimum values. A level measurement is the input. If the value of e exceeds the maximum allowable value, the controller output is 100%, and the level control valve, which can be a block valve rather than a throttling-type control valve, is opened to allow liquid to leave the vessel. When the value of e drops below the minimum allowable value, the controller output is 0% and the level control valve closes, as shown in Figure 8.2.

For noncritical applications, the level measurement may be made using one or two level switches rather than a full level reading instrument. In a single switch system, the measurement and controller functions are integrated together in a single device. For this type of application, a level switch should be specified that contains an internal band gap function that can be adjusted externally. When the level exceeds the high limit of the switch, the switch generates a 100% output. This value remains the output value until the switch reads a level below the low limit of the switch. At the lower limit, the switch generates an output of 0%. This scheme is shown in Figure 8.3a.

In a two switch system, both switches read the level measurement of the vessel, but are typically installed at different heights in the vessel. The outputs from the two switches are routed to a calculation block where their values are added together. When the level is higher than the setpoint of both switches, each switch sends a value of "1" to the calculation block. When the controller receives a value that is greater than 1.1 from the calculation block, the output value of the controller is set to 100%. When the level is lower than the setpoint of both switches, each switch sends a value of "0" to the calculation block. When the controller receives a value that is less than 0.9 from the calculation block, the output value of the controller is set to 0%.

Controller Theory

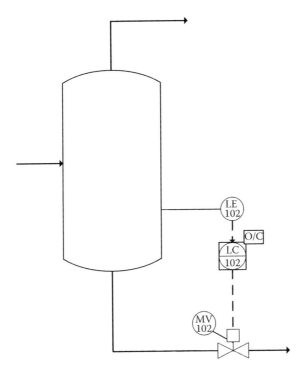

FIGURE 8.2
A typical on/off level controller.

When the level is in between the setpoint values of the two switches, the calculation block will receive a value of "0" from the upper switch and "1" from the lower switch. The calculation block sends a value of 1.0 to the controller. The controller keeps the same output value as long as the value of "1" is received. This special type of *on/off* controller is sometimes given other names, such as an *open/hold/close* (O/H/C) controller. This scheme is shown in Figure 8.3b.

8.2 PID Control

The proportional-integral-derivative (PID) controller is the workhorse of the process regulatory control domain. The vast majority of all controllers use this algorithm. Unless a different type of controller is specified, you can assume a PID controller is being utilized.

In any feedback controller, the objective is to minimize the difference between the measured parameter and a desired value for that parameter,

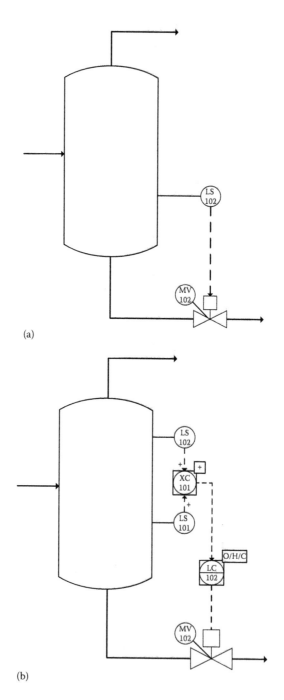

FIGURE 8.3
A typical on/off level controller with (a) a single level switch and (b) a two switch arrangement. O/H/C: denotes an open, hold, close type on/off controller.

known as the setpoint. This difference is known as the "error," which we denote in this textbook as e. If we could freeze time, this would be easy. We would calculate the error, determine how much to change the state of the control variable (CV), and send that information, known as the output, to the CV. After the CV was changed, we could verify that the system was now at the desired condition by calculating a new error. If exact control is not required, we could use an on/off controller, as described in Section 8.1.

But in the real world, time and the process do not stand still while our controller is determining the correct output for the CV. Instead, the system is dynamic and constantly changing in response to earlier CV adjustments and to changes in the process itself. Mathematically, we can express this as:

$$e(t) = MV_{sp} - MV_I(t) \tag{8.1}$$

where
  $e(t)$ is the error at time t
  MV_{sp} is the controller setpoint
  $MV_I(t)$ is the input value of the measurement value at time "t"

Note: if the setpoint is adjusted externally, such as the setpoint for the slave controller in a cascade loop, then MV_{sp} is also a function of time and would more accurately be written as $MV_{sp}(t)$.

First, let's consider a brand-new feedback control loop. If the loop is well designed, then the target measurement value would be at the 50% point of the range of the measurement instrument. Further, this measurement would correspond to the 50% point of the range of the control variable. The ideal value of the output to the control variable at its midpoint range is known as the *output bias* or simply the *bias*. Now let's put the control loop in service. If the loop operates exactly as designed and the process is at the exact design operating conditions, the error will be zero and therefore:

$$CV_o = CV_{ob} \tag{8.2}$$

where
  CV_o is the output from the controller to the control variable
  CV_{ob} is the output bias

In reality, there will always be some error, so the goal of the control algorithm is to determine how to change CV_o such that the impact of that error is minimized.

The most commonly used controller algorithm used in time-dependent controllers is the PID controller, although sometimes the derivative (PI control) and less often the proportional components (DI control) are not enabled. These controllers use a combination of three types of responses to calculate the output, CV_o, which is sent to the CV. Let's look at each of these separately before we put them together.

8.2.1 Proportional Response Control

In proportional response control, the change to the output is proportional to $e(t)$:

$$CV_o = CV_{ob} + K_p * e(t) \tag{8.3}$$

where K_p is the constant of proportionality, known as the *controller gain constant* or simply the *gain*.

The gain determines how sensitive the controller output is to the measurement error for that specific control loop. By adjusting the gain, the output can be made more or less responsive to the measurement error. Note, by changing the sign on the gain, the direction the controller output changes based on changes in the error can be reversed.

Here's an example of how this controller algorithm works. Consider the heat exchanger control scheme shown in Figure 4.2. In this example, a very hot process fluid is being cooled by generating steam from a boiler feed water stream. The process temperature is measured and sent to a temperature controller. The controller output is used to adjust the valve position of a CV on the boiler water feed to the exchanger. A low-low liquid level safety control loop is used to *air-fail-open* (AFO) this CV if the liquid level in the heat exchanger is below the specified level.

Now let's apply Equation 8.3 to this control scheme. The bias is programmed into the controller algorithm when the control loop is initially configured. The error is the setpoint value minus the actual value. If the temperature is too high, $e(t)$ will be negative. If the temperature is too low, $e(t)$ will be positive. We want the CV to open further if the fluid is too hot, letting in more fluid. Conversely we want the CV to reduce the opening when the fluid is too cold so that less utility fluid is fed to the heat exchanger.

Since the CV is designated AFO, it is a direct-acting valve (it fails to the 100% position when the output has a value of 0 or "fail"). When the controller output increases, higher air pressure pushes down on the valve stem, reducing the opening area and reducing the flow rate of the fluid (see Section 10.7 for more details on how throttling CVs work). Therefore, we want to leave the sign on K_p positive. This will result in $e(t) * K_p < 0$ when $e(t)$ is a negative number (the fluid is too hot). The controller output will be the bias value minus the measurement error (Equation 8.3). So when $e(t)$ is a larger negative number (the fluid is getting hotter), the value of the output, CV_o, will go down. A lower output value tells a direct-acting (AFO) CV to open further when the fluid gets hotter, increasing the flow of cooling water into the heat exchanger and decreasing the fluid temperature. When $e(t)$ is a smaller negative or positive number (the fluid is getting cooler), the value of the output will go up. A higher output value tells an AFO CV to close further when the fluid gets colder, decreasing the flow of cooling water into the heat exchanger.

If the valve were an AFC-type valve (air-fail-close; reverse-acting), a higher controller output would lead to a higher air pressure, which would move the stem further out of the valve body, leading to a greater valve opening area. Then we would want to use a negative sign on K_p so that a higher temperature (leading to a negative value for $e(t)$) would lead to a positive value for CV_o. A positive CV_o to a reverse-acting valve will tell the control valve to open further while a negative CV_o will tell this type of valve to close further.

Some controllers use an adjustment parameter known as the *proportional band*, PB, instead of controller gain. The two are related. For a dimensionless K_p:

$$PB = \frac{100\%}{K_p} \qquad (8.4)$$

Note: if K_p is given in measurement units, then PB is in inverse measurement units expressed as a percentage.

Equation 8.3 has no provision for the physical limitations of the process, the measurement variable instrument, or the CV. When the system hits one of these limits, the controller output will no longer make the expected change in the system. This is known as *controller saturation*. In process applications, the low value alarm must always be set at or, preferably, higher than the controller's lower saturation limit and the high value alarm must always be set at, or preferably, lower than the controller's higher saturation limit. This insures that the control operator is alerted to deviations from the desired operation before the controller's utility is nullified (no changes occur outside of the saturation limits of the controller).

Another limitation of Equation 8.3 is the inability of proportional control to hold the process at the desired steady-state value unless this value happens to be exactly at the value reflected by the controller bias. Note that when $e(t)$ goes to zero, Equation 8.3 will output the bias value to the CV. But if the current desired measurement value setpoint does not correspond to the CV position of the bias, the process will not be at the desired steady-state value. This steady-state error is known as the *offset error* or simply the *offset*. In order to eliminate the offset and allow our system to reach any desired steady-state condition, we add the integral controller response to our controller.

8.2.2 Integral Control

In integral control, the output is adjusted away from the bias by using a term that is calculated by integrating the value of the error over a specified time interval:

$$CV_o = CV_{ob} + \frac{1}{\tau} \int_0^{t_m} e(t) \, dt \qquad (8.5)$$

where
  CV_o is the controller output
  CV_{ob} is the controller bias
  τ is an adjustable constant known as the integral or *reset time*
  $t_m - 0$ is the integration time interval
  $e(t)$ is the measurement error as defined in Equation 8.1

Adding this integrating term to our controller provides two important features. First, it eliminates the offset error. When $e(t_x)$ is zero, CV_o will still be different from CV_{ob} since no single value of $e(t)$ is used to calculate the output. The only time when CV_o will equal CV_{ob} is when the current desired measurement value setpoint exactly corresponds to the CV position of the bias. Thus, even if $e(t_x)$ goes to zero, the controller will still drive the process to the desired steady-state value.

The other important feature is the dampening action inherent in integrating the error function over a time interval. A proportional and/or derivative controller without an integral term would make sudden, large changes to the controller output, which introduces dynamics into the process and may even make the process go unstable. The integral term smooths these changes out so that the controller output changes gradually over time. How gradually the change occurs is usually tuned using the integral time constant τ.

This dampening feature is also the reason why a pure integral controller is not employed. A pure integral controller would be sluggish, since the controller output will not have an appreciable change to an upset or change in the process until the error function change is sufficient to affect the integrated term. Therefore, the responsiveness of proportional control is needed so that the controller will begin to react immediately to a change. But the integral control will insure that the controller does not overreact to the change. The instrument technician or control engineer who sets up and commissions the control loop will adjust the relative values of the gain (K_p) and reset time (τ) so that the controller will have the best balance of these two features for the specific control application.

The impact of this balance is shown in Figure 8.4. If the proportional term is too large compared to the integral term, the controller will overreact to a change in the input. This is known as an underdamped controller. When the contribution of the integral term is too large compared to the proportional

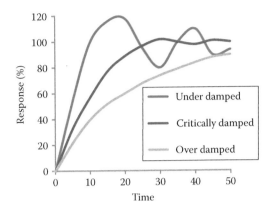

FIGURE 8.4
Impact of loop tuning on the response of a PI controller.

term, the controller will never reach the desired value. This type of controller is known as an overdamped controller. When the two tuning constants are at the correct balance, known as critically damped, the response will slightly overshoot the desired result but will quickly oscillate to the desire value.

In Section 8.2.1, we saw that the proportional controller could reach a point where the controller action was constrained by a physical limitation of the system, a condition known as saturation. Saturation is an even bigger problem for the integral controller. When a sustained large error occurs, the controller can get stuck in saturation for most of the integral time interval even after the error decreases. This is known as *reset windup* or *integral windup*. To counter this, commercial controllers have a feature known as *anti-reset windup* that disables the integral term from the controller output calculation whenever the controller is at or beyond the saturation point.

While anti-reset windup is useful for continuous processes, it is critical for PID controllers used for one or more steps within a batch sequence (either a batch process or a semi-batch unit operation embedded in a continuous process; see Chapter 1 for an explanation of these processes). When the batch sequence is on a step that doesn't use the controller, it will accumulate a sustained large error until the sequence changes to the step where the controller is being used. Without anti-reset windup, that step would not react quickly enough to the new conditions of the process. This same thing occurs if the controller setpoint is changed significantly between two steps in a batch sequence.

Combining the proportional and integral controllers together (Equations 8.2 and 8.5) we have the PI control equation:

$$CV_o = CV_{ob} + K_{pi}\left(e(t) + \frac{1}{\tau}\int_0^{t_m} e(t)dt\right) \qquad (8.6)$$

where K_{pi} is the gain term of a PI controller, which includes the gain term of the proportional controller modified by part of the reset time from the integral controller.

Note that since the gain and reset time are adjustable constants, it doesn't matter whether the gain term is embedded with the proportional term or pulled outside of the summation of the two controller terms, as shown in Equation 8.6. However, it is easier to generate the corresponding transfer functions in the "s" domain (using Laplace transforms) if the equation is arranged as per Equation 8.6. See Appendix A for further details on transfer functions.

8.2.3 Derivative Control

For some process situations, waiting to respond to the error in the measurement will result in negative consequences for the process. For example, if we are processing a food product that is temperature sensitive and a heat exchanger used to cool the food down does not react quickly enough to an increase in throughput, a significant amount of the food product could be spoiled by remaining at too high of a temperature before the controller increases the cooling fluid rate.

Derivative control was developed to handle these types of situations. By measuring the rate of change of the error rather than the magnitude of the error, the controller can react to changes in conditions faster. Consider the error function shown in Figure 8.5. By measuring the rate of change, control action can anticipate increases or decreases in the error before they become significant. This information can then be used to mitigate the impact of the change by making adjustments to the controller output that anticipate the error.

Thus, the derivative controller has an equation of the form:

$$CV_o = CV_{ob} + \tau_D \frac{de(t)}{dt} \tag{8.7}$$

where τ_D is the derivative time constant often known as simply the *derivative time*.

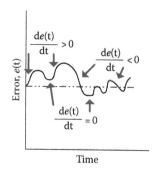

FIGURE 8.5
Demonstrating the derivative of the error function.

Controller Theory

As with proportional control, derivative control has a steady-state offset error. In this case, whenever the rate of change of the error function passes through a minima or maxima, the derivative value is zero and the controller output will be the value of the bias. Further, the offset is the difference between the measurement value corresponding to the measurement value setpoint and the measurement value corresponding to the CV at the bias value. Therefore, derivative control is rarely used without integral control.

8.2.4 Combined PID Control Algorithm

Proportional, integral, and derivative control equations are commonly combined in one of the following two forms:

$$\text{Parallel:} \quad CV_o = CV_{ob} + K_{PID}\left[e(t) + \frac{1}{\tau}\int_0^{t_M} e(t)dt + \tau_D \frac{de(t)}{dt}\right] \quad (8.8)$$

$$\text{Series:} \quad CV_o = CV_{ob} + K_{PID}\left[\left(e(t) + \frac{1}{\tau}\int_0^{t_M} e(t)dt\right)\left(e(t) + \tau_D \frac{de(t)}{dt}\right)\right] \quad (8.9)$$

where
  CV_o is the controller output
  CV_{ob} is the controller bias
  K_{PID} is the gain constant
  τ is the reset time
  τ_D is the derivative time

The full PID controller has three separate tuning constants that can be adjusted to provide the best response for the controller to each specific control application. Modern controllers have loop tuning algorithms built into their software and the instrument technician, control engineer, or process engineer simply chooses one and lets the controller tune itself. The values of the constants from the controller's automatic tuning functions sometimes need minor adjustments that can be made manually. An introduction to loop tuning can be found in Appendix B.

PID controllers have a number of advantages, including the following:

- Corrective action begins as soon as the measured/control variable deviates from setpoint.
- No system analysis is necessary to implement a controller.

- It is versatile and robust.
- Direct information is received about how well the controller is performing.

However, PID controllers have some limitations, such as the following:

- Corrective action begins only after the measured/control variable deviates from the setpoint, although this can be mitigated somewhat by including feed forward control in the control scheme.
- There is no predictive capability to anticipate disturbances, except for that provided by the derivative term. Again this can be mitigated somewhat by including feed forward control in the control scheme.
- For large time constant processes, PID control will not react quickly and can therefore allow large and frequent disturbances to occur; the process may never reach steady state. Again this can be mitigated somewhat by including feed forward or cascade control in the control scheme.
- If the desired measurement variable cannot be measured online, direct feedback control cannot be used. Many times, inferential control can be used to overcome this deficiency (see Section 7.2).

There are many texts available that discuss PID controller theory in much greater detail. For the purposes of most process engineers, it is enough to have an understanding of the basic theory of the PID controller and the purpose of each of the controller tuning constants.

Some of the most common modifications or alternatives to straight PID control are discussed in the next few sections.

8.3 Modifications to Minimize Derivative Kick

Equations 8.8 and 8.9 can be modified to minimize interactions between the terms, to provide filters to minimize the impact of derivative and/or proportional controller kick (the large response of the controller to a sudden rapid change in the measurement setpoint or process conditions), or other factors. For example, if derivative kick is a concern, the term:

$$\frac{dMV_I(t)}{dt} \qquad (8.10)$$

can be used instead of:

$$\frac{de(t)}{dt} \tag{8.11}$$

yielding, for example,

$$CV_o = CV_{ob} + K_{PID}\left[e(t) + \frac{1}{\tau}\int_0^{t_M} e(t)dt + \tau_D \frac{dM_I(t)}{dt}\right] \tag{8.12}$$

Since the measurement value is much larger than the error value, the rate of change of the actual measurement value is often much lower for the same process change compared to the rate of change of the error in the measurement value (the difference between two large numbers compared to the difference between two small numbers). Since using the measurement value serves to dampen the derivative term, it should only be used where there are concerns in derivative response to large changes in the process.

8.4 Nonlinear Control Strategies for PID Controllers

PID controllers are based on the assumption that the relationship between the measurement variable and the independent variable in the control loop is linear. But the response of a measurement variable to a change in its independent variable is never truly linear. Nonlinearities are due to the dynamics of the control elements as well as the dynamics of the process itself.

Most of the time, these nonlinearities are minor and linear controls work well. But if there is a concern about the nonlinearities, one way to mitigate their effect is to operate in a relatively narrow range around the design operating point. If the range is narrow compared to the nonlinearities, a linear correlation within this narrow range will provide acceptable operation (Figure 8.6).

However, there are occasions where the nonlinearities are significant and the operating range is too wide for stable operation of a traditional PID controller. There are three common strategies employed for these situations:

- Modified PID control algorithms
- Tuning parameter scheduling
- Nonlinear transformations of input or output variables

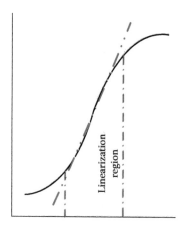

FIGURE 8.6
Truncating to the linear region of a nonlinear function.

8.4.1 Modified PID Control Algorithms

Equation 8.8, the full PID algorithm, can be rearranged to:

$$CV_o - CV_{ob} = K_{PID}\left[e(t) + \frac{1}{\tau}\int_0^{t_M} e(t)dt + \tau_D \frac{de(t)}{dt}\right] \quad (8.13)$$

The linearity in this equation comes from the fact that each term in the algorithm uses a simple difference calculation (Equation 8.1) for the error term. A simple way to introduce a nonlinear response into the equation is to replace the gain constant, K_{PID}, with a function. The simplest of these is the *error-squared* controller. Let K_{sq} be the substituted gain into Equation 8.13, which is calculated by:

$$K_{sq}(t) = K_{PID}\left(1 + \alpha|e(t)|\right) \quad (8.14)$$

where α is an adjustable constant.
Then Equation 8.13 becomes:

$$CV_o - CV_{ob} = K_{sq(t)}\left[e(t) + \frac{1}{\tau}\int_0^{t_M} e(t)dt + \tau_D \frac{de(t)}{dt}\right] \quad (8.15)$$

Now $e(t)$ enters into the overall algorithm twice, once in the gain term and once in each of the P-I-D terms, resulting in a response that is now proportional to the square of the error. The absolute value of the error is used in the gain term (Equation 8.14) to avoid changing the sign of the output.

Controller Theory

The most common application of an error-squared controller is for surge drum level control when it is desirable to make a larger output action near the high and low limits than at the midpoint. One of the roles of the surge drum is to dampen dynamics from the process. Usually, the exact level is not important as long as a reasonable level is maintained. With the error-squared controller, the output action is small when the level is near the target value. As the level changes farther away from the desired point (in either direction), the output action increases by the square of the error. So the controller is responsive at large errors and nonresponsive at small errors. This is ideal for a surge drum.

8.4.2 Tuning Parameter Scheduling

We can now extend this technique further using a method known as *tuning parameter scheduling*, also known as *controller parameter scheduling*. Since the vast majority of applications adjust the gain in the PID control algorithm, it is also widely known as *gain scheduling*.

Recall from Section 8.2 that a PID controller has three adjustable parameters: K_{PID} is the gain constant, τ is the reset time, and τ_D is the derivative time. These are also known as the *tuning parameters*, because the values are typically adjusted to tune the control loop for maximum stability during commissioning. For the vast majority of control loops, the original loop tuning will provide stable control for the life or most of the life of that control loop. However, there are certain cases where the process conditions change sufficiently over time such that the control loop will become unstable by overreacting to the dynamics of the process. In other cases, the process conditions change sufficiently over time such that the control loop will become sluggish and the response is too slow for good control. Where either of these cases is anticipated, tuning parameter scheduling can be used to insure good control under widely different conditions.

For a scheduled tuning parameter, the value of the parameter changes based on the value of a reference variable, which is typically known as the *scheduling variable*. There are three common methods:

1. Vary a controller parameter (usually the gain) continuously with changes in a scheduling variable (usually the control variable)
2. Change the settings for a controller parameter (usually the gain) for different operating regions of control or for different magnitudes of the error
3. Interpolate from current settings based on the value of a scheduling variable (usually the control variable)

Consider the reaction system described in Section 7.3 and shown in Figure 7.13 with a high select controller. This type of control scheme is adequate if there is a relatively sharp change in the fluid temperature as the reaction front (the region where most of the reaction is occurring) moves

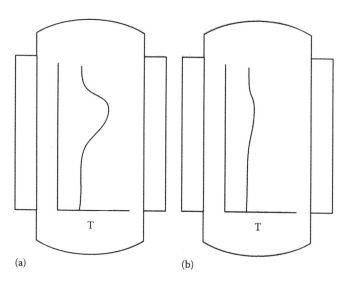

FIGURE 8.7
The temperature profile through a reactor subject to uniform catalyst deactivation at (a) initial operation and (b) near end of life operation. The temperature controller may be too sluggish under condition (b) for stable operation if tuned under condition (a).

through the reactor. However, sometimes the catalyst deactivation is more uniform (see Figure 8.7). When this occurs, the temperature profile tends to flatten out. If the controller is tuned for the case of fresh catalyst when the profile has a distinct bulge, then over time, the controller action will become sluggish and may not react quick enough to changes in the reactor.

In this case, we'd like to change the tuning of the gain constant so that the controller is more responsive to changes in the selected temperature in the reactor as the catalyst ages. If the suite of temperature measurements used in Figure 7.13 were employed, then we could calculate the difference between the maximum and minimum temperature measurements in the reactor and normalize this value using the setpoint for the associated temperature controller. This temperature difference would represent the scheduling variable, SV(t), for the gain schedule for the controller:

$$SV(t) = \frac{MAX[TI101, TI102, TI103, TI104, TI105] - MIN[TI101, TI102, TI103, TI104, TI105]}{TC100_{SP}}$$
(8.16)

Since the value of the scheduling variable will change gradually over the lifetime of the catalyst, method 1 could be used; the value of the gain could increase as the value of the scheduling variable decreases:

$$K_{PID} = \alpha \frac{K_{PID,O}}{SV(t)}$$
(8.17)

where
- α is an adjustable tuning constant to adjust the rate of change in the gain based on the value of the scheduling variable
- $K_{PID,O}$ is the original gain value

Now, let's add a complication to this example. What if the catalyst can be periodically regenerated back to its near-original activity? In this case, there will be a step change in the shape of the temperature profile each time the catalyst is regenerated. In this case, it might be better to use method 2. The maximum temperature difference will occur when the catalyst is new. A slightly smaller temperature difference will occur when the catalyst is regenerated. These two values could define region 1, when the controller is the least responsive to changes in the reactor maximum temperature.

As the catalyst ages, the temperature difference decreases. At a given point, the difference will be so low that it triggers an automatic or manual (via an alarm) regeneration cycle. We can divide the range between region 1 and the regeneration point into a series of regions; for this example let's use two. Then region 2 would be the temperature difference range from the low point of region 1 to a point roughly halfway between that point and the regeneration point. Region 3 would be the temperature difference range from the low point of region 2 to the regeneration point. If the temperature controller were also used as part of the regeneration cycle, a region 4 (or even more regions) could be defined that changes any or all of the tuning parameters during the regeneration process.

This scenario is depicted in Table 8.1 where ΔT could be the same as $SV(t)$ in Equation 8.16, ΔT_{high} would be the temperature difference immediately after regeneration, ΔT_{mid} is the approximate average between the temperature difference immediately after regeneration and the minimum acceptable temperature difference, and ΔT_{low} is the minimum acceptable temperature difference (that point at which conversion in the reactor drops below acceptable levels, indicating the catalyst should be regenerated). $K_{PID,O}$ is the original gain constant; $K_{PID,Regen}$ is the gain constant during the regeneration step; and α_1 and α_2 are adjustable tuning constants used to modify the gain depending upon the region.

TABLE 8.1

An Example of Regional Gain Scheduling

Region	Range	K_{PID}
1	$\Delta T > \Delta T_{high}$	$K_{PID} = K_{PID,O}$
2	$\Delta T_{mid} > \Delta T > \Delta T_{high}$	$K_{PID} = \alpha_1 * K_{PID,O}$
3	$\Delta T_{low} > \Delta T > \Delta T_{mid}$	$K_{PID} = \alpha_2 * K_{PID,O}$
4	$\Delta T_{low} > \Delta T$	$K_{PID} = K_{PID,Regen}$

This leads us to another common application of tuning parameter scheduling, batch and semi-batch processes. Consider the simple batch blending system described in Section 2.11. During the second step in the sequence, the heating coil is energized and the fluid is heated until it reaches the target temperature. At that point, the temperature must be maintained at that temperature. If the same PID controller is used for both functions, then the controls will be more stable and reactive if the tuning parameters change once the target temperature is reached. Method 2, using changes based on regions of the fluid temperature value, would be the appropriate choice here.

8.4.3 Nonlinear Transformations of Input or Output Variables

The concept for these controllers is essentially the same as the concept introduced in Section 7.1. In that section we described how an intermediate calculation is performed to take the square root of the inlet signal from Bernoulli-type flowmeters. Most control system software incorporates this intermediate calculation right in the configuration options for the controller. We now extend this principle to other nonlinear parameters.

If the range over which the measurement variable changes is on the order of multiple orders of magnitude, then another controller, the *logarithmic controller*, can be used. This controller is most commonly used for an analyzer control loop when the composition varies exponentially with changes in the process. This is the case for many reactions where the reaction rate varies exponentially with temperature. Instead of using Equation 8.1 to calculate the error, a modified equation is used:

$$e(t) = \ln[MV_{sp}] - \ln[MV_I(t)] \qquad (8.18)$$

This will dampen the output, avoiding an unstable control for these types of systems. However, it will act sluggishly if the changes are not orders of magnitude in range. Therefore, pairing this controller with a more reactive slave controller in a cascade loop is usually recommended.

Now let's generalize this concept. When the measurement variable changes significantly and nonlinearly over time, the measurement variable function may be estimated using the best mathematical model:

$$MV_{calc}(t) = f[MV_1(t)] \qquad (8.19)$$

This model may be analytically determined, like the exponential function of the Arrhenius model, or empirically determined. The error function is then calculated using the inverse of the function defined for Equation 8.19:

$$e(t) = MV_{sp} - f^{-1}[MV_I(t)] \qquad (8.20)$$

Controller Theory

For the Bernoulli-type flowmeter, MV is the pressure difference term and f(t) is the kinetic energy term, v^2, leading to:

$$MV_{calc} = K(P_{in} - P_{out})^2 \tag{8.21}$$

Since the cross-sectional area of the pipe where the flowmeter is located is constant, the volumetric flow rate in the pipe is proportional to the inverse function, which is the square root of the measurement variable—the pressure difference:

$$e(t) = MV_{sp} - K'\sqrt{(P_{in} - P_{out})} \tag{8.22}$$

In the composition controller described above, MV is the temperature from the reaction. The function is the Arrhenius equation, which relates the concentration to the temperature:

$$MV_{calc} = Ae^{-MV_1(t)/RT(t)} \tag{8.23}$$

The inverse function for an exponential function is a logarithmic function, yielding the logarithmic controller:

$$e(t) = MV_{sp} - A'\ln\frac{-MV_1(t)}{RT(t)} \tag{8.24}$$

Other functions relating the target variable to the measurement variable may be best represented by a parabola, a hyperbola, or a cubic equation.

Empirical curve fitting can also be used. Various forms of the empirical equation may be used, but the most common is:

$$f(t) = a(MV) + b(MV)^2 + c(MV)^3 + \ldots n(MV)^n \tag{8.25}$$

Note: MV = MV(t) in this equation and a, b, and c are adjustable constants, often known as the weighting factors.

8.5 Adaptive Controllers

All of the controllers described in Sections 8.2 through 8.4 use the basic PID controller paradigm as their foundation. In this section, we move away from the PID paradigm and learn about controllers that are not based on this foundation.

There was a time when many in the automation industry believed that *adaptive controllers*, also known as *heuristic controllers*, would replace PID controllers as the primary type of controller used in the process industries. While the demise of the PID controller proved to be false, the use of adaptive controllers has become a widely applied method of determining the controller output value in single input/single output controllers.

An adaptive controller combines the features of the tuning parameter scheduling controller and the generalized form of the nonlinear controller but with one significant difference: the PID algorithm is replaced by a more generalized empirical model. Adaptive controllers are particularly useful when process dynamics and/or operating conditions are unknown prior to loop tuning. In fact, many modern stand-alone controllers will use an adaptive control algorithm only during the initial tuning phase to determine the best tuning constants for a PID controller. Once the learning or tuning mode is completed, the controller switches to PID mode. This is known as a *self-tuning* or *self-adaptive* controller.

Adaptive controllers are most commonly used in temperature controllers where the input temperature measurement comes from one or more thermocouples. Thermocouple signals are particularly susceptible to interference from electromagnetic radiation, generating "noisy" signals. It is impossible to shield out all of this type of interference in a process plant. In this application, the adaptive controller switches rapidly between reading the thermocouple signal and reading a signal that is only made up of the interference. It then cancels out the interference signal from the thermocouple signal, ideally leaving only the actual reading (Figure 8.8). This corrected reading is then used to evaluate the error from the setpoint.

Another good application of the adaptive controller is for situations where the process conditions gradually change over time. Examples of this include heat exchanger fouling and reactor catalyst deactivation. The adaptive controller can adjust its tuning to accommodate these changes to provide the best controller response throughout the life of the unit operation.

Adaptive controllers are also useful when there are frequent changes in feedstock conditions, product specifications, or in batch applications. Another application is for unit operations that are impacted by diurnal or seasonal changes in the environment, such as air temperature for air coolers or cooling water temperature. Instead of having to manually retune a PID controller with every change or having to preset a series of tuning constants using tuning parameter scheduling, the adaptive controller automatically self-tunes to the new conditions.

Finally, when the relationship between a dependent and an independent variable is inherently extremely nonlinear, an adaptive controller can be used. The controller is not limited by the linear PID algorithms and thus can more flexibly adapt to nonlinear changing conditions.

Controller Theory

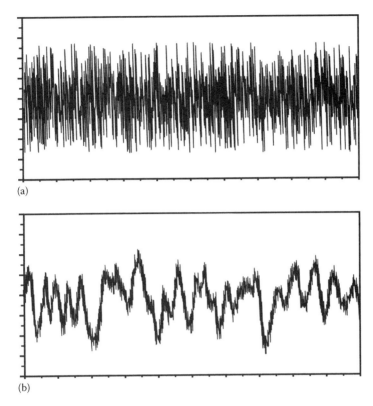

FIGURE 8.8
Adaptive controllers can be used to isolate a measurement signal in a noisy environment (a) to provide useful information (b).

Problems

8.1 Consider the process control scheme shown in Figure 8.1. In certain very high temperature applications, there is concern that the temperature at the surface of the heating tubes will be above its safe limit. In these cases, a temperature sensor is attached to the outside surface of the tube.

   8.1.1 Why is the sensor attached to the outside of the tube rather than the inside of the tube? Does it matter where along the length of the tube the sensor is attached?

   8.1.2 One way to use this sensor reading is to use a high select control scheme. Show how Figure 8.1 would be modified for this type of control scheme.

8.1.3 Another way to use this sensor is as a separate safety control system. The temperature sensor is attached to a digital switch inserted in the temperature control loop. When the value of high limit of the tube temperature is reached, the switch value changes from "0" to "1." This value is used to interrupt the power to the heater when the temperature is above the tube temperature limit. Show how Figure 8.1 would be modified for this type of control scheme.

8.2 Add LAHH and LALL safety control loops to the control scheme shown in Figure 8.2.

8.3 Add LAHH and LALL alarm safety control loops to the control schemes shown in Figure 8.3a and b.

8.4 When the concentration of polar compounds, such as water, exceeds the saturation point in an organic liquid, an aqueous phase is formed in addition to the primary organic phase. This very small aqueous phase is often removed in process drums downstream of the point where the water phase forms. For careful separation, a water boot is attached to the low point of the drum. The interface level in the drum is measured and an on/off controller is used to drain the boot.

8.4.1 Draw this process and control scheme.

8.4.2 What are the relative densities of the organic and water phase (i.e., which is higher, which is lower) when this type of system is employed?

8.4.3 Draw the comparable process and control scheme if the relative densities were reversed.

8.5 Describe how each of the three terms contributes to control in a PID controller.

8.6 If you want to make a PID controller more responsive to process changes, which tuning constant should you adjust? Should the constant be made smaller or larger?

8.7 If you want to make a PID controller more reactive to process changes, which tuning constant should you adjust? Should the constant be made smaller or larger?

8.8 If you want to smooth out the action of a PID controller, which tuning constant should you adjust? Should the constant be made smaller or larger?

8.9 In a PID controller, the relationship between the measurement variable and the control variable is assumed to be linear as shown in Equation 8.13. If, instead, the relationship is exponential, how would you modify the basic equation to linearize the relationship?

9

Higher-Level Automation Techniques

All of the process automation functions described in Chapters 2 through 8 can be performed using simple, highly reliable regulatory control blocks. However, to maximize the efficiency and profitability of the process, we may need to layer more advanced automation functions on top of this basic level. These functions are performed using applications (apps) written and stored in computers with more power, but less reliability, than the regulatory control system. We will introduce the most common of these techniques in this chapter. The features contained in these applications vary depending upon the supplier and are continually being improved. Therefore, we will only include an introduction to the concepts that each technique represents. Details of specific apps are best learned directly from the application supplier.

9.1 Plant Automation Concepts

In order to understand good plant automation system (PAS) design, we must consider the PAS from an integrated perspective. To do this, we need to recognize that all of the functionality of the PAS must contribute to achieving the overall objectives of the facility in which the process resides. For any process facility, these will usually be based on the following three general objectives:

1. Maximizing profit
2. Meeting acceptable health and safety standards
3. Minimizing environmental impact

From these general objectives, we can build more specific objectives for any real process enterprise. Each of the automation elements included in the PAS should be focused on supporting the overall plant objectives. We can categorize the automation elements that may be included in the PAS using a layered "functional" model as shown in Figure 9.1.

The highest-level functions are those that set the direction and review progress for the entire facility. Next are functions that serve to optimize the

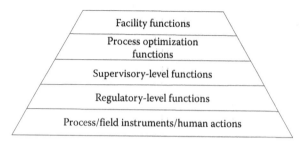

FIGURE 9.1
A hierarchy of automation elements in a process facility.

overall operation of the entire process operation. Supervisory-level functions are advanced functions that occur at the process unit level. Regulatory-level functions monitor and control the direct process using the basic process and field instrumentation level.

One of the key features of process plants is the presence of real-time processes. These processes may be steady-state, dynamic, or batch processes. While the processes are real time and continuous, modern process control systems utilize discrete measurements. Thus, with real-time processes we must be concerned with issues such as data resolution, information priority, system criticality, reliability, and controllability/tuning. Therefore, it is useful to understand the way information is passed between the various layers of the PAS and also the timing of information transfer. We can do this by expanding the functional model into a "data/information" model. In this model we can categorize the following layers in the process enterprise (Figure 9.2):

Enterprise management (process enterprise, business management) *layer*: Average accuracy, high integrity, weeks to years of resolution, high degrees of freedom, and low priority. Examples of the kinds of operational data/information that are inputs and outputs (I/O) for this layer are as follows: it passes production targets/schedules to the plant-wide layer and process enterprise profitability to corporate systems; it receives performance data from the production management layer and corporate directives/opportunity information from corporate systems.

Production management (process optimization/plant-wide management) *layer*: Medium accuracy, high integrity, hours to weeks of resolution, and moderate priority. Examples of the kinds of operational data/information I/O for this layer are as follows: it passes setpoint/constraints to the process management layer and performance data to the enterprise management layer; it receives operating data from

Higher-Level Automation Techniques

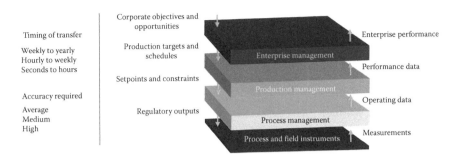

FIGURE 9.2
A typical data/information transfer model for the various layers of a plant automation system.

the process management layer and production targets/schedules from the enterprise management layer.

Process management (supervisory/process unit management) *layer*: High accuracy, seconds-to-hours resolution, high priority, and low degrees of freedom. Examples of the kinds of operational data/information I/O for this layer are as follows: it passes regulatory outputs to the process and field instruments layer and operating data to the production management layer; it receives measurements from the process and field instruments layer and setpoints/constraints from the production management layer.

Process and field instruments (regulatory control) *layer*: Very high accuracy, milliseconds-to-minutes resolution, very high priority, and zero or low degrees of freedom. Examples of the kinds of operational data/information I/O for this layer are as follows: it passes measurements to the process management layer and receives regulatory outputs from the process management layer.

Within each layer of the PAS are applications and supporting software and hardware that perform specific functions that support the PAS objectives. There are many options available and each control system vendor has their preferred configuration. Starting from the lowest layer of the PAS functional model and moving upward, a general listing of some of the applications that may be included in the PAS are as follows (Figure 9.3).

9.1.1 Field Instrumentation

Objective: Directly measure the values of dependent variables and directly manipulate independent variables in order to achieve the process enterprise objectives in the most efficient manner.

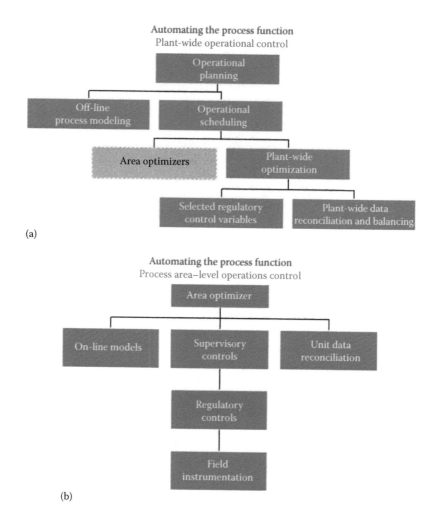

FIGURE 9.3
Process automation applications/layers for (a) the entire plant and (b) an individual process area. The light shaded box shows where the individual process area systems, as shown in (b) fit within the plant-wide system, shown in (a).

Global description: The most common dependent variables measured, and the speed of information dynamics, are:

- Pressure. Incompressible fluid flow—speed of sound (milliseconds)
- Temperature—slightly faster than the velocity of fluid flow (seconds)
- Compressible fluid flow. Compositon from direct analysis—velocity of fluid flow (seconds)

- Level—on order of the residence time of the reservoir (seconds to minutes)
- Composition from Sampled Analysis—measuring device dependent (tens of seconds to hours)

The most common independent variables manipulated are throttling-type control valves (CVs). The next most common is electrical current to a variable speed motor. Refer to Chapters 2 and 10 for further details on measurements in processes.

9.1.2 Regulatory Controls: The Backbone of Real-Time Process Controls

Objective: Maintain the process conditions at some optimum steady-state condition in order to achieve the process enterprise objectives in the most efficient manner.

Global description: A subset of all the potential dependent variables in the process is measured at specific time intervals. The time interval is determined based on the capabilities of the field instrumentation and of the regulatory control system components. The time interval must be short enough to be responsive during upsets. The results of these measurements are compared to the setpoint. Based on the comparison of the measurement variable to its setpoint, a suite of independent (control) variables, most commonly control valves, are manipulated in order to bring the measurement variable values closer to the setpoint.

The most common function set in the regulatory control layer is the single parameter feedback control loop. A dependent variable is measured using a field instrument and the measured value is compared to the desired value (setpoint). The error between the measured value and setpoint is used to determine the adjustment to a single independent variable (the controller output). The measurement/control variable combination is chosen such that a change in the output affects the measurement variable. The two variables may be taken from the same process stream or from two related process points.

See Chapters 2 through 8 for further details.

9.1.3 Supervisory (Advanced Process) Controls

Objective: Implement control strategies that cannot be implemented without intermediate manipulation of direct measurement data used to determine changes to independent variable setpoints.

Global description: A control strategy that involves the manipulation of the direct measurements rather than their direct use. Supervisory controls are indirect. Measurements are assessed by the control scheme and then actions are taken to change the setpoints of regulatory controllers.

See Sections 7.1 and 9.2 for further details.

9.1.4 Online Models

Used to infer properties for supervisory control strategies. These models are especially useful for predicting compositions, since analytical measurements may be expensive and slow. Online models are also used for performance assessment such as heat exchanger fouling, reactor performance/catalyst activity monitoring, and rotating equipment efficiency. Models must be simple enough to achieve acceptable response times. In general, update times should at least be as fast as a sampling type analyzer (see Section 10.5 for a description of analyzers)—on the order of seconds to minutes.

See Sections 7.2 and 9.3 for further details.

9.1.5 Area Data Reconciliation

Used to identify bad measurements automatically, to predict the true value of a bad measurement, and to allow generation of mass and energy balances. When included with a fully integrated PAS, maintenance work tickets can be automatically generated for repair of bad instruments. By replacing bad measurement values with predictions, area data reconciliators allow supervisory controls to remain online even when one or more measurements are not available. Further, this reduces the need to place control loops and control strategies on manual control, which improves plant safety. Mass and energy balance information is used by area and plant-wide optimizers.

See Section 9.3.1 for further details.

9.1.6 Area Optimizer

Used to set process objective targets for supervisory control strategies within a discrete process area (these are often known as unit optimizers, referring to process units, not unit operations). Area optimizers are often closely coupled with multivariable controllers which are part of the supervisory controls layer. Area optimizers are best justified for processes that change process objectives frequently. The optimizer can be used to determine the most efficient path to move from one set of operating objectives to another. The optimizer can also help to efficiently recover from a major process upset and can help to optimize catalyst life, minimize energy consumption, and minimize environmental impacts.

See Section 9.3.2 for further details.

9.1.7 Plant-Wide Operational Control Applications

Operational planning, operational scheduling, off-line process modeling, plant-wide optimization, and plant-wide data reconciliation.

See Section 9.3 for further details.

9.2 Advanced Process Controls

The term advanced process control or APC is used for a wide variety of automation functions. Sometimes, even cascade control loops are included in this category. For our purposes, we define an APC as a control strategy that uses a control computer (except for sequential logic control, which is a special category of control) to convert one or more inputs into one or more outputs. Because of the more complicated nature of APC, it is best used to provide external setpoints to regulatory control loops rather than to provide direct outputs to control variables.

Control computers allow the process control engineer/technician to program APCs using higher-level languages such as Basic, FORTRAN, C, or C++. The code used in the APC is often known as the *control script*. In order to keep APC as responsive as possible, control scripts should be as simple as possible and computationally efficient. In this text, we will use a simplified form of Basic to depict examples of control script. Further information about the script can be found here and in Appendix C.

We have previously seen examples of how to execute controls using calculation blocks in regulatory controls (see Section 7.1). However, an APC uses computer code within a control computer instead of calculation blocks and thus can be used for more complicated calculations. For some applications, an APC is preferred to a regulatory control because it is more computationally efficient and less expensive to use a control script instead of a series of calculation blocks.

For example, Figure 7.7 shows how to calculate the effective overall heat transfer coefficient for a heat exchanger using six calculation blocks in regulatory control in order to monitor the fouling or scaling of a heat exchanger. This same objective can be achieved more efficiently in an APC. The control script for an APC to calculate the overall heat transfer coefficient in order to monitor heat exchanger fouling or scaling, when the process fluid is being cooled, might look like the following (details on the script used in this text can be found in Appendix C):

1. LET THH = TI-101(t)
2. LET HC = TI-102(t)
3. LET TCC = TI-104(t)
4. LET TCH=TI-103(t)
5. LET FC = FI-101(t)
6. DELT1 = THH – TCH
7. DELT2 = THC – TCC
8. LT = ln(DELT1)-ln(DELT2)
9. LMTD = (DELT1-DELT2)/LT

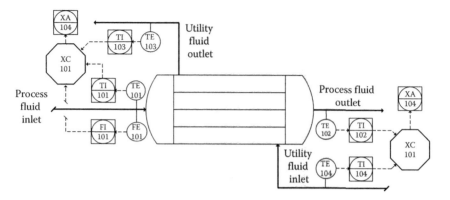

FIGURE 9.4
System to monitor the overall heat transfer in a heat exchanger prone to fouling or scaling where the temperature of one or more of the inlet streams varies widely using a MISO APC instead of calculation blocks. *Note:* the computer symbol was duplicated to make the drawing easier to read; it does not imply two separate APC functions. XC-101 calculates the overall heat transfer coefficient. This coefficient is compared to an externally determined minimum acceptable value. When the value drops below this value, an alarm, XA-104, is activated to alert the operator to take the heat exchanger out of service for cleaning.

10. M(t) = FC*AXC*RHOC *AXC is the cross section of the pipe where *FI-101 is *measured; RHOC is the density of the FI-101 fluid; assumed *to be a constant
11. Q(t) = M(t)*CPC*(TCH-TCC) * CPC is the density of the cooling fluid
12. U(t) = Q(t)/A/LMTD *A is the surface area of the heat exchanger
13. If U(t).LT.UMIN THEN ALARM=1 ELSE ALARM=0
14. LET XA-104(t)=ALARM *alarms when U falls below the minimum *acceptable value
15. GO TO 1

The details of the control script are typically documented separately from the process unit's drawings. On a P&ID using this type of control, the scheme might be depicted as shown in Figure 9.4, although it is equally common not to include any indication of an APC on P&IDs at all.

9.2.1 Multiple Input/Single Output Controls (MISO)

MISO controls are APCs that have multiple inputs that are used in some manner in order to generate the remote setpoint for a single controller. Consider a process scheme where a hot process fluid is used in a cross exchanger to provide the energy necessary to vaporize the reboiler return stream for a separation column (distillation, stripper, etc.), as shown in Figure 9.5. In a traditional reboiler, the energy/unit of steam used to provide the reboiler energy is

Higher-Level Automation Techniques 271

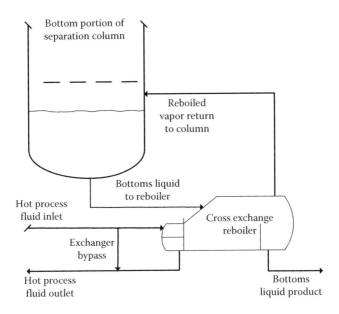

FIGURE 9.5
The bottom portion of a separation column using a cross exchanger type partial reboiler.

essentially constant. Thus, a change in steam flow rate provides a predictable change in the hot side energy flux from the steam to the process fluid.

However, for the scheme provided in Figure 9.5, the energy/unit of hot side process fluid may vary due to upstream process conditions (e.g., the temperature of the hot process fluid inlet stream could change or the heat capacity of the hot process fluid inlet stream could change due to a change in fluid composition). If these changes are substantial or frequent, the operation of the distillation system can be impacted (reduced or excessive reboil can change the vapor traffic in the column and can also have short-term impacts on the bottoms product composition).

For such a system, it may be better to set the rate of flow of the hot fluid into the reboiler (i.e., adjust the CV on the bypass line) to vary heat flux for the hot side fluid. Recall from Chapter 5 that this type of system has two independent (control) variables: the flow of hot side fluid into/out of the exchanger and the fraction of the inlet cold side liquid that leaves as bottoms liquid product (with the rest being vaporized and returned to the column as vapor).

The two key measurement variables are the level of the cold side liquid in the reboiler and the light key composition in the bottoms liquid product. So let's configure the level control loop as a cascade loop where the reboiler level controller provides an ESP to a bottoms liquid product flow controller that adjusts a CV on the bottoms liquid product stream. The bottoms product analyzer controller can then be configured to provide a setpoint to an APC "heat flux" controller that we will denote as q_{sp}. Then:

$$q_{sp} = F_{hot}(T_{h,in} - T_{h,out})C_p \qquad (9.1)$$

Rearranging for F_{hot}:

$$F_{hot} = \frac{q_{sp}}{(T_{h,in} - T_{h,out})C_p} \qquad (9.2)$$

where
F_{hot} is the flow rate of the hot fluid
$T_{h,in}$ is the inlet temperature of the hot side fluid
$T_{h,out}$ is the outlet temperature of the hot side fluid
C_p is the average heat capacity of the hot side fluid

This strategy can be implemented using a MISO APC. Using the instrument tag numbers from Figure 9.6, the control script might look like the following:

1. LET QSP = AC-201(t)
2. LET THI = TI-301(t)
3. LET THO = TI-302(t)
4. FHSP = QSP/(THH-THO)/CP
5. LET APC-200(t) = FHSP

This might be depicted on a P&ID or process control scheme as shown in Figure 9.6.

9.2.2 Fuzzy Logic Controllers

Sometimes, there are process steps where an exact condition, as defined by one or more measurement variables, is not required or even desired. In these cases, a more qualitative control is preferred. The *fuzzy logic controller*, also known as the *decision tree controller*, was developed for these circumstances. In a fuzzy logic controller, inputs are evaluated against one or more criteria using inequalities. For example, an input might be compared in a control script such as:

1. LET MV1 = T101(t)
2. LET MV1SP = T101(SP)
3. LET MV2 = T102(t)
4. LET MV2SP = T102(SP)

Higher-Level Automation Techniques

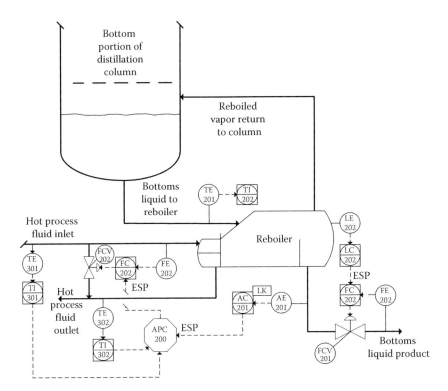

FIGURE 9.6
Depicting a MISO APC to use heat flux as the measurement variable for the intermediate control loop in a nested cascade scheme for a cross exchanger type partial reboiler. LK denotes light key.

5. IF MV1.LT.MV1SP GO TO 4
6. e(t) = 0.5*(MV2-MV2SP)
7. GO TO 5
8. e(t) = −0.25*(MV2-MV2SP)
9. GET CVO = f(e) *calls the controller function to calculate the value *of CVO
10. LET T102(CVO) = CVO

For this example, the value of one input, denoted as MV1, is used to determine how the controller will address the error between the measurement input, denoted as MV2, and the measurement setpoint, MV2SP. If MV1 is greater than its setpoint, then 50% of the error between MV2 and its setpoint will be used in the control algorithm to calculate the output from the controller to the control variable. If MV1 is less than or equal to its setpoint, then 25% of the negative value of the error between MV2 and its setpoint will be used in the control algorithm.

TABLE 9.1

Classification of Treatment Regions within a Metal Curing Furnace

Region	Temperature Range	Required Action	Comments
Cold	$0-T_{cl,min}$	Increase temperature quickly	$T_{cl,min}$ is the minimum temperature for the cool region
Cool	$T_{cl,min}-T_{cl,max}$	Increase temperature slowly	$T_{cl,max}$ is the maximum temperature where an increase in temperature is required
Ideal	$T_{cl,max}-T_{wm,min}$	None	$T_{wm,min}$ is the maximum temperature where no increase is required
Warm	$T_{wm,min}-T_{wm,max}$	Decrease temperature slowly	$T_{wm,max}$ is the maximum temperature for the warm region
Hot	$T_{wm,max}-\infty$	Decrease temperature quickly	

Fuzzy logic controllers are most commonly used in manufacturing applications where multiple replicates of the same item are being processed. No two replicates will be exactly identical, so a qualitative control strategy allows automation of manufacturing tolerances to be embedded in the process.

Let's consider one of the circumstances where this controller might be preferred in a process industry application. In the process of producing metal alloys in a metal refinery, the solid metal solution often must be raised to a temperature that allows the ions in the metal solution to reorganize into its lowest energy state. As long as the metal stays within a certain temperature range for the proper length of time, the objective of this process step will be achieved. So, consider such a process where the acceptable temperature range is divided into five unequal temperature range regions labeled: "cold," "cool," "ideal," "warm," and "hot," as shown in Table 9.1. Any temperature within the curing furnace in the "cool," "ideal," or "warm" regions is acceptable. Temperatures below "cool" are too "cold" and temperatures above "warm" are too "hot." In order to avoid getting too cold or getting too hot, the controls should begin to adjust the heat input to the curing furnace when the temperature approaches one of these limits. Therefore, our fuzzy logic controller might use a decision sequence such as the following:

Step	Decision	Comment
1.	LET TF=TI-100(t)	* where TI-100 measures the curing * furnace temperature
2.	If TF.LT.TCLMIN GO TO 6	* go to max adjust if temp is cold
3.	If TF.LT.TCLMAX GO TO 8	* go to min adjust if temp is cool
4.	If TF.LT.TWMMIN GO TO 12	* the temperature is ideal, do nothing

(*Continued*)

Step	Decision	Comment
5.	If TF.LT.TWMMAX GO TO 10	* go to min adjust if temp is warm; * otherwise max adjust for hot
6.	$e(t) = 2*(TF-TIAVG)$	* the temperature is either too cold or * too hot, use twice the difference * between the current temperature and * the average temperature in the ideal * range as the error function value in * the output calculation
7.	GO TO 14	
8.	$e(t) = (TF-TCLMAX)/(TCLMAX-TCLMIN)$	* the temperature is cool, use the * relative error from ideal as the error * function value in the output * calculation
9.	GO TO 14	
10.	$e(t) = (TF-TWMIN)/(TWMMAX-TWMMIN)$	* the temperature is warm, use the * relative error from ideal as the * error function value in the output * calculation
11.	GO TO 14	
12.	$e(t) = 0$	* the temperature is in the ideal range, * no adjustment is required
13.	$CV_o = CV_{ob} + K_{PID}\left[e(t) + \dfrac{1}{\tau}\int_0^{t_M} e(t)dt + \tau_D \dfrac{de(t)}{dt}\right]$	* calculate the output based on the * error
14.	GO TO 1	

9.2.3 Multiple Input/Multiple Output APC

The next level of complexity is to use multiple inputs to calculate multiple outputs. This is known as a MIMO APC. Consider the reactor control scheme shown in Figure 6.5. The optimum condition is based on the reaction temperature, the concentration of the reactants, and the residence time at the reaction conditions. The reaction pressure may also be important for some reactions. However, the optimum temperature and residence time may change based on the relative concentration of the reactants in the reactor.

Let's assume that laboratory or pilot plant testing has been used to generate a four-dimensional relationship function between optimum reactor temperature, residence time, and the ratio of reactants A and B in the reactor (Figure 9.7). A MIMO APC could be used to determine the optimum point on this relationship function from multiple inputs and then generate external setpoints for both TC-201 and FC-101 (FI-101 would be converted to a

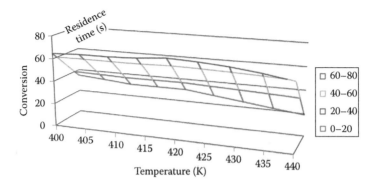

FIGURE 9.7
A 3D correlation plot to select the best temperature and residence time combination to maximize conversion for a single reactant ratio. The actual database would be four dimensions, adding the reactant ratio variable (equivalent to a series of 3D plots at different ratios).

controller and a CV added to the inlet or outlet lines; note that this might affect upstream unit operations, so a more comprehensive analysis of the overall process would need to be performed to determine how to do this without causing other problems).

9.2.4 Multivariable Controls

If the fuzzy logic controller concept is extended to a MIMO APC paradigm, the result is the *multivariable control system*, or MVC. Variations of this type of control system are known as a *Model Predictive Control System* or a *Dynamic Matrix Control System*.

Consider the gas absorption control scheme shown in Figure 5.14. In this system, the concentration of solute in the lean solvent (AC-300) is used to set the steam rate to the reboiler (TCV-300), and the concentration of solute in the solute-lean outlet gas (AC-101) sets the ratio used to control the flow rate of solvent into the absorber (FC-201) and thus in the entire recycle loop. In this scheme, the concentration of solvent in the solvent fluid stream (typically water) is based on a manually input setpoint (AC-203) and would usually be based on vendor recommendations or laboratory/pilot system tests.

In a real system, there is an optimum balance between the solvent concentration and the solute concentration in the absorber solvent feed stream. The balance considers the costs of regeneration, the rate of solvent degradation due to the regeneration process, pumping costs in the transfer pump, the costs associated with the trim cooler, and the loss of solvent in both the solute-lean gas outlet and recovered solute gas streams. In addition, some of the measurement variables, like the solvent chemical concentration in the lean solvent stream, only have to be within a range of acceptable values, not at a specific setpoint. For these measurement variables, inequalities and fuzzy logic are the best way to determine outputs to associated control variables. Clearly, this is too complicated for regulatory control. In fact, there are so many interacting variables that it would be difficult to design a MIMO APC that could calculate the overall costs and generate multiple external setpoints to move the overall control scheme to its optimum point (the ESPs would be for AC-101, AC-201, AC-300, TC-201, and TC-301 in Figure 5.14).

The only reasonable way to optimize this system is with an empirical or model-based control system that can assess the dynamics of all of the variables simultaneously. Usually, a specialty vendor is used to implement their version of an MVC to any given application. However, it is useful to have an overview of the basic concepts involved and the capabilities of such a system so that you are prepared to work with a specialist from the MVC vendor and thus to maximize the effectiveness of the MVC controls applied.

A typical MVC consists of an objective function, a process or predictor module, a controller module, and a feedback or residual module. The predictor module can be based on: (1) a process-based model, similar to a process simulator, (2) a dynamic model that replicates the dynamics of the process and the regulatory control system, or (3) a purely empirical mathematical matrix-based model. Regardless of the model type involved, the MVC uses some type of objective function, often economic, and determines the optimum values for the control variables that maximize the objective function. This vector of optimum values is then used by the controller module to generate a suite of outputs that are sent as external setpoints to key regulatory control variables associated with the process of interest. The controller module can be based solely on PID control algorithms, empirical (adaptive) control algorithms, or a combination.

Instead of just calculating a single step, most MVCs will determine every time step that will take the process from its current state to its optimum state using the predictor module. In this way, not only does the MVC identify the optimum state, it also determines the optimum series of steps that reaches this optimum state, which is known as the event or prediction horizon. The MVC will adjust the vector of outputs based on multiple predicted steps in the event horizon. The steps used are known as the control horizon. Usually, only a specified portion of the entire prediction horizon is used, say the next 5 min, as predictions beyond this point are less accurate and may become invalid as additional process variations occur. The adjusted vector of outputs is sent to the process' regulatory control variables as ESPs.

In order to insure that the MVC prediction module is accurate, the MVC uses the predictor module to determine the value of the suite of input variables, both measurement and disturbance, that matches the values for the control variables predicted by the output ESPs generated by the controller module. The feedback module generates an error vector based on the difference between the actual measured variables and their predicted values. The MVC uses this information to improve the predictor module.

This prediction/evaluation process is repeated after each MVC step, insuring that the MVC output moves toward the most recent optimum in the most efficient manner. Using this methodology, the MVC can move the process from one set of conditions to another distinctly different set of conditions, along the optimum trajectory.

The most commonly used MVC system uses a mathematical matrix-based process model and an empirical control model, so this system will be used in the discussion that follows. However, keep in mind that other variations are also commonly used and newer, advanced systems are sure to develop in the future.

Recall the adaptive controllers described in Section 8.4. These controllers determined their optimum tuning parameters by testing the actual control loop behavior. In MVC, this same principle is extended to a whole suite of controllers. The designer specifies all of the measurement, disturbance, and control variables associated with the overall control scheme and inputs their regulatory tags into the MVC's control matrix database. One unique feature of an MVC is that you can overspecify the system. That is, you can specify more measurement variables than control variables. The MVC will then build internal dependencies to generate pseudo or intermediate variables to satisfy the mathematical equations of the process model.

Once the regulatory controls are commissioned, the MVC is put into learning/tuning mode. The MVC makes small (usually less than 5%) changes to control and disturbance variables and evaluates the overall response of the measurement variables to these changes. It then uses this evaluation to develop tuning/weighting factors for use in simulated adaptive controllers which then generate ESPs for the MVCs control variables. The MVC continues making these changes to the simulated adaptive controllers until it has fully explored the dynamics of the entire process/process control scheme. The MVC is then put into control mode.

Back when MVC systems were first developed, the learning mode could take up to 6 months, since the MVC perturbed each measurement/disturbance variable combination individually, requiring thousands of tests. The use of more sophisticated design-of-experiment type methodologies has reduced the number of tests and learning time to a matter of a few weeks in current MVC systems.

MVC systems may require dozens or even hundreds of input variables. However, there is a good chance that one or more of these inputs will be unavailable at any given time. Therefore, the MVC must have the ability to operate with an incomplete set of input variables. If a data reconciliation system is also employed (see Section 9.3), data reconciliation techniques can be used to provide the best available predicted measurement value to replace the measurement from an out-of-service instrument. When these are not available, the system may simply use the last available value. More commonly, the MVC will reassess the number of input and output variables available in the matrix and adjust the number of internally generated equations to balance the matrix. If the system becomes underspecified (too few measurement variables for the number of control variables), the MVC may remove one or more control variables from the matrix, leaving the associated controller ESPs at the last available values.

9.3 Plant-Wide and Process Area Automation

9.3.1 Data Reconciliation

Online data reconciliation apps provide services and information that can be used by regulatory and supervisory controls as well as by optimizers and data generation apps. The data reconciliation system reduces the impact of instrument failures on the PAS and therefore increases both the availability and reliability of measurements. This in turn insures that the online time for the controls at every level of sophistication is maximized and thus increases the profitability of the process plant. It also provides mass and energy balance information that can be used to assess process performance and assist in determining the optimum operating state of the facility.

There are typically two basic features of a data reconciler. Its first feature is to compare current measurement values to historical values. The historical values might be the average and standard deviation for the last hour, shift, or week. If the current measurement differs from the average by more than one or two standard deviations, the app can send an alarm to the control operator and suggest a substitute value for the measurement based on the average historical value. The operator then decides whether to use the original measurement or the suggested replacement.

The second feature is to calculate mass and energy balances. By overspecifying the number of measurements used in the balance calculation, the app can then determine if one of the measurements is "bad," where "bad" means the measurement is inconsistent with the rest of the measurements in the balance calculation. When a bad measurement is detected, the app can send an alarm to the control operator and suggest a substitute value for the measurement based on a balance calculation that excludes that measurement. The operator then decides whether to use the original measurement or the suggested replacement.

If the balance isn't overspecified, the data reconciliation app can suggest a replacement value using historical data and can highlight the bad measurement and send an alarm to the control operator. The PAS may also have the capability to run an online or off-line process model that can be used by the data reconciliation app to determine a suggested replacement value. This capability may be invoked automatically to provide the control operator with a suggested value or may need to be run manually, depending upon the sophistication of the automation system.

In addition to using mass and energy balances to detect bad measurements, data reconciliation apps can also be used by optimizers and MVCs to determine the state of the process and to suggest process changes to move toward the process' optimum operating state. They can also be used to generate routine reports for operations, engineering, and planners to assist in human evaluation of the process.

9.3.2 Plant-Wide and Process Area Optimization

Real-time optimizers (RTOs) are used to set process objective targets for supervisory control strategies. Plant-wide optimizers do this for key selected controllers and control strategies across the entire plant, while process area optimizers focus on a specific process area or a complex suite of unit operations. A real-time optimizer is often closely coupled with MVCs to allow more sophisticated objective function calculations to be performed than can be accomplished within the MVC app. Area optimizers are best justified for processes that change process objectives frequently as the optimizer can be used to determine the most efficient path to move from one set of operating objectives to another. The optimizer can also help to efficiently recover from a major process upset and can help to optimize catalyst life, minimize energy consumption, and minimize environmental impacts.

The goal of the RTO is to determine the vector of controller setpoints/supervisory targets, $T(x,t)$, that maximizes an objective function, $OBJ(T)$, while satisfying a matrix of measurement value constraints, $C(y,t)$. The objective function can be linear (the most common), nonlinear, or empirical. The goal is to maximize revenue producing elements while minimizing cost elements. Weighting factors can also be used to bias the function for certain targets compared to other targets. For example, if the plant is operating below its maximum capacity due to sales limits, the weighting factors for revenue elements can be reduced and/or the weighting factors for cost elements increased so that the RTO drives the process toward efficiency rather than toward productivity.

A typical optimizer script might look something like this:

$OBJ(t) = a_1 OB_1(t) + a_2 OB_2(t) + \ldots$
$OB_1(t) = \text{MAXIMIZE}(\text{FC-100}(t) * \text{PRODVAL1})$ *example of a revenue
$OB_2(t) = \text{MINIMIZE}(\text{FC-200}(t) * \text{HPSTMVAL})$ *example of a cost
.
.
.
$C(1,t) = \text{AC-100 .LE. PROD1HKCONC}$ *example of a specification
                                              *constraint

C(2,t) = SC-303 .GT. PMP1LOW.AND.SC-303.LE.PMP1HIGH
       *example of an equipment
       *operating range constraint

.
.
.

As you can see, the number of relationships within the RTO can get extensive and a large number of variables may be used in the calculations. Fortunately, the RTO only has to update the target matrix at reasonable time intervals—hourly, per shift, daily, and so on—which coupled with the computing power of today's computers allows very complicated RTOs to be enacted.

To get the maximum benefit out of a process area optimizer, a process engineer should prepare an optimizer objectives statement that the programmers can use to construct the optimizer. The act of preparing this statement requires a deep analysis and understanding of the process and it is not unusual for improvements in the regulatory or APC controls or even in the process itself to come out of this analysis.

9.3.3 Planning and Scheduling

A planning app is used to determine the overall inputs and outputs to the plant. Inputs would be items like raw material supply schedules and inventories while outputs would be product delivery commitments and inventories. Cost data are also needed, including utility costs, raw material and product values, and so on. Constraints, such as any emission limits on the plant's environmental or operating permit, are also considered. The planning app uses all of this information to determine target inputs and outputs of the plant such as the target amounts of each product that should be produced. The planning app provides information used to schedule shipments to/from the plant and predicts the inventory of materials and good in the plant's storage areas.

A scheduling app is used to determine the throughput for each process area in order to meet the plant's production targets. It provides the production targets that will be used by the operations staff and automation system to determine the inputs and outputs that will be generated in order to satisfy the production targets provided by the planning software and/or planning department staff. This information is typically generated once per shift, once per day, or even once per week.

Although planning apps and scheduling apps can be used separately, they are often coupled together in a single coordinated system. Scheduling software often includes a linear optimization routine that can be used for optimization. However, if an RTO is available, the planning app can use the

RTO to perform plant-wide optimization. The output of the planning and scheduling system will be used by process area optimizers and/or MVCs to determine the most profitable way to operate the process.

9.3.4 Enterprise and Supply Chain Management

A logical extension of planning and scheduling apps is the enterprise management system, also known as enterprise resource planning/management, business management systems, or supply chain management systems. An enterprise management system integrates a number of business functions together to increase coordination of these functions. For a process plant these functions include the functions described above for planning and scheduling, plus inventory management, purchasing, and shipping. It may also include resource allocation (staffing, etc.) and accounting functions such as billing and payments.

When the enterprise management system is tied into the plant's automation system, it can operate in real time even though its functions may be managed at a remote corporate headquarters. The system can track business resources and the status of business commitments.

9.4 Higher-Level Automation of Batch Processes and Semi-Batch Unit Operations

The development of higher-level automation tools for batch processes and semi-batch unit operations has lagged the comparable development for continuous processes. As a consequence, you may need to custom design/program APC and other higher-level applications for your specific process. Part of the problem is the dynamic nature of batch processing. If the process doesn't remain in the same step in the batch sequence long enough, it may not be possible to determine the optimal condition. However, for batch processes that are repeated frequently and for semi-batch process steps, it may be possible to use information from multiple iterations of the sequence to gather the data needed for optimization. In this subsection, we'll provide some ideas of the types of higher-level automation that might be useful for these types of processes. This description is in no way comprehensive and you may be able to come up with many other powerful and useful techniques of your own.

MISO and MIMO control schemes can sometimes improve the efficiency of batch processes and many of the concepts introduced in Sections 9.1 and 9.2 can be applied to individual steps within a batch sequence. Consider the semi-batch control scheme shown in Figure 5.21 for a fixed bed adsorption system. The length of time the bed can remain in process service depends upon the capacity of the bed to adsorb the solute. However, as the bed ages, this capacity reduces. Thus, the original design conditions are usually based on a design bed capacity that is closer to the end-of-life condition than the beginning-of-life condition. As the scheme is specified in Figure 5.21, the length of the process step will start out much longer than that in the original design and gradually get shorter and shorter over time.

During regeneration, a fixed quantity of a hot inert gas is heated to a target temperature and passed through the bed at a target flow rate until the solute has been desorbed out of the bed to below a given concentration. In the scheme shown in Figure 5.21, all of the targets (i.e., setpoints) for these variables are manually specified by the control operator, usually based on the original design conditions. When the bed is new, these conditions will likely regenerate the bed quickly, and the bed will sit in idle mode waiting for the next bed to require regeneration. It might be better to regenerate the saturated bed over a longer period of time when the bed is new. The economic optimum state for this system will be based on the costs to compress the inert gas, the costs to heat the inert gas, and the degradation costs to the adsorbent due to the regeneration step. An MVC using an optimization function (either internal to the MVC or as a separate area optimizer app) can be used for such a system because the overall process is continuous and cyclic and therefore provides sufficient time and data to implement the MVC. Even if a sophisticated MVC is not specified, a simplified model-based control scheme coupled with an area optimizer could be used to accomplish this same objective.

Let's look at another example. Consider the typical bacteria population growth curve shown in Figure 6.9 and the control scheme for a typical semi-batch biological reactor system shown in Figure 6.11. By dividing the growth curve up into specific regions, a sequence of steps can be developed that moves the process through the life of the colony in order to accomplish the objective of the bioreactor. The scheme in Figure 6.11 assumes that the dynamics of the system within each step will be linear for the entire sequence. With proper tuning, a stable control scheme can be applied based on this assumption. However, the overall system will not react optimally to upsets or changes because the dynamics will not be linear for the entire sequence. For example, the second region shown in Figure 6.9 is the exponential growth phase. Changes to the process in this phase will most likely react exponentially rather than linearly. To properly specify each controller for each sequence, a complete dynamic analysis of the process and the control system would need to be performed (note, this topic is introduced in Chapter 12). In some of the sequences, a nonlinear controller, such as a logarithmic controller (see Section 8.3 for details of nonlinear PID controllers), might be the better choice for one or more steps in the sequence.

As this last example demonstrates, it is possible to have a batch sequence where different types of controllers would be best for different steps in the sequence. A simple way to handle this is to have multiple regulatory controllers connected through an internal hand switch. The SLC sends a signal to the hand switch at each step, telling the switch which controller's output to send to the control variable. Dynamic testing, either after commissioning, or using a model, is used to specify the controller type for each step. The disadvantage of such a system is that multiple controllers need to be specified, which may be expensive depending upon if each represents a distinct physical controller or if they are just virtual computer-based controllers.

Another way to accomplish this is to use a model or heuristic controller. The model could include each of the functional forms that are optimum for each step in the sequence:

$$e'(t) = a*e(t) + b*\log(e) + c*e^{-n} + \ldots \tag{9.3}$$

where $e'(t)$ is the error value sent to the controller, while a, b, and c are on/off weighting factors specified by the SLC at each step of the sequence. Note: the SLC would also update the tuning constants to the controller algorithm (PID, etc.) for each step of the sequence.

For many biological processes, the steps in the sequence are long enough that a heuristic or self-tuning controller can be specified instead. In this case, the controller tests the system dynamics at the beginning of each step in the sequence and then generates the optimum tuning constants for that step.

Problems

9.1 Write an APC control script to replace the calculation block scheme(s) shown in:

9.1.1 Figure 7.1

9.1.2 Figure 7.5

9.2 In the Preimo Foods company Honey-baked turkey plant, turkeys are first preweighed and sorted by size. Those birds with a precook weight of between 5.5 and 7 kg are then based with a honey-lemon-water solution in batches of 20. A temperature sensor is inserted into the breast of each turkey. The batch is then conveyed into an oven where they are slowly baked to a target temperature of 76°C. However, every bird must reach a minimum temperature of 74°C to be considered fully cooked. None of the birds should reach a temperature of 80°C or they will become dry and less edible. When all of the birds have reached the minimum temperature and at least 50% of the birds reach a temperature of 75.5°C, the batch cooking step is stopped and the birds are conveyed to the cooling unit. However, if any of the birds exceed 80°C, then the cooking cycle is halted. If all of the birds are above the minimum, they are conveyed to the cooling unit. If one or more of the birds are below the minimum, an alarm sounds and the operators are notified which birds have not reached the minimum (so they can remove these from the conveyor before it sends the birds into the cooling unit).

9.2.1 Develop a logic flowchart for this process.

9.2.2 Develop a fuzzy logic control script that could be used for this process.

9.3 Describe a MIMO APC that would adjust the residence time in a fixed bed reactor by adjusting the outlet pressure for a gas-phase reactor system based on the yield and conversion of raw materials A and B into desired product C. Write a typical control script that might be used for this APC.

9.4 For the control scheme shown in Figure 5.16, identify the dependent and independent variables that could be useful for an MVC designed for this system.

9.5 For the control scheme shown in Figure 5.16, what additional measurements would you add for a data reconciliation system?

9.6 For the control scheme shown in Figure 5.16, write optimization statements that could be used to design a unit optimizer for this system.

9.7 Find the website for a chemical plant. Describe the specific overall objectives for that process plant that correspond to the three general objectives of any process facility: maximizing profit, meeting acceptable safety and health criteria, and minimizing environmental impact.

9.8 Consider the gas absorption system shown in Figure 5.14. Sometimes, when the solute is absorbed into the solvent, the mixture has a large heat of solution, raising the temperature of the solvent sharply at the point of maximum absorption. As the solvent moves farther down the absorber, the temperature drops as heat is transferred to the gas the solvent is contacting. This can result in a fairly significant temperature bulge at the point of maximum absorption. A simple, indirect way to optimize the solvent flow rate is to keep the temperature bulge at a specific point within the absorber. The bulge shape can be replicated using a parabola. If the column is outfitted with temperature sensors along the length of the column, a parabolic function can be fitted to the temperature profile. The flow rate of the solvent can then be adjusted to keep the maxima of the parabola at the desired location in the column.

Develop an APC control script to perform this control. Replicate the control scheme in Figure 5.14 and add the temperature elements and indicators necessary and use these indicator identifiers in your control script.

10

Instrumentation (Types and Capabilities)

The process control system relies on field instruments to provide measurement data and to manipulate the process. In this chapter, we introduce the most common field instruments that have been developed to provide these functions.

10.1 Pressure

Pressure is the easiest and most responsive parameter to measure in the process plant. Devices with accuracy adequate for the vast majority of process applications are economical and require very little maintenance. As such, it is usually justified to include pressure measurements in most process streams, even when not needed for controls. It would only take using a pressure measurement to troubleshoot and determine the cause of one or two upsets over the life of the instrument to justify its purchase.

Pressure measuring devices used for gas and liquid fluids are commonly referred to as pressure gauges. The most common of these can be classified as force-summing devices. In these types of gauges, an elastic element, the force-summing device, is exposed to the pressure force of interest. The elastic element changes shape due to this force. The sensor in the gauge then measures the shape change and converts it into a displacement that correlates to the pressure of the fluid. Of the wide variety of force-summing devices, the most common are Bourdon tubes and diaphragms. Bourdon tubes provide large displacement motion, which is useful in mechanical pressure gauges, while diaphragms yield a lower level of motion and are thus typically limited to electromechanical sensors.

In a Bourdon tube pressure gauge (Figure 10.1), a spring is the elastic element that changes shape with changes in the force exerted upon it (pressure is just force per normal surface area). Mechanical linkages convert the spring's motion to that of a pointer. Adjustment springs and linkages are designed to allow calibration for zero, linearity, and span. In some gauges, instead of linking the spring's motion to a pointer, the motion of the spring, or more commonly of a diaphragm, can be linked to a differential transformer to convert the motion into an electronic signal. Other gauges employ a motion amplifying mechanism tied to the wiper

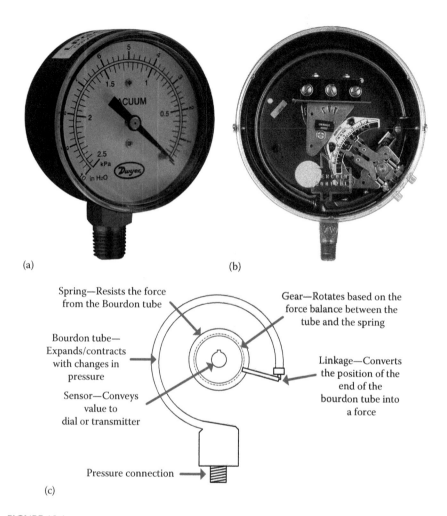

FIGURE 10.1
A Bourdon-tube-type pressure gauge: (a) local dial gauge type, (b) internals, (c) schematic. (a and b: Images courtesy of Dwyer Instruments Inc.)

of a potentiometer. Bourdon tube gauges are a good choice when the pressure reading is for local, manual use only.

Electromechanical pressure sensors (Figure 10.2), or pressure transducers, convert motion generated by an elastic element, usually a diaphragm, into an electrical signal. These sensors are the most common devices used when the pressure measurement is being used in a control system. The electrical output is directly proportional to the applied pressure over the gauge's pressure range. For rapidly changing—dynamic—pressure measurement, the frequency characteristics of the transducer are an important consideration.

FIGURE 10.2
An electromechanical pressure sensor. (Courtesy of Dwyer Instruments Inc.)

Strain gauge transducers are based on metal or silicon semiconductor strain gauges (Figure 10.3). The gauge material can be sputtered onto a diaphragm, diffused into a silicon diaphragm structure, or configured as a spring or a bellows assembly. Strain gauges are made of materials that exhibit significant electrical resistance change when strained. This change is the sum of three effects: (1) changes due to changes in the length of a conductor, (2) changes in the cross-sectional area of the conductor due to the change in its length (resistance change that is approximately proportional to a change in area),

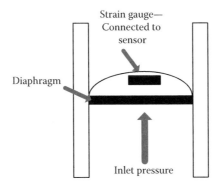

FIGURE 10.3
A strain gauge transducer for pressure measurement.

and (3) changes in the piezoresistive or "bulk" resistivity of the conductor material due to the level of strain it is undergoing.

Manometers provide highly accurate measurement of the pressure by comparing the level in one arm of a "U" tube to that in the other arm (Figure 10.4). The manometer contains a fixed quantity of a high-density fluid (mercury used to be used, but fortunately has been replaced by less toxic substances). The space above one arm of the U tube is attached to the process fluid while the other arm is attached to a reference source (often just the atmosphere). The difference in the pressure force from the two measured fluids results in a change in the relative levels of the two manometer fluids. This change can be correlated to the pressure of the fluid. While highly accurate and inexpensive to install, the manometer requires a level sensor that can translate the fluid level change into an electronic or digital signal. They also require more maintenance than many of the other methods described above. Manometers are commonly used in lab- and pilot-scale applications but less commonly in industrial applications.

Pressure gauges and sensors are usually configured to measure the pressure in one of three ways: the absolute pressure relative to a vacuum, the gauge pressure relative to the surrounding atmosphere, and the differential pressure between two sensing points. Absolute pressure measurement has units of kPa (kiloPascals, which should always be based on a vacuum) in the

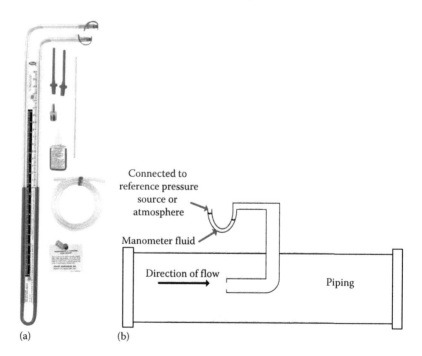

FIGURE 10.4
Manometers can be configured for (a) pressure measurement or (b) flow measurement. (a: Courtesy of Dwyer Instruments Inc.)

SI system or PSIA (pounds per square inch, absolute) or ATM (atmospheres, which should always be based on a vacuum) in the FPS system (also known as the Imperial system in the United Kingdom or the "common" system in North America). Gauge pressure measurement has units of kPa_g (this is rarely used) or barg in the SI system or PSIG in the FPS system. Since differential pressure is the comparison of two sensors, it is independent of any reference pressure, so the units are kPa or PSI (with no "a" or "g" designator).

From a controls perspective, bar and ATM units do not provide sufficient resolution for use by control operators and process engineers. Therefore, the units of kPa and PSIA are preferred. To accurately convert from gauge pressure to absolute pressure, the local barometric pressure is required, although the pressure at sea level, 101 kPa or 14.7 PSIA, is often used. As long as these are consistently used for every pressure measurement (in other words, absolute and gauge instruments aren't mixed together), the absolute error should not matter in the control of the process.

10.2 Flow/Mass

The flow rate of gas and liquid fluids can be measured in a wide variety of ways. The most common class of flowmeters are the Bernoulli-type flowmeters. These meters are based on the Bernoulli relationship:

$$\frac{P}{\rho} + \frac{V^2}{2} + gZ = \text{Constant} \tag{10.1}$$

If one of the terms in Equation 10.1 is held constant, then there is a direct relationship between the other two terms. The most common flowmeter is the orifice meter. A fluid is passed through a horizontal section of piping containing a pair of pipe flanges. Inserted in the pipe flanges is a piece of metal with an orifice directly in its centerline. Resistance to flow caused by the smaller flow area in the orifice reduces the pressure in the fluid. The back edge of the orifice is typically beveled to reduce the flow disturbances caused as the fluid is forced through the smaller opening. The pressures of the fluid both upstream and downstream of the orifice are measured. Because the orifice meter is at a constant vertical height, the term gZ becomes a constant and Equation 10.1 can be rearranged to:

$$\Delta P \propto v^2 \quad \text{or} \quad v \propto \sqrt{P} \tag{10.2}$$

Since the cross-sectional area of the pipe is known, the volumetric flow rate can be calculated directly:

$$G = v * A_x = kA_x\sqrt{P} \tag{10.3}$$

Orifice meters (Figure 10.5) are simple, inexpensive, and reliable. They can be used for both gas and liquid fluids. Their accuracy in real systems is on the order of ±2%–5% depending upon the pressure, corrosivity, and cleanliness of the fluid. The primary disadvantage is the pressure drop required, which increases the costs to move the fluid through the process. Thus, an orifice meter is not an energy-efficient flowmeter.

The venturi meter (Figure 10.6) uses the same principle as the orifice meter but is a more accurate and energy-efficient alternative. Unfortunately,

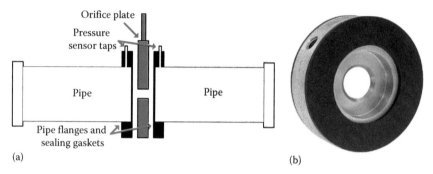

FIGURE 10.5
An orifice-type flowmeter. (a) Schematic and (b) image. This is the most common type used in the process industries. (Product Image courtesy of Dwyer Instruments Inc.)

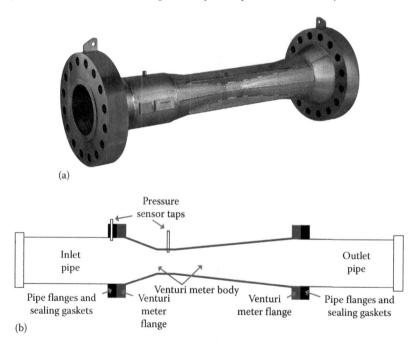

FIGURE 10.6
A venturi-type flowmeter. (a) Image and (b) schematic. (Image courtesy of Badger Meter, Inc.)

venturi meters are also more expensive and take up much more room, especially for very large diameter piping systems. In a venturi meter, the fluid passes through a "neck," where the inside diameter of the pipe is gradually reduced. As the diameter decreases, the velocity of the fluid increases. Based on Equation 10.2, the pressure must decrease when the velocity increases. At the minimum diameter point, the "throat," the piping diameter begins to increase. The pipe diameter after the throat increases much faster than the rate of narrowing in the neck and most of the original pressure in the fluid is recovered. The pressure is measured at two points along the meter, usually at the entrance to the meter and at the throat. The change in pressure is then related to the fluid velocity using Equation 10.3.

Manometers, which were described in Section 10.1, can also be used as a Bernoulli-type flowmeter. The manometer leads are attached to two points along a horizontal pipe. The difference in the pressure between the two points, due to the pressure drop in the pipe, results in a difference in the level in the manometer fluid, which can be correlated to fluid velocity. The meter is essentially the same as the version used for pressure measurement (Figure 10.4).

Another class of flowmeters takes advantage of the physics of fluid flow. The most common of these is the Coriolis or mass flowmeter (Figure 10.7). When a fluid flows in a pipe or tube that is rotating or vibrating, a change in direction of the flow field will generate a Coriolis force due to the inertia of the fluid. So in a Coriolis flowmeter the fluid flows through a curved section of tubing that is either rotating around an axis or vibrating. The curved section will deflect in a predictable manner based on changes in the inertial

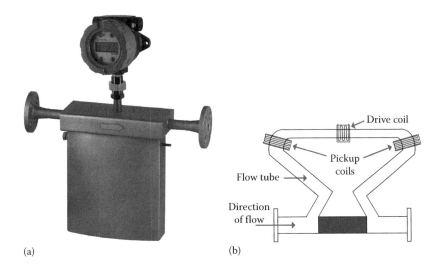

FIGURE 10.7
A Coriolis-type mass flowmeter. (a) Image and (b) schematic. (Product Image courtesy of Badger Meter, Inc.)

momentum (and therefore the mass flow rate) of the fluid. The deflections can be measured and correlated to the mass flow rate. The cost of these meters have decreased significantly in recent years, and they are now the second most common type of flowmeter after the orifice meter.

The pitot tube meter is another Bernoulli-type flowmeter. These meters are most commonly used to measure the flow rate of low pressure gases at low flow rates, a condition that is difficult to measure with orifice or Bernoulli meters. In a pitot tube flowmeter, a tube with an opening at the tip is inserted into the flow field. The tip is the stationary (zero velocity) point of the flow (Figure 10.8). When pressure at the tip is compared to the static

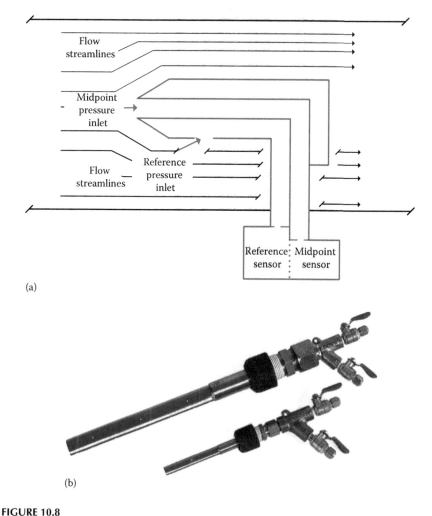

FIGURE 10.8
Pitot tube flowmeters are often used for very low flow, low pressure gas flow measurements. (a) Schematic and (b) image. (Product Image courtesy of Badger Meter, Inc.)

Instrumentation (Types and Capabilities) 295

FIGURE 10.9
Rotameter-type flowmeter. (Image courtesy of Dwyer Instruments Inc.)

pressure of the fluid, the difference can be used with Equation 10.3 to calculate the flow velocity.

The rotameter (Figure 10.9) is another Bernoulli-type flowmeter. However, in this meter, the pressure term is kept constant and the velocity term is related to the potential energy term. In a rotameter, the fluid is routed upward where it impacts a float. The float's position is the sum of the upward acting drag force generated by the fluid's momentum and the downward acting force of gravity. With the pressure term a constant, Equation 10.1 can be rearranged to correlate the fluid velocity to the change in potential energy, as measured by the position of a float of known weight, size, and shape.

Rotameters are very accurate. For applications where the volumetric flow is low but the pressure is sufficient to allow for small diameter piping, they are also relatively inexpensive, which is why they are very commonly used in laboratories and pilot plants. However, the costs increase substantially as the volumetric flow rate measured is increased. They also cannot be used in low flow, low pressure applications (where pitot tubes are usually used

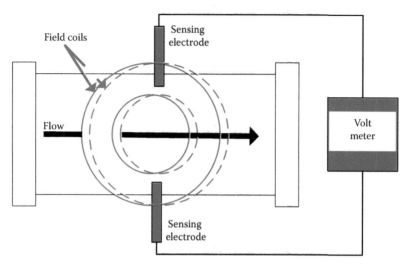

FIGURE 10.10
An electromagnetic-type flowmeter.

instead). A rotameter installed in a pipe of larger than around 5 cm or 2 in. in diameter is rare.

Another type of meter that takes advantage of the physics of fluid flow is the magnetic or electromagnetic flowmeter (Figure 10.10). Faraday's law of electromagnetic induction states that a voltage will be induced when a conductive fluid moves through a magnetic field. In the magnetic flowmeter, the magnetic field is created by energized coils outside the flow tube and the voltage is measured using electrodes mounted in the pipe wall. The voltage produced is directly proportional to the flow rate.

Magnetic flowmeters have a relatively high power consumption and can only be used for electrically conductive fluids such as water (note: there are always dissolved solids in water that make it conductive). These types of meters are particularly valuable where direct contact with the fluid (such as a radioactive or highly toxic fluid) must be avoided.

The ultrasonic flowmeter (Figure 10.11) exploits the Doppler effect, which is the change in the frequency of a reflected sound wave due to the velocity and direction of fluid flow. So in an ultrasonic flowmeter, the frequency from the fluid moving toward a transducer is compared to the frequency when the fluid is moving away from a transducer. The frequency difference is equal to the reflected frequency minus the originating frequency and can be used to calculate the velocity of the fluid.

Another class of flowmeters is the direct mechanical flowmeter class. The most common are paddle and turbine flowmeters (Figure 10.12). As the fluid passes through the meter, it impacts a paddle or the blades of a turbine, causing the paddle or turbine to rotate. The rate of rotation is directly proportional to the mass flow rate through the meter.

Instrumentation (Types and Capabilities) 297

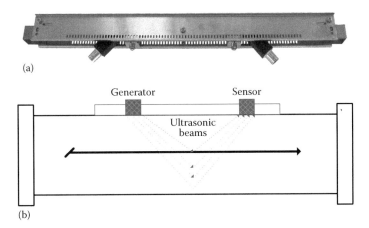

FIGURE 10.11
An ultrasonic-type flowmeter (a) Image and (b) Schematic. (Product Image courtesy of Dwyer Instruments Inc.)

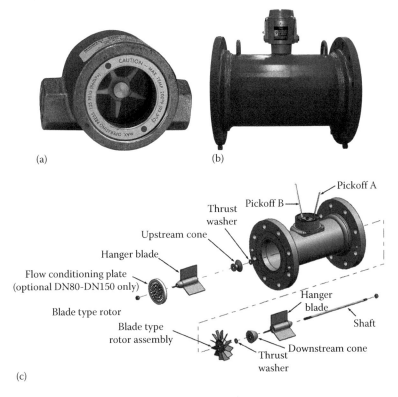

FIGURE 10.12
Direct mechanical flowmeters: (a) paddle meter image, (b) turbine meter image, (c) turbine meter schematic. (a: Image courtesy of Dwyer Instruments Inc.; b and c: Image and schematic of a Daniels Series 1500 Liquid Turbine Meter courtesy of Emerson Automation Solutions.)

Another type of mechanical flowmeter is the positive-displacement flowmeter. In this meter, the fluid is separated into segments of known volume, which allows semi-batch operation of the meter in a continuous flow stream. Fluid feeds into a chamber until it is full, and the time it takes to fill the chamber is measured. The total volumetric flow rate can then be calculated from the rate at which the volume is filled.

Some versions also measure the displacement rate from the chamber by measuring how fast the fluid displaces a known volume out of the chamber. These meters are relatively expensive and so are usually only applied for special applications. For example, they are a good choice for measuring very low flows at low pressures and fluids with a high viscosity, applications that are often very difficult to measure by the more commonly used meters.

Using a completely different principle, vortex flowmeters, also known as vortex shedding flowmeters or oscillatory flowmeters, measure the vibrations of the downstream vortexes caused by a barrier in the moving stream. The vortex vibration frequency is related to the velocity of the fluid and can be used to calculate the flow rate (Figure 10.13).

Weigh-in-motion systems (Figure 10.14) are a common method used to measure solids flow. Solids continuously flow into a container or onto a small conveyor located on a weigh scale or load cell and continuously flow out the bottom (or off the end) at the setpoint flow rate. In another version, the flow rate out/off is allowed to vary with the solids flow rate into/onto the platform/container. Changes in the weight of the material in the container or on the platform are used to determine the mass flow rate of the solids. They can also be used to adjust the rate of flow of the conveyor or other device directing the solids into the container. In another version of these systems, a container located on a weigh scale is filled with the solid. The rate to fill the container to a target weight is used to determine the flow rate of the solids stream.

If the solid is entrained in a gas, wave sensors can be used to detect the particle density in the stream. These types of sensors are described in

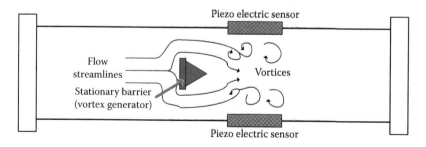

FIGURE 10.13
A vortex-type flowmeter.

FIGURE 10.14
A weigh-in-motion mass flowmeter for solids measurement. (Image courtesy of Vande Berg Scales.)

Section 10.4 for level control. Combining the particle density measurement with an independent gas flow measurement provides a measurement of the mass of solids flowing in the stream. This same type of system can be used when solids free fall from a hopper through a pipe using a transport fluid.

This description is far from complete. There are many additional types of flowmeters for specific applications. However, the ones described here represent the vast majority of those employed in the process industries.

10.3 Temperature

There are a number of different types of sensors for temperature measurement. Each of these infers a change in temperature or a difference in temperature compared to a reference source using a temperature-dependent physical property. The four most common types used in the process industries are described here.

The most common type of temperature sensor is the thermocouple. As shown in Figure 10.15, two metals with different electromotive force (emf) potentials are joined together at one end. This creates a galvanic circuit and the voltage induced by the circuit can be measured at the open end. As temperature changes, the emf at both the open and joined or closed ends change.

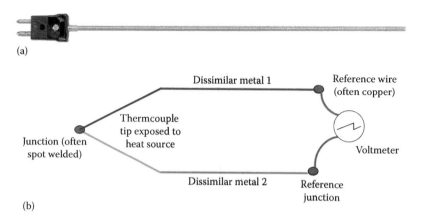

FIGURE 10.15
(a) Image and (b) schematic showing how a thermocouple measures temperature. (a: Product Image courtesy of Dwyer Instruments Inc.)

The open end is held at a constant reference temperature so that the only change will be at the closed end, which is in contact with the fluid.

Different types of wires are used depending upon the range of temperatures that the thermocouple will experience. The most common type is the "k" type which has a sensitivity of 41 µV/°C and can be used in the temperature range of −200°C to +1350°C. The k type thermocouple has one wire made of chromel and the other wire made of alumel.

To avoid contacting the thermocouple wires with the fluid, the thermocouple can be inserted into a sheath, known as a thermowell. As long as the thermowell has sufficient thermal conductivity, the loss of responsiveness to fluid temperature changes is minimized.

The primary limitation of thermocouples is the low voltage of the generated signal, on the order of microvolts. Interference from externally generated emf can make it difficult to read the true signal from the signal noise. Therefore, adaptive controllers are recommended when using thermocouples in noisy environments (basically any in-plant application) as these controllers have a better capability to cancel out noise from low-voltage signals. In real process applications, the accuracy of a temperature measurement from a properly configured and calibrated thermocouple system will be on the order of ±1°C–2°C or 2°F–5°F.

When more accurate or reliable temperature measurement is required, a resistance temperature device (RTD) may be specified (Figure 10.16). RTDs capitalize on the fact that the electrical resistance of a metal changes as its temperature changes. A typical RTD consists of a platinum wire wrapped around a mandrel and covered with a protective coating of glass or ceramic.

Thermistors are similar to RTDs except the temperature measurement is based on resistance change in a ceramic semiconductor instead of a metal.

Instrumentation (Types and Capabilities)

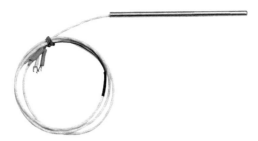

FIGURE 10.16
A resistance temperature device (RTD). (Image courtesy of Dwyer Instruments Inc.)

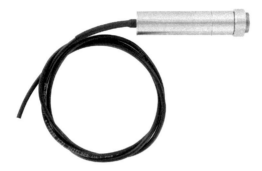

FIGURE 10.17
An infrared sensor for temperature measurement. (Image courtesy of Dwyer Instruments Inc.)

The resistance-temperature relationship of a thermistor is negative and highly nonlinear. Thermistor vendors incorporate special design features to mitigate the consequences of nonlinearity for these sensors.

Infrared sensors (Figure 10.17) measure the strength of the radiation waves in the infrared spectrum emitted by a surface in contact with the fluid. As the temperature increases, the signal strength and frequency of infrared radiation from the surface increases. Infrared sensors are more expensive and usually only used for special applications such as extremely high temperature or for fluids where direct contact by a thermowell must be avoided.

10.4 Level

There are a wide variety of methods for measuring the level of a liquid or the interface between two liquid phases in a vessel. The simplest is visual observation of the level in a sight glass. However, this method isn't very useful in process control and there are many other limitations, including pressure and fouling.

The most commonly applied class of automatic level measurement sensors are those based on the density (i.e., specific gravity) difference between the liquid being measured and the fluid in the space above the target liquid, which may be a gas or a lighter density liquid. Of these, the simplest is the float. A float is a solid object having a specific gravity between those of the target process fluid and the fluid above. The object will float at the surface, accurately following its rises and falls. The level of the float is based on its physical location, which can provide an absolute reading or activate a switch. Float systems may use mechanical components such as cables or conductive tapes (Figure 10.18) to communicate level. A more recent development is the magnet-equipped float so that the position of the float can be determined remotely.

Conductive tapes or straps can also be used without the float. The tape hangs in the tank from the top to the bottom (or low-low level point) of the tank and is grounded to the tank wall, forming a galvanic cell. The electrical resistance in the tape is measured. If the tank contains a conductive liquid, it will short-circuit the portion of the tape below the liquid level, changing the resistance of the cell and thus the current and/or voltage generated. Measurement of this change can be correlated to the level in the tank.

A similar level sensor is the capacitance sensor. The advantage of this sensor over the conductive sensor is that it can be used for nonconductive liquids as the sensor measures capacitance, which depends upon the fluids dielectric constant, rather than its electrical conductivity. Oils have dielectric constants from 1.8 to 5, pure glycol is 37, while aqueous solutions are between 50 and 80. The capacitance of a body of liquid varies with liquid level. As the fluid level rises and fills more of the space between the plates in the sensor, the overall capacitance rises proportionately. A capacitance bridge electrical circuit is used to measure the overall capacitance and provide a continuous level measurement.

Hydrostatic head measurements have also been widely used to infer level. Hydrostatic devices include bubblers, differential-pressure transmitters, and

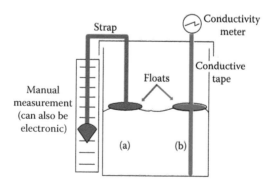

FIGURE 10.18

Level measurement using (a) float position measurement and (b) a conductive tape and float system.

displacers. These devices are Bernoulli-type sensors where the velocity term is zero. Because the Bernoulli equation assumes a constant density of the fluid, changes in the density due to temperature or composition changes will affect the accuracy of these types of sensors. In a bubbler (Figure 10.19), a "dip" tube is used to sense the pressure at the low level point in the vessel. The pressure will vary with a change in the height of the liquid above the measurement point. The same thing can be accomplished by directly measuring the pressure at the bottom and top of the vessel using one of the pressure measurement devices described in Section 10.1 and calculating the differential pressure directly. Changes in the pressure are measured and these are then correlated to changes in the height of the liquid.

Displacers (Figure 10.20) work on Archimedes' principle. A column of solid material with a density greater than the process fluid (the displacer) is suspended in the vessel from the low level to the high level points. The column displaces a volume of fluid equal to the column's cross-sectional area multiplied by the process fluid level on the displacer. A buoyant force equal to this displaced volume multiplied by the process fluid density pushes upward on the displacer, reducing the force needed to support it against the pull of gravity. This force, which can be measured, varies with the liquid's level.

For highly accurate measurement of large volume liquid levels, a wave-based measurement sensor can be used (Figure 10.21). These sensors typically operate by measuring the distance between the liquid level and a reference point at a sensor or transmitter near the top of the vessel. A pulse wave is generated at the reference point, which travels through either the vapor space or a conductor, reflects off the liquid surface, and returns to a pickup at the reference point. An electronic timing circuit measures the total travel time, which is correlated to the distance to the surface of the fluid and thus to the liquid level. Different devices employ different kinds of wave pulses, with the most common being RADAR (microwaves), ultrasonic waves, or photons (light).

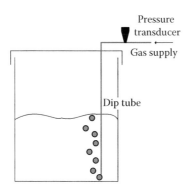

FIGURE 10.19
A bubble-type level measuring device measures the pressure required to push bubbles through a fluid. As the liquid level increases, the force required increases.

FIGURE 10.20
A displacer-type level measurement device. (Image courtesy of Dwyer Instruments Inc.)

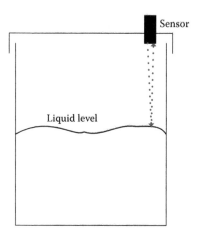

FIGURE 10.21
A wave-based level sensor. The wave could be RADAR, ultrasound, or light.

Another variation of the wave-based sensor is the piezomagnetic sensor. The speed of a torsional wave along a wire is used to determine the position of magnets embedded in a float. These devices can be highly accurate and work for both total level and interface levels. Some models will measure both the total and interface levels simultaneously. They can be substantially more expensive than other methods, although the costs have been decreasing steadily with technology improvements.

10.5 Composition

Knowing real-time changes in the composition of a process fluid is probably the most valuable information that can be used for process control. Unfortunately, composition is usually the most complicated, maintenance-intensive, and expensive measurements in the regulatory control system. Online analyzers can be divided into two basic categories: direct process measurements and sampled process measurements.

An online analyzer has a sensor (Figure 10.22) that is directly in contact with the process fluid. The most common are pH, conductivity, and gas partial pressure sensors, such as oxygen. The first two are indirect measurements

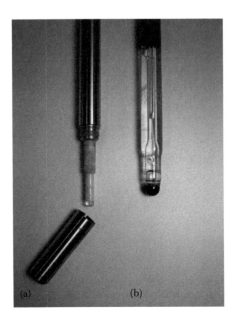

FIGURE 10.22
Examples of sensors for a direct measurement type online analyzer: (a) dissolved oxygen and (b) pH.

that must be correlated to composition. The third type is used to measure the partial pressure of a specific gas phase compound within the process fluid with the most common being hydrogen, carbon monoxide, and oxygen. Even these analyzers do not directly measure the compound of interest; instead, they isolate and assess an effect that would be caused uniquely (or nearly uniquely) by that compound. For example, oxygen is quantified by measuring changes in a physical property, such as the electrical conductivity, of a fluid based on its oxygen content. Specific details can be obtained from the supplier.

The advantage of online analyzers is that they have a responsiveness that is similar to flow measurements. Their disadvantage is the accuracy of the indirect methods incorporated into the online system. Further, they usually must be designed to operate in areas requiring intrinsically safe or explosion-proof devices, which may limit the type of analysis that can be performed or substantially increase the complexity and/or cost of the analyzer.

Sampled process analyzers allow for more complicated and sophisticated analytical methods at the cost of responsiveness and complexity. A sample is drawn from the process fluid and passed through a sample conditioning system (Figure 10.23) to remove any unwanted material from the fluid and to bring the fluid to a standard temperature condition. The fluid is then routed to one or more analyzers to determine the concentration(s) of interest. These analyzers are often known as continuous emission monitors (CEMs), although they are not limited to streams that are being discharged. A wide variety of techniques are employed in these analyzers, including infrared detectors and gas chromatography.

FIGURE 10.23
For sampled type online analyzers, a slip stream of the process fluid is routed through a sample conditioning system such as is shown here. (Image courtesy of LT Industries/Process NIR Analyzers.)

There is a delay from the time the sample leaves the process stream until the concentration reading from that sample is reported. This delay can be on the order of tens of seconds, minutes, or tens of minutes. As such, sampled analyses should be considered nonresponsive measurement variables. When used in regulatory control loops, it should almost always serve as the master control in a cascade control system.

10.6 Vibration

When a mass rotates around a centerline, imperfections to the radial distribution of mass will cause the mass to wobble or vibrate. For centrifugal or axial pumps and compressors, minor changes in the fluid will also contribute to the radial vibration as well as induce axial vibration. Electrical motors may also experience these small effects due to the load attached to the motor. So while vibration is not usually used directly in process control, vibration measuring instruments are commonly used in process plants and the results of these measurements reported via the human interface of the control system.

A rotating body will generate both electrical and magnetic fields, known as the excitation fields. Thus, a common way to measure vibration is to measure changes in one of the properties in either of these electrical fields. For example, it can be shown that the amplitude of the electrical field frequency directly correlates to the amplitude of the vibration and this property forms the basis for the vibrometer. An accelerometer measures changes in the amplitude caused by changes in the speed of the vibration and uses this to measure vibration. A tachometer matches the frequency of the excitation electrical field to a standard resonance frequency. When the frequencies match, the amplitude of the vibration will match the amplitude of the resonance frequency and can thus be used to quantify the vibration.

Similar sensors measure the strength of the induced magnetic field, which is then correlated to vibration. These devices are highly accurate but more expensive than their electrical analogs.

10.7 Throttling Control Valves

Throttling valves, also known as control valves, are used to meter or control the flow of a fluid as it passes through a pipe. It is the most common independent variable that we manipulate in process plants to control the

FIGURE 10.24
A pneumatically actuated throttling control valve.

process conditions. Most control valves have three components: the valve body (also known just as the valve), the valve actuator, and the valve positioner (Figure 10.24).

The valve body holds the mechanism that adjusts the opening through which the fluid flows. The most common types are reciprocating and rotary globe valves. These types are precise (<2% accuracy) and easy to automate. High-performance versions of ball valves and butterfly valves are also available, which under acceptable process conditions perform very well. Diaphragm valves, surprisingly enough, have excellent throttling characteristics.

Reciprocating globe valves (Figure 10.25) are the most rugged, and usually the most expensive, particularly in the larger sizes. The torturous path through this type of valve provides excellent energy conversion of the process fluid resulting in accurate, repeatable control. Severe service, noise, and cavitation control trims are available in this valve style.

The most common designs are sliding-stem globe and angle valves. A globe valve has an opening that forms a seat. A beveled plug or disk moves up or down within the seat to vary the opening through which the fluid can flow. The plug is connected to a stem that is moved up or down by the actuator.

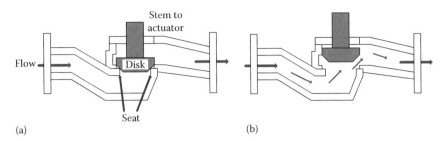

FIGURE 10.25
Globe-valve-type control valve in the (a) closed and (b) open positions.

In an angle valve, the fluid changes direction as it passes through the seat, which increases the accuracy of the plug position for many applications. In these types of valves, the positioner is a sliding stem, which is attached to the actuator and to the plug.

Rotary globe valves consist of cammed plug and segmented ball designs. They have similar characteristics to reciprocating globe valves, except that they have the benefit of rotary motion. Rotary valves do not have a tortuous path, but they have significantly more capacity and turn-down than reciprocating globe valves. For a ball-type valve, a hole is drilled through the ball. When the ball is in the open position, the hole is aligned to the direction of flow. When the ball is in the closed position, the hole is completely isolated from the fluid by the valve seats. The fluid passage can be throttled by adjusting the fraction of the ball opening that is available to the fluid. In the cammed plug, there are divots in the side of the plug that can be rotated to let fluid pass through the seat or to block the flow.

The most common type of actuator is the pneumatic diaphragm/spring type (Figure 10.26). Air is used to apply pressure to one side of a diaphragm. A spring applies pressure to the other side. The force balance determines the degree of deflection of the diaphragm. The valve stem is connected to the diaphragm. The air pressure is regulated based on the output value of the controller. As the air pressure changes the diaphragm deflection changes, moving the stem and therefore adjusting the opening in the valve body.

In a direct acting valve, as pressure increases, the stem is pushed down into the valve seat and the opening is decreased. If there is no air pressure on the diaphragm, the stem is raised to its maximum height. This is known as an air-fail-open (AFO) valve. In a reverse acting valve, as pressure increases, the stem is raised up out of the seat and the opening is increased. If there is no air pressure, the stem is pushed down into the seat. This is known as an air-fail-closed (AFC) valve. This type of actuator is almost always used for reciprocating type valves.

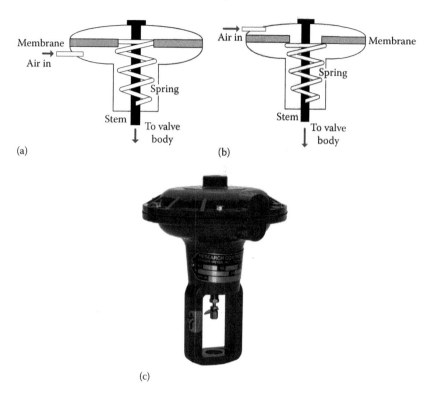

FIGURE 10.26
(a) Reverse acting (AFC), (b) direct acting (AFO) pneumatically driven actuators, and (c) a typical actuator body. (Image courtesy of Badger Meter, Inc.)

10.8 Speed Control Systems

The most common type of speed controller used today is the electronic speed controller (ESC), which is an electronic circuit that varies an electric motor's speed based on the controller output. A common type of ESC varies the switching rate of a network of field effect transistors. The rapid switching of the transistors changes the frequency of an AC supply to the motor. The use of multiple transistors allows for more accurate and smoother control, especially at lower speeds. Another variation uses transistors as relays to open and close the electrical supply circuit to the motor. This type will work for both AC and DC type motors.

A mechanical speed controller uses a resistive coil and moving arm to vary the current supplied to the motor and thus the motor speed.

10.9 Remotely Operated Block Valves

Block valves typically operate in two positions: fully open and fully closed. Valves made specifically for on/off service are designed with tight reliable shutoff in the closed position and little restriction in the open position. Ball valves, gate valves, butterfly valves, diaphragm valves, and plug valves are the most commonly used isolation valves. It should be noted that the closed position of these valves (and indeed of any valve) will not guarantee 100% isolation of the upstream fluid. All valves leak and the degree of leakage increases with use due to wear on the valve components.

Remotely operated ball valves are the most common type of remotely operated block valve (Figure 10.27). A ball is inserted between the seats in the valve body. There is a hole drilled through the ball. When the ball is in the open position, the hole is aligned to the direction of flow. When the ball is in the closed position, the hole is completely isolated from the fluid by the valve seats. These valves provide tight shutoff and high capacity with just a quarter-turn to operate. Ball valves can be easily actuated with pneumatic and electric actuators. In this application, the hole in the ball usually matches the diameter of the fluid channel in the valve body (a fully ported valve), which reduces the pressure drop through the valve in the fully open position.

Butterfly valves (Figure 10.28) are typically used for large pipe diameters where ball valves are not practical. In this valve, a disk that matches the diameter of the opening in the valve body is attached to a stem at its

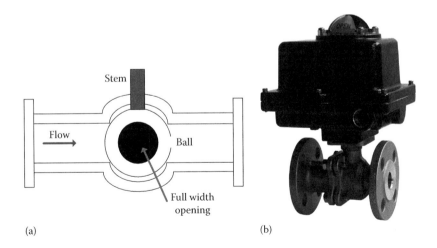

FIGURE 10.27
A remotely operated ball valve. (a) Schematic and (b) image. The hole is full width in a block valve, but smaller when used as a throttling control valve. When the hole is turned into the direction of the flow, the valve is open. (Image courtesy of Dwyer Instruments Inc.)

FIGURE 10.28
A remotely operated butterfly valve. These are typically used for very large diameter pipes. (Image courtesy of Dwyer Instruments Inc.)

centerline. By rotating the stem, the disk is rotated within the valve body. At fully closed, the edges of the disk are pressed against the valve seat, sealing the opening.

Gate valves (Figure 10.29) have a sliding disk (gate) that reciprocates into and out of the valve port. Gate valves are an ideal isolation valve for high pressure drop and high-temperature applications where operation is infrequent. Multiturn electric actuators are typically required to automate gate valves; however, long stroke pneumatic and electrohydraulic actuators are also available. Gate valves are less responsive than ball valves and so are not typically used for routine on/off control applications.

Plug valves are similar to ball valves, except instead of a spherical element, a cylindrical element is used as the internal restriction. Plug valves are typically more expensive than ball valves, but they are inherently more rugged as well. The plug is guided by a sleeve that acts as the sealing member.

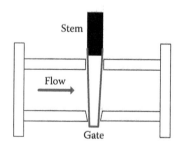

FIGURE 10.29
A remotely operated gate valve. These are typically used for very high pressure drop or temperature applications.

Plug valves require more torque to operate than ball valves but are easily automated with quarter-turn actuators. Plug valves are also available in three-way and 180° configurations.

10.10 Pressure Relief Valves

Pressure relief valves are one of the primary safety system devices in the process plant. Pressure relief valves must be installed on all vessels plus all piping downstream of a motive force device (pump, compressor, etc.) that can be remotely isolated from the nearest relief valve, and blocked in sections of liquid lines that may be exposed to heat. Lines containing liquids that vaporize at room temperatures or lower (such as liquid nitrogen) must also be protected. There are two primary types: the spring-loaded angle valve and the rupture disk.

The most reliable valve is the spring-loaded relief valve (Figure 10.30). The vapor space of the process vessel or piping is connected to a vertical inlet that is blocked with a disk. It is held in place by a spring. The amount of pressure required to hold the spring in place is adjustable within certain limits, which allows the relieving pressure to be calibrated and tested. When the pressure

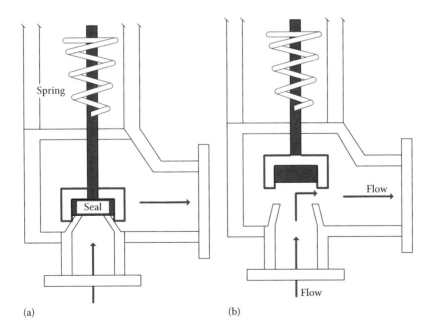

FIGURE 10.30
A spring-loaded pressure relief valve in both (a) normal (closed) and (b) event (open) positions.

holding the disk is exceeded, the disk is lifted out of its seat and fluid from the pressure vessel or piping is discharged around a bend and horizontally out of the valve into the relief system piping. In some versions, a bellows is used to hold the disk instead of a spring. These types are most common for large diameter valves.

The relief system piping must be carefully designed as the fluid will leave the valve at the speed of sound, generating substantial momentum. When the pressure in the protected vessel/piping drops below the relieving pressure, the disk will reseat. Most valves have a slight offset between lifting and seating pressure to minimize valve chattering (where the valve cycles open, close, open, close). In addition, back pressure from the relief system piping can add to the pressure required to open the valve compared to the shop-tested pressure. Therefore, the discharge piping must be designed to minimize pressure drop, even at very high flow rates and estimated back pressure should be calculated and taken into account when the relief valve's relieving pressure setting is specified. A special type of bellows valve, the balanced bellows valve, is not affected by back pressure. The spring metal in the bellows is also protected from the process fluid, reducing potential sticking problems due to corrosion.

A duplicate valve is usually installed as a spare so that the relief valve can be taken out of service on a routine schedule (most commonly once every 6 months or so) and shop tested. If there is a change in phase from liquid to vapor, two operating relief valves are required with an additional valve installed as a backup.

A lower cost protection device is the rupture disk (Figure 10.31). The disk consists of a thin sheet of metal that is scored so that when it fails, it will fail along a known path and the metal will be moved out of the flow path. The disk is typically installed between a pair of flanges. The thickness of the metal and the depth of the scores determine the bursting pressure of the disk. Once the disk fails, it cannot be closed or reused. Therefore, the fluid in the vessel or piping protected will depressurize until it reaches the back pressure of the relief system piping. Further, the disk must then be replaced before

FIGURE 10.31
A rupture disk. These should only be used for noncritical service since they cannot be tested prior to installation to insure proper use.

Instrumentation (Types and Capabilities)   315

the protected operation is placed back in service. Another disadvantage of rupture disks is that they cannot be directly tested. Instead, an identical valve must be tested in its place. Due to these limitations, rupture disks should only be used for noncritical service, such as a vessel filled with a nontoxic, nonflammable, low-value fluid.

Problems

10.1 Which of the following measurement devices would be considered responsive and which would be considered to be nonresponsive from a controls perspective (note: you are not determining the responsiveness of the process being measured, just the instrument doing the measuring)?

    10.1.1 A strain-gauge-based pressure sensor
    10.1.2 A venturi-type flow sensor
    10.1.3 A pitot-tube-type flow sensor
    10.1.4 A thermocouple-type temperature sensor
    10.1.5 A Bourdon tube pressure sensor
    10.1.6 A float-type level sensor
    10.1.7 A Coriolis flow sensor
    10.1.8 A capacitance level sensor
    10.1.9 A direct measurement composition sensor
    10.1.10 An orifice flow sensor
    10.1.11 A radar-type level sensor
    10.1.12 An indirect measurement composition sensor

10.2 What is the primary use for a vibration sensor in process control?

10.3 Explain how a Bourdon-tube-type pressure measurement element works.

10.4 What are the most common reference sources for pressure measurements?

10.5 What is one valid reason why you might specify a venturi-type flowmeter instead of an orifice-type flowmeter?

10.6 Explain how a Coriolis-type flowmeter works.

10.7 Explain how a thermocouple works.

10.8 Explain how a bubble-type level measurement device works.

10.9 What are the two basic classes of online composition measurement devices?

11
Automation and Control System Projects

This chapter provides an overview of key concepts for formulating and executing automation and control system projects. The first section focuses on various concepts that should be considered during specification and design phases of the project, while the second section provides guidelines for project execution.

11.1 Specification and Design Concepts

In order to maximize the productivity and efficiency of the process enterprise* (the summation of all activities associated with a process plant), an integrated design is recommended. Even if your project is just a partial retrofit, developing a comprehensive integrated design outline will help to align the project to the overall objectives of the process enterprise and provide justification for the addition of features whose purpose might not otherwise be so obvious.

To consider the plant automation system (PAS) from an integrated perspective, we need to recognize that all of the functionality of the PAS must contribute to achieving the overall objectives of the facility. These are based on the following:

- Maximizing profit
- Meeting acceptable safety and health criteria
- Minimizing environmental impact

From these general objectives, we can build more specific objectives for any real process enterprise (see Chapter 9 for an example of more specific objectives for the process portion of the process enterprise). Once the objectives for the enterprise have been identified, we use a top-down approach to designing the system: design from the top down but build from the bottom up. That is, each successive layer in the automation hierarchy is designed to support the higher layer. However, when it comes time to build the PAS, we must start at the most basic layer and work upward.

So, what are these layers in the PAS that we must design and build? The answer is more complex than it would first appear because the PAS is

---

* A glossary of terms is provided at the end of this chapter.

multidimensional and the layers vary depending upon which aspect of the PAS is under consideration. Let's start with a view of the PAS that is most readily aligned to the topic material we have covered in Chapters 2 through 10. For most process plants, the model depicted in Figure 11.1 can be used as a "functional" view of the PAS. These layers were defined and discussed in Chapter 9. While this model represents a functional view of the operational aspects of the PAS, it doesn't reflect how the computer software and hardware are related to each other to provide the functions at each layer of the PAS.

Figure 11.2 represents a "physical" view of the PAS that shows the various layers that support any specific function within the PAS. The top

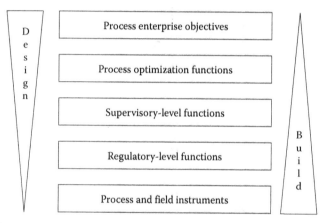

FIGURE 11.1
A functional view of the operational section of the plant automation system; the section of the system that has been the focus of this textbook.

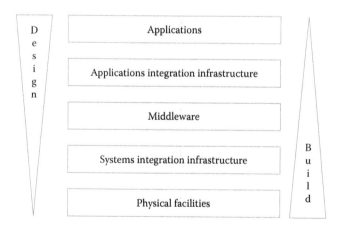

FIGURE 11.2
A physical view of the computer software and hardware necessary to provide and support the functions of the plant automation system.

layer consists of all the *applications* that provide the functions depicted in Figure 11.1. These include the techniques used to measure dependent variables such as flow, temperature, and so on, as described in Chapter 10. These also include the controller modules, the advanced control strategies, and the higher-level vendor-supplied applications such as a real-time optimizer (see Chapter 9). Further details on PAS applications are provided in Section 11.1.2.

The *applications integration infrastructure* is also known as the *application wrappers*. This layer includes all of the computer software that allows the individual applications to interact with everything else in the automation system. In the personal computing domain, programs such as Windows, Linux, and Google Chrome publish protocols that application designers can use so that their products integrate with computing systems using their operating systems. The suite of these protocols used by the application designer constitutes the application wrapper for that particular application's computer program. This same concept can be applied to the more complex and more diverse suite of applications in the plant automation system.

Middleware are computer programs that provide utility services that can be used by other computer programs within the PAS. Examples include graphics building tools, cut-and-paste functions, spell checkers, formatting functions, and so on.

The *systems integration infrastructure* represents the base-level computer software necessary to support the operation of the PAS. These include drivers for devices, communications protocols, and operating systems. The most fundamental of these are permanently burned onto computer chips in the form of firmware.

All of these layers reside within the *physical facilities* of the PAS. These include processors, monitors, printers, storage devices, servers, interconnecting wires, and even the building that houses the PAS equipment. These are discussed in more detail in Section 11.1.3.

In addition to these five physical layers, there are two other important layers that bound all five of these layers: the *human user interface layer* and the *security layer*. We will discuss these in Sections 11.1.3 and 11.1.4, respectively.

11.1.1 Applications

The applications layer represents the complete suite of process enterprise functions that must be accomplished to achieve the process enterprise objectives. These applications are typically divided into categories that correspond to the organizational chart of the personnel working in the facility. A typical "applications model" of the process enterprise is shown in Figure 11.3. In a comprehensive plant automation system, our vision must extend beyond the real-time control of the process (the operations portion) and incorporate all the supporting functions that allow the process enterprise to function.

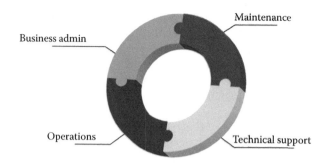

FIGURE 11.3
The applications and supporting layers in an integrated plant automation system can be classified according to their general functions, such as the overview classification shown here.

These other functions can be divided into business/administrative functions, maintenance functions, and technical support functions.

Business/administrative functions include manpower planning, accounting, purchasing, sales, and management. Automation of many of these functions may be provided by central corporate systems, such as payroll management or via an enterprise resource management system (see Chapter 9).

Maintenance is the subset of functions and resources required to maintain, repair, or physically modify the physical facilities of the process enterprise. Most major process plants utilize a maintenance management system to log, track, and facilitate the execution of day-to-day maintenance activities. These systems will maintain spare parts and operating equipment inventories as well as maintenance tools and equipment. Work orders and work permits can also be coordinated by these systems.

Technical support includes engineering, laboratory services, drawing management, training schedules, and so on. The plant automation support resources are often coordinated out of technical support, although they can also be managed out of the maintenance system.

It is also useful to understand the way information is passed between the various layers of the PAS applications and the timing of information transfer. This is defined using a "data/information" model such as the one shown in Figure 9.1 and described in the accompanying text. Note that this model can also be extended to the other application areas: business, maintenance, and technical support. The details of those areas are outside the scope of this book, which focuses solely on the process area. In addition, data/information transfer occurs between the application functional areas, usually at the same layer level in the data/information model.

11.1.1.1 Objectives-Based Controls: Changing the Process Control Paradigm

The traditional process control paradigm is *parameter-based* and can be summarized as a five-step process:

1. Measure-dependent process parameters such as pressure, temperature, flow rate, level, and composition.
2. Off-line, determine the desired parameter setpoint that will meet the plant's operating plan and operating targets; this is often done by operator feedback experience.
3. Manually specify the desired operating point (the setpoint) for each independent process parameter.
4. Manually link each dependent process parameter to an independent variable such as a control valve.
5. Manually tune the response of each independent process parameter to the linked measured value of the dependent process parameter.

We followed this paradigm in Chapters 2 through 7, although we went beyond these simple steps by using cascade loops and external setpoints. While all efficient operating plants attempt to meet the overall process enterprise objectives with plant operations, the parameter-based control paradigm is focused on controlling and assessing performance based on the ability of the operations function (the control system and operators) to operate to a suite of parameters. Therefore, this approach is limited in the ability to integrate the entire plant automation system so that the facility's ability to meet its objectives is optimized.

With the more sophisticated controls described in Chapters 2 through 7 and especially the techniques introduced in Chapter 9, we began to move beyond the parameter-based paradigm toward a better paradigm, one based on objectives. An *objectives-based* paradigm might be described as follows:

1. Determine the long- and short-term operating objectives of the plant.
2. Automatically plan and schedule plant-wide operating objectives and unit-level operating objectives.
3. Automatically specify and adjust independent process parameter targets to most efficiently meet the operating objectives.
4. Use measured dependent process parameters to assess how well the operating objectives are being achieved.

An objectives-based control system does not replace the parameter-based system. The traditional system is still required as the base level of control for most continuous processes. However, the objectives-based control system incorporates objective-based functionality that allows the PAS to automatically specify and adjust independent process parameter targets so that the plant will most efficiently meet the long- and short-term objectives of the process enterprise. Therefore, a project to install regulatory control is improved if the following factors are incorporated into the design methodology:

1. Regulatory controls are designed as an integrated part of an entire PAS.
2. The requirements for regulatory control functionality are analyzed and described functionally. This approach can minimize obsolescence.
3. Since regulatory control is the foundation for all operations functions of the PAS, we should overdesign this portion of the facility to avoid generating inadvertent information bottlenecks that limit access to the substantial profitability improvements available from the PAS functionality provided by higher-layer applications.

11.1.2 Physical Automation Systems

The physical automation system equipment and associated basic software and firmware are known as the systems integration infrastructure. These elements constitute the most fundamental level of the PAS and include the primary hardware and software environments on which the PAS applications and applications integration infrastructure reside and execute. Included are the following:

- Buildings, computer rooms, marshaling facilities, and utilities
- Data highways and other physical networks
- Network gateways, routers, and bridges
- Servers
- Peripherals
- Standards-based communication protocols
- Operating software

To deliver a well-designed PAS, the systems integration infrastructure should be the last portion of the PAS designed and the first portion of the PAS built (design from the top, build from the bottom). This can make implementation a challenge. For large projects this challenge can be mitigated by using a phased implementation plan.

In general, plan on overdesigning the systems integration infrastructure. The degree to which you should overdesign should be directly related to how labor intensive or intrusive it would be to expand capacity in the future. For example, data highways should include at least 100% extra capacity since it is both expensive and intrusive to route new data cables after the process units have been constructed. By contrast, servers should be supplied that just meet capacity since they will be replaced in just a few years with more powerful units.

In the long run, it is cost effective to invest in good ergonomic facilities for intensive PAS users, particularly control board operator workstations. This also improves plant safety.

11.1.3 Human User Interfaces

The effectiveness of the PAS is directly related to the quality of the human user interface (HUI) functionality provided. A well-designed, effective HUI layer is necessary to leverage the substantial investment made in applications. This is similar to when you first start using a new computer program on your PC. If the human interface is easy and intuitive, you are able to quickly utilize the full power and functionality of the program. However, if the HUI isn't easy and intuitive, it can take a long time just to master the program's basics.

From a functional viewpoint, the interaction that any given user will have with any given PAS function can be categorized as *intensive* or *casual*. An *intensive user* is proficient in the application executing the function of interest. Interaction with this function on a routine basis is part of the intensive user's job requirements. The intensive user's scope is limited to a small subset of the functions within the PAS, usually functions supplied by one or two applications. This user needs to have access to most or all of the capabilities that the application provides and has the incentive in terms of time/training to learn and understand the application's specific environment. For example, the maintenance planner uses the maintenance management system intensively but rarely needs to access any of the functionality provided by the process control system.

A *casual user* is not proficient in the application executing the function of interest. Interaction with this function happens rarely. The casual user normally interacts with this function as part of a larger, more general scope that may include a wide subset of the functions within the PAS. This user does not have the incentive to learn and understand the application's specific environment. Consider a maintenance planner who will occasionally access functions provided by the enterprise resource planning system (ERP). For example, the planner might need to know the labor rates that should be applied to work performed by maintenance personnel. The planner usually doesn't need to know specific individual user salaries, just an average value that covers the salaries, fringe benefits, and other overhead expenses of all of the workers within a given job category, say "welders." This information is generated in the ERP from information collected and managed by the overall enterprise administrative staff (accountants).

The HUI needs of these two classes of users are very different. The intensive user's needs are typically best supplied by dedicated human user interfaces (DHUI). Every commercial off-the-shelf (COTS) software package provides a

DHUI for the operation and maintenance of that application. These interfaces provide the power and accessibility required by the intensive user. There are two limitations to COTS DHUIs:

1. It is usually difficult to integrate information from external applications into the interface.
2. The user must operate within the paradigm specified by the developer of the interface. This includes terminology, ergonomics, and so on.

Casual users are not well served by DHUIs. Instead, access through a "universal" human user interface (UHUI) is preferred.

The UHUI provides a common interface that can be used to access information from the entire PAS. A well-designed UHUI will have the following features:

- Allow predefined information sets from multiple applications
- Allow custom information set and command requests in an intuitive, easy-to-understand language
- Provide meaningful information displays by including a screen/graphics builder capable of building text and graphic screens. Information must be integrated within a single PAS window.
- Provide the means for creating and managing role-focused environments discriminated by user, role, location, and workstation type.
- Coordinate with the security layer to guarantee PAS integrity against access by unauthorized and/or external systems, applications, and users.

11.1.4 Physical and Systems Security

When a well-designed PAS is implemented, new and revised work processes arise that quickly render the PAS critical to the operation of the plant. The PAS gives greater access to information and more power to efficiently conduct the business of the enterprise. As a result, it is imperative that the **security layer** be given the proper level of attention during the design, budgeting, and implementation of the PAS.

For PAS to be effective there must be a balance between protection from undesirable events (sabotage, fire, unauthorized access) and accessibility (easy access to authorized PAS functionality). It has been shown that oversecure systems translate into undersecure systems. That is, if the security system employed is too cumbersome, personnel will find ways to circumvent it even when protection is in their best interests.

A reasonable security layer can be provided with the PAS. The ideal PAS security system utilizes a primary security access system—a login ID and password—for normally sensitive security levels. This security concept is equivalent to the security system the banking industry uses to protect your money. If it's safe enough for an ATM, it's safe enough for a process plant! However, just as your banking records are not available to the general public, there are specific applications and information contained in the PAS that must also be controlled to additional levels of security.

Functionally, the security layer can be divided into two main classes: *physical security* and *access security*. Physical security deals with the control of the physical equipment of the PAS. Some good design concepts are as follows:

- Access to the rooms containing computer servers, peripherals, communications equipment, and data switches should be controlled. Only authorized employees should be allowed to unlock the doors to these rooms using their ID/password. Authorization should be location specific. The security system should generate, index, and archive a time-stamped log identifying when authorized personnel enter a secured location.
- Multiple clustered servers, located in geographically separate rooms reduce the risk of total system loss due to a fire or sabotage. Databases of sensitive information should be physically separate from the storage location for normal data.
- The security system should be housed in a separate server, enclosed in a dual-tier (one must go through two separate doors and use two separate passwords) physically secure room. The security server should be redundant and should not be included in the main server cluster. Interaction between the main PAS network and the security network should be by a secure network gateway.
- Control board operator workspaces should have a controlled access barrier. Authorization should be location specific, but in the event of a crisis, authorized management personnel (such as a shift superintendent or the unit supervisor) should be able to override access barrier control (typically by electronically unlocking the barrier/door).
- Workstations that can directly manipulate the process plant should require physical access—a card reader containing the user ID plus a password. Access should be location specific. A workstation may access data from all units of the process plant but only manipulate control valves, and so on, for a specific portion.
- Copies of all PAS software and archived backups of all PAS data stores should be maintained at a remote, offsite location or at a cloud-based data storage facility.
- Secure, callback modems should be used for remote system access. Alternately, a secure website can be utilized. Certain system functions should not be available for remote system access under any circumstances (such as payroll records, personnel information, process control, etc.).

Access security is provided so that access to the applications, information, data, and computer programs residing on the PAS computers is restricted. Some of the features of this security category are:

- A database should be maintained within the security system specifying the exact accessibility of each person who will interact with the PAS including those located at remote locations (such as corporate support groups).
- A standard library of accessibility should be developed and maintained for each employee category.
- Accessibility should include time limits for temporary assignments.
- Access to this database should be maintained at the highest level of security within the PAS (either dual- or tertiary-tier security). Accessibility to certain portions of the PAS environment should be limited to specific physical workstations. The entire PAS environment should be identified by security level. A typical process plant system would have the following levels:
  - *Normal security*: single-tier security; access to authorized PAS environment from a single login ID/single password combination.
  - *Confidential security*: dual-tier security; access to authorized PAS environment requires a single login ID but two separate passwords.
  - *Critical security*: dual-tier security; access to authorized PAS environment requires the physical insertion of a card containing the login ID plus a second login (either physical or via keyboard) and two separate passwords.

11.1.5 Automating Batch Processes

It is important to recognize some distinct differences when controlling batch processes compared to continuous processes. These are summarized in Table 11.1. In continuous processes, the primary regulatory control is the feedback control loop. In batch processes, the primary regulatory control is a sequential logic control loop (often known as PLC control).

Some important design considerations for batch processes are:

1. Batch process automation has a hierarchy similar to continuous processes.
2. Originally, batch process control was developed on a *time-based* paradigm. But like all processes, optimum performance is achieved when an *objectives-based* paradigm is adopted. Objective-based automation of batch processes can be more challenging than continuous processes because of the time-dependent, sequential nature of most

TABLE 11.1

Important Differences between Continuous and Batch Processes

Continuous	Batch
Independent variable setpoints are time independent and relatively constant over time.	Independent variable setpoints are time dependent and may change frequently or dramatically with time.
Steady state is the ideal operating state: time independent.	The operating state is dynamic and "ideal" may drastically change over time.
Operating states change smoothly and gradually.	Operation is sequential by nature. Operating states tend to change in step changes.
History can be utilized immediately to assess performance and optimize operation.	History cannot be used immediately. Cyclical processes can use history to assess performance and optimize operation. For noncyclical processes, higher-level analysis is required to use history.

batch processes. It is important to recognize batch unit operations that are incorporated into a continuous process. The automation of these unit operations needs to be designed using *batch processing* techniques rather than *continuous processing* techniques.

3. Batch processes can be optimized in real time when dependent parameters can be measured and analyzed with a time resolution at least one order of magnitude faster than the effective timescale of the process step being undertaken.

4. Supervisory control strategies can be effective in optimizing batch processes that are cyclical by nature. In these situations, history learned from earlier cycles can be used to improve the next cycle(s). See Chapter 9 for further details.

5. Online models and unit optimizers can be used effectively for many batch processes. Small parameter perturbations can be used to assess performance at fast time resolutions within the model and then used to extrapolate a new "model optimum" set of conditions.

6. If almost all unit operations within the process plant are batch processes, PLC-based regulatory control systems can be used in lieu of DCS-based regulatory controls.

7. Higher-level automation functions are not as well developed for batch processes compared to continuous processes. You may need custom software to achieve comparable results.

11.1.6 Safety Automation Systems

The topic of process safety deals with the techniques, facilities, and systems we use to minimize the risk of injury or death to people—both in the plant and in the community—resulting from our process plant.

11.1.6.1 Basic Principles

To make our process as safe as possible, we try to minimize three major areas of risk:

1. Fire
2. Detonations (explosions and implosions)
3. Hazardous/toxic material release

In addition to achieving these objectives, we want to design our safety systems so that they:

- Minimize adverse environmental impact
- Minimize loss of assets—materials and facilities
- Minimize costs

Fires require three things (Figure 11.4):

1. A fuel
2. An oxidizer
3. An ignition source

If we can eliminate ANY of these three, we eliminate the fire. We normally try to eliminate two of the three. Internal to our process we usually eliminate or minimize the ignition source and the oxidizer (although there are exceptions). Sometimes we WANT fire within the process (such as a fired heater). In these cases, we isolate the fire in a controlled manner so that we burn what we want to burn and nothing else. External to our process, we usually minimize the ignition source and the fuel (we can't eliminate the oxidizer—it's in the air that surrounds us).

For many hydrocarbons and other chemicals in the process there is a temperature at which no oxidizer is required for ignition. This temperature is known as the auto-ignition temperature or auto-ignition point. The most

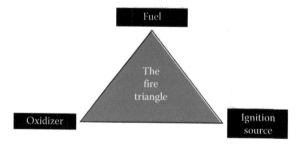

FIGURE 11.4
The fire triangle.

"famous" auto-igniting substance is phosphorus, which has an auto-ignition point below room temperature.

Detonation occurs when there is a sufficient buildup of both the fuel and the oxidizer before exposure to an ignition source such that when ignited, the combustion process goes out of control.

To aid evaluation of process risks from fire and detonation, the U.S. National Fire Protection Association (NFPA) has developed a classification system known as the NFPA area classification system. Similar classification systems have been developed for most countries, with an International Standards Organization (ISO) version as the most universally applied version. The area classification diagram is used to help insure that plant layouts conform to these requirements.

Under the NFPA specification, the first classification is based on the probability of having a flammable atmosphere:

- Division 1: event probability > 1 h/1.1 years
- Division 2: 1 h/1.1 years < event probability < 1 h/114 years
- Nonhazardous event probability < 1 h/114 years

By knowing your risk of having an explosive mixture present, you can take the correct, "best" approach—where best is a combination of safety, risk, mitigation of cost, and system complexity—to minimize the chance of fire or detonation.

In Division 1 areas, the strategy is to always prevent the formation of an explosive mixture by:

- 100% sealed equipment; so-called *explosion-proof* equipment
- Continuous dilution—forced ventilation to insure that the atmosphere is always below the LEL (lower explosive limit) of the fuel source
- Elimination of ignition sources

In Division 2 areas, the strategy is to minimize the probability of forming an explosive mixture and to minimize the probability of having an ignition source present where an explosive mixture can form. However since the probabilities are much lower than in division 1 areas, this strategy is executed with less expensive techniques than in division 1 areas:

- Increase distance from release source to any possible ignition source, thereby using dilution to reduce risk.
- Require all electrical connections be in sealed housings (eliminates ignition from arcing).
- Using *intrinsically safe* enclosures—these must passively avoid the formation of explosive mixtures by ventilation, sealing, or both.

Familiarity with the area classification concept is important in plant automation projects because it impacts the specifications needed for field instruments.

11.1.6.2 Hazards Analysis

Coupled with the area classification concept is the concept of *contingency*. Usually the process plant is designed for a single contingency event. That is, planning systems and facilities such that if one thing goes wrong no safety incidences will occur. But we must design such that any possible thing can be the one thing that goes wrong. Usually the most rigorous is an electrical failure.

Sometimes we design for double contingency. That is, any two things go wrong at a time without a safety incident. We use double contingency when the consequences of having any single incident are high, so we want to insure that the probability of having a significant consequence is very, very low. In very rare cases, a triple contingency design is required such as the radioactive side of nuclear power plant processes. That's why those plants are so expensive to build.

The area classification concept is all about estimating the probability of having a flammable or explosive environment and then minimizing the consequences if such an environment occurs. This concept can be extended to all aspects of the process. A comprehensive evaluation of the probability and consequences of unplanned (i.e., bad) events that is performed at the unit operations level is known as a *preliminary hazards analysis*. When a comparable evaluation is performed at the individual component level (i.e., every valve, fitting, pipe, instrument, piece of equipment, etc.), it is known as a *HAZOP analysis*. The most common outcome of these analyses is a set of recommendations leading to changes in the instrumentation, controls, and safety automation aspects of the plant automation system.

During these analyses, the probability and consequence of an event that impacts health, safety, or the environment is qualitatively evaluated. When the combination of probability and consequence reaches a certain level, steps must be taken to reduce the risk and/or impact. This is shown conceptually in Figure 11.5. The safety automation system is one of the most important aspects of the process plant to accomplish this objective.

11.1.6.3 Process Safety Layers

One concept that helps us visualize how to make our plant safe is to think about process safety in layers:

1. Innermost layer: the best way to maximize process safety is to make it inherently safe wherever possible. Sound, conservative design criteria focused on minimizing the possibility of experiencing an unsafe

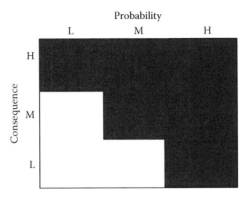

FIGURE 11.5
Qualitative classification of event probability and consequence during hazards analyses. If an event lands in the gray shaded area, steps must be taken to minimize the probability and/or consequences associated with the event.

condition should be employed. Both the process and the technology employed should be studied sufficiently and understood to recognize potential hazards. The process design should then minimize these hazards.

2. If we cannot make a process step inherently safe, the next best strategy is to use dedicated instrumentation and controls for protection. This suite of instrumentation and controls is known as the safety automation system. We have introduced many of these strategies throughout the previous chapters, such as including high-high level systems.
3. The next best strategy is to use process instrumentation and controls for protection. It is important to recognize that the primary purpose of the process control system is to adjust independent variables to optimize the operation of the plant. Its role as a safety layer is secondary to this purpose. Therefore, using this layer to replace the secondary safety automation system layer is discouraged.
4. Finally, the least attractive way to make our plant safe is to use operating procedures to insure safe operation. But, since people make mistakes, we never want to base our process design on this layer. These procedures should only be used for secondary safety protection or as a temporary measure until a more permanent solution can be implemented.

11.1.6.4 Emergency Automation Systems

The most damaging safety incidences involve fires, explosions, and inadvertent material releases. In many process plants, the units are tightly integrated with minimum surge and inventory storage capacity separating the process functions. As a result, an incident in one operating unit can propagate into upset conditions throughout the process. Suddenly, an isolated upset becomes a plant-wide problem. To minimize this problem, we want to design our plants in a manner that allows isolation of release points.

There are three different and distinct types of safety systems. All are commonly referred to as emergency shutdown systems:

1. Emergency isolation system (EIS)
2. Emergency shutdown system (ESD)
3. Equipment protection system (EPS)

The *emergency isolation system* (EIS) is the main technique used in the safety systems of process plants. An emergency isolation valve(s) is closed to isolate a piece of equipment or unit operation from the remainder of the process. EIS is usually invoked in response to measured process parameters such as level, pressure, flow, or temperature. We have already learned some of the ways how this technique is applied in Chapters 3 through 7. For example, in Figure 2.25 when the level in a knockout drum reaches the low-low level alarm point, a signal is sent to an independent controller, which in turn sends a signal to a remotely operated block valve to block the flow of liquid leaving the drum. This safety control loop is part of the EIS.

The EIS includes more comprehensive actions beyond simple single pipe isolation. A properly designed EIS usually involves a degree of sequential logic to avoid propagating the upset to other units. The EIS may redirect the flow of fluids away from the area of upset/fire to an intermediate storage area or may put entire groups of unit operations into a continuous recycle loop. Another feature of EIS is speed of response. Sequential logic processors utilized in EIS should have the highest scan and processing times available (currently on the order of 50 ms or less). The dynamics of isolation valves must also be considered.

The *emergency shutdown system* is a more complicated function than EIS. In ESD, an entire process system (collection of unit operations) is taken out of service. By its nature, this action usually impacts other process units in the facility significantly. Complex sequential logic must be executed rapidly to correctly and safely shutdown the process without damage to the equipment or undue impact to the environment. ESD should only be employed in process units where EIS is not feasible. In some processes, ESD is a second stage function that is invoked if EIS is ineffective in containing the event. Each process unit should function independently and shutdown safely in the event of a unit's ESD being activated. Keep the ESD systems decoupled.

Equipment protection systems are used to protect specific high-value equipment from damage. The most common application is the protection of important rotating equipment such as compressors, turbines, and high-speed pumps.

The goal is to allow the equipment to operate in the safe, stable region while still optimizing performance criteria (energy consumption, maintenance costs, etc.). Unfortunately, most equipment vendors label their equipment protection systems as ESDs. However, these systems are not typically designed to directly minimize health, safety, and environmental impacts. Their only safety protection is through the indirect benefits that may accrue if the integrity of the equipment is protected.

Some techniques useful in the design of these emergency safety systems are:

- Provide emergency isolation valves for each discrete unit operation.
- Do not use removal of motive power as an isolation technique. Instead, use physical isolation with valves to block flow in pipes, and then interlock the pump motor control to the EIS valve.
- Incorporate intermediate recycles into the process system whenever feasible. The safest way to manage an upstream or downstream unit operation when a significant upset occurs is to isolate the unit and maintain the basic operating parameters using recycle.
- When recycle is not feasible, route intermediates to predesignated storage vessels. This may generate contaminated streams, but may avoid loss of valuable material due to flaring or release.

11.1.6.5 Redundancy

Redundancy involves the duplication of a component in the safety system for one of two reasons:

1. *Availability*: improve the availability of the functions provided by the component
2. *Assurance*: insure the correct operation of the component (typically the accuracy of the component) in performing its function

These two functions are very different, and it is important to know which function you are trying to achieve when you design for redundancy.

The relationship between redundancy and availability is as follows:

1. No redundancy: there is only a single component available. If this component fails, the functionality provided by the component is not available until the component is restored. This level is also referred to as nonredundant.
2. Dual redundancy: there are two identical components available, one in active operation and one in standby or tracking mode. If one component fails, the second unit provides the functionality. This level is sometimes referred to as redundant or dual redundant.

3. Triple redundancy: there are three identical components available—one in active operation and two in standby or tracking mode. This is often termed triplicated redundancy, tri-redundant, or triplexed redundancy.

The relationship between redundancy and assurance is as follows:

1. No redundancy: improper operation must be inferred by comparing current operation to history or through the use of data reconciliation techniques.
2. Dual redundancy: there are two identical components available. Depending on the function, one or both may be active. For process instruments, a comparison can be made between the two readings. If both readings are the same within expected tolerances, the measurement is believed. If the readings are different, the measurement is rejected. Unfortunately, there is no way to tell which reading is correct. Thus, if the measurement is used for a safety device, the safety system must assume that the worst-case reading is correct. As a result, processes that use dual redundant safety measurement assurance criteria have a lower availability factor than those with no redundancy.
3. Triple redundancy: there are three identical components available. For process instruments, a two-out-of-three voting logic is usually used for assurance. If two of the three readings are the same within expected tolerances, the measurement from these readings is believed and the third measurement is rejected. Triple redundant design provides a higher level of assurance and availability than either of the other two levels. As a result, well-designed process safety systems now utilize triple redundant design criteria for most safety system components.

When designing a safety system's redundancy, remember that the overall level of assurance and availability of a suite of components will be same as the lowest level of any of the individual components that comprise this suite.

11.2 Guidelines for Automation Projects

PAS projects are different from traditional process plant projects because they involve a combination of hardware and software. Further, the technologies used in these systems continue to improve and become more powerful on a routine basis. As such, plant automation system (PAS) projects should incorporate techniques to minimize technology obsolescence.

11.2.1 Project Life Cycles

A typical life cycle for a plant automation project involves seven steps:

1. Scoping: describes what "finish" is. In this phase, we define a preliminary set of functional requirements. The deliverable is a functional specification document (FSD).
2. Planning: identifies how to get to "finish." In this phase, we define the schedule, resources, and project execution strategies required. The deliverable is the project execution plan.
3. Analysis: completely describes, in functional terms, what will be provided. In this phase, we determine the full functional requirements of the PAS. The deliverables are a detailed functional specification document and a preliminary test plan.
4. Design: physically describes what will be provided. In this phase, the design of the physical system is completed and procurement specifications are developed. The deliverables are the solution design document, purchase orders, a detailed test plan, a quality assurance plan, and a training plan.
5. Test and build: is a prototype of what will be provided. A small-scale model of the PAS is built offsite so that we can procure and test software and perform offsite system configuration activities. The deliverable is certified PAS software, partially configured for implementation.
6. Implementation: provides the functionality on-site. We install, finish configuration, and test the PAS components. Deliverables include certified, installed, and configured PAS components, fully trained intensive users, plus PAS support personnel (operators, systems support personnel, maintenance personnel), and complete documentation.
7. Commission: places the PAS functionality in service. We begin using the PAS in the process enterprise and correct problems uncovered during operation.

11.2.2 How to Proceed with Automation Design for Process Plants

Step 1: Identify the business objectives of the enterprise. These are specific objectives based on the three general objectives listed in Section 11.1, above.

Step 2: Identify the plant-wide process objectives required to support the business objectives. These are usually product- and product-specification-related objectives as well as efficiency-related objectives such as energy consumption, and so on. Examples are:

- Maximize the production of three main products based on the most economical allocation given the current contract commitments, net profit for each product, and availability of raw materials.

- Minimize the consumption of raw materials required to meet a specific contract product supply commitment.
- Minimize the consumption of utilities.
- Minimize the generation of waste streams.

Step 3: Identify unit operation objectives to support the plant-wide process objectives. These should be used to generate the objective function for the unit optimizer.

Step 4: Specify the unit operations applications required to support the unit objectives.
- Identify unit data reconciliation requirements.
- Identify online models.
- Identify multivariable controller (MVC) applications and specify the independent and dependent variables associated with each MVC.
- Specify the functionality of other advanced control applications. Simple applications are documented on the P&IDs. More sophisticated applications are documented on supplemental drawings.
    - Identify all of your process objectives.
    - Determine the dependent variables that most closely measure your objectives.
    - Select an independent variable for each dependent variable. They should be related.
    - If you have additional important dependent variables, you must construct dependencies for them, usually by using cascade strategies.

Step 5: Identify the regulatory controls required to support all of the above layers in the PAS operational application model. These are documented on the P&IDs.

Step 6: Identify the field instruments necessary for the regulatory controls and data reconciliation applications. These are documented on the P&IDs.

Glossary

Administration: the subset of functions and resources required to run the process enterprise. Typical functions include plant management, personnel, timekeeping, accounting, reporting, and training.

Application Wrapper: a software program that allows a COTS application to interact and integrate into a PAS.

Application: a computer program that automates a specific suite of process enterprise functions.

Applications Systems Infrastructure: the set of hardware and software resources required to support the operation of applications within a PAS.

Automation: taking a sequence of operations that were performed manually and making the performance automatic.

COTS: "commercial off the shelf" refers to a general computer program marketed by a third party vendor to multiple users. The opposite of COTS is custom software.

Data: knowledge in the form that it was generated.

Discrete: data or information that is not continuous. All digitally based process control systems manipulate discrete data/information.

Function: defines the relationship between one or more independent variables and one or more dependent variables. A process enterprise is a suite of functions designed to achieve a specific set of objectives.

Information: knowledge in a form that is useable for making decisions.

Integration: coupling two separate and distinct operators (applications, programs, communications systems, databases, etc.) in some manner. In a PAS, integration occurs on many different levels.

Maintenance: the subset of functions and resources required to maintain, repair, or physically modify the equipment of the process plant. Often, only a portion of the maintenance functional area is company owned/employed. Contractor, supplier, and vendor resources may be used to achieve many of the objectives of the maintenance functional area.

Middleware: one or more computer programs that perform utility services in an environment that integrates PAS. Windows is an example of a personal computer middleware product.

Obsolescence: the loss of usefulness or relevancy of a resource.

Open Systems: a computer program that is structured to allow the functionality of the program to interact easily with unrelated computer programs.

Operations: the subset of functions and resources required to execute the normal operation of the process plant. The exact composition of operations varies somewhat from plant to plant. Operations is a subset of the process plant enterprise.

Physical Facilities: the noncomputing technology resources required to implement a PAS. These resources include buildings, HVAC, electric utilities, and cabling.

Plug and play: an integration design concept in which the goal is to allow any specific computer program to be replaced with a different program of similar functionality without impacting the surrounding integrated computing system.

Process Plant Enterprise or Process Enterprise or Process Facility: the complete set of functions and resources (people, equipment, materials, etc.) required to execute the objectives of the process plant.

Process Plant: an integrated series of unit operations designed to convert a suite of raw materials into a suite of products.

Real Time: data generated by a continuous measurement or information provided to an independent variable operating continuously on a process condition. Note that processes may contain continuous process conditions but the data and information of a digitally based process control system are always discrete, although they can be "virtually" real time.

Systems Integration Infrastructure: the set of hardware, software, and physical facility resources required to support the operation of a PAS. This is a lower level of infrastructure than the applications systems infrastructure.

Task: the specific implementation of a function. In a process enterprise, tasks are used to implement the functions designed into the enterprise.

Technical Support: a subset of functions and resources that support the operations and maintenance areas of the process plant. Typical functions include troubleshooting problems, determining process improvements, monitoring operator performance, and laboratory services. The source of technical support varies widely depending on the process plant and Company policy. Small, simple plants may have no local support. A large integrated petrochemical plant may have a substantial in-plant support staff.

Transactions: the subset of functions and resources required to transfer resources to/from the process enterprise. Typical functions include product marketing, raw material supply coordination, contracting, and procurement.

Problems

11.1 For each of the following components of PAS, identify where it resides (which layer) of the PAS data/informational model shown in Figure 9.1:

   11.1.1 The laboratory management system
   11.1.2 A flowmeter
   11.1.3 A temperature controller
   11.1.4 A work permit routing procedure
   11.1.5 The plant manager's force deployment report

Automation and Control System Projects

11.1.6 A routine to optimize the blending of products at the tank farm

11.1.7 A cascade control scheme to smooth out the flow out of a tank using level control cascade to flow control

11.2 For each of the following components of PAS, identify where it resides (which layer) of the PAS physical model shown in Figure 11.2.

11.2.1 The laboratory management system

11.2.2 The software that allows the lab management system to communicate with the other software in the PAS

11.2.3 A temperature controller

11.2.4 Windows operating software

11.2.5 The graphics building program used to build operator displays

11.2.6 The communications routine that determines the priority and routing of data throughout the PAS data highway

11.2.7 The central control room

11.3 For each of the following applications, identify where it resides in the Applications Layer Model shown in Figure 11.3:

11.3.1 The laboratory management system

11.3.2 The process control system

11.3.3 A materials inventory system

11.3.4 The payroll system

11.4 From the Internet find the overall description of a process plant. Develop specific facility objectives for that process plant based on the three global objectives: maximize profit, minimize environmental impact, and meet required safety criteria.

11.5 Describe the two types of human user interfaces.

11.6 Describe the two classes of security associated with the plant automation system.

11.7 What is a detonation?

11.8 What is the relationship between the probability of a safety-related event and the consequences of a safety-related event?

11.9 What is an event contingency? What is the most common contingency level used in the design of process facilities?

11.10 Where does the safety automation system fall in terms of the concept of process safety layers? Where does the process control system fall?

11.11 When should operating procedures be used instead of an automated safety function for facility protection?

11.12 What are the three types of safety systems?

11.13 Redundancy is used in process plant equipment and instrumentation for two distinct purposes: availability and assurance. Redundancy for availability decreases the probability that a given piece of equipment or instrument is out of service. Redundancy for assurance decreases the probability that the measurement or action is incorrect. For each of the following scenarios, identify whether redundancy is added for availability, assurance, or both. Also, identify the contingency level and redundancy level of the design.

  11.13.1 Normally, 100% backup units are supplied for all shell and tube heat exchangers.

  11.13.2 On ASME Code 8 vessels (those classified as boilers—a pressurized vessel in which water is converted into steam), there are three 100% capacity relief valves: two in operation and one car-sealed and blocked closed.

  11.13.3 On a critical product transfer tank, two identical level indicators measure the liquid level. The signals from these indicators are sent to a redundant distributed control system where the operator can see both readings.

  11.13.4 In order to avoid a runaway reaction in a reactor vessel, the temperature in the overhead gas-phase product line is measured with three separate thermocouples (TCs). The signals from these three TCs are routed independently to three separate input cards on a triplex programmable logic controller (PLC) used for the unit's safety automation system. If the temperature is too high as determined using 2/3 voting logic, the PLC will initiate a process unit shutdown.

  11.13.5 The signals from the TCs in #4 are also routed to a redundant distributed control system where the operator can see all three readings.

11.14 Please refer to Figure 11.6, showing the relief valve installation for a certain pressure vessel.

  11.14.1 What is the protection contingency level?

  11.14.2 What is redundancy level provided for availability of operation?

11.15 What does it mean to be inherently safe?

11.16 When should an operating procedure be used as part of the plant's overall safety strategy?

11.17 What does it mean if a process area has a safety area classification (as defined by the NFPA area classification diagram) of Division 2?

11.18 What does it mean if a process area has a safety area classification (as defined by the NFPA area classification diagram) of Division 1?

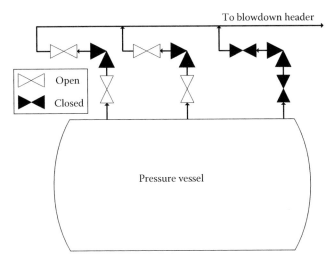

FIGURE 11.6
Relief valve protection of a pressure vessel.

11.19 Redundancy is used in process plant equipment and instrumentation for two distinct purposes: availability and assurance. Redundancy for availability decreases the probability that a given piece of equipment or instrument is out of service. Redundancy for assurance decreases the probability that the measurement or action is incorrect. For each of the following scenarios, identify whether redundancy is added for availability, assurance, or both. Also, identify the contingency level and redundancy level of the design.

11.19.1 Four firewater pumps are installed: two in one location and two in another. At each location, one pump has a motor driver and one pump has a diesel engine driver.

11.19.2 On process vessels operating below atmospheric pressure, there are two 100% capacity vacuum breaker valves: one in operation and one car-sealed and blocked closed.

11.19.3 On key process vessels, a separate temperature indicator loop is often installed in parallel to the measurement element used in a temperature control loop.

12

Process Dynamic Analysis

Chapters 2 through 8 focused on the design of basic, stable control strategies for the regulatory level of the plant automation system, while Chapter 9 introduced more complicated automation systems and Chapter 10 provided an overview of field instruments. In Chapter 11, concepts and methods for control projects were introduced. After a process plant and its associated plant automation system is designed, built, and commissioned, problems may arise during operation. Some of these problems may manifest themselves as unstable operation or in the form of unstable control of the process. When these types of problems occur, analyzing the dynamic behavior of the process can be useful in identifying the cause and solutions to the instability. This chapter provides an introduction to these types of analyses.

Our window into how well a process is performing is through the measurement of process parameters (e.g., temperature, flow, pressure). The overall behavior that can be measured (and therefore controlled) in a real-time process is a combination of the behavior of the process itself and the behavior of the instrumentation and controls applied to that process. These two sets of behavior are not independent, as the process responds to changes from the control system and the control system responds to changes in the process. After all, this coupled behavior forms the basis of process control.

At the most complicated level, a multivariable controller (MVC) attempts to model an entire interrelated portion of a process plus the associated regulatory controls. When these are present, the MVC can be used in an off-line mode to analyze the process. However, it is more common to conduct process analysis by assembling models of those unit operations in the process that are most likely to be contributing to the instability, along with models of each measurement device, controller, and control variable device (e.g., control valves). These models can be mass/energy balance-based models similar to those used in process simulators or heuristically developed mathematical descriptors of the dynamic behavior of those processes.

The advantage of the heuristically developed model is that the interactions between model elements usually do an excellent job of replicating the interactions of the process and its control system. Special training may be required to employ a software package that develops real-time heuristic models. Once the software is understood, the models can efficiently simulate the responses of the process to various types of upsets, which can then be used to formulate solutions to reduce process instability. The biggest disadvantage of these models is that the process and its controls are essentially a "black box." Thus, it is much harder to define the root cause of a process instability or to investigate substantial process changes that could, in the end, debottleneck the process or provide other similar benefits as well as mitigate the process instability.

More insight into the process and the control system can be obtained using models based on mass and energy balances. These models may be more difficult to develop, especially if the control system dynamics and the interactions need to be accurately modeled and they tend to only be used for relatively simple systems. Describing how to develop such models is often the largest component of process control textbooks because the steps involved in developing these models can, of themselves, help the process engineer to gain a better understanding for the operation of real-time processes. Too often, however, this extended discussion of process modeling comes at the expense of information related to the actual control of process unit operations. We have chosen to address this topic later and to focus on model development rather than on analytical methods and solutions to these models. Further details are available in a wide variety of textbooks focused on process dynamics and control.

The remainder of this chapter will be devoted to examples of how these types of models can be developed.

12.1 Dynamic Models for Common Unit Operations

In this section, we will use a few simple, commonly used unit operations to illustrate how dynamics models are developed. These models will then be combined with models from Section 12.2 in Section 12.3 to illustrate how these types of models can be used to evaluate the overall dynamics of both the process and its regulatory controls.

12.1.1 Modeling a Liquid Knockout Drum: A Dynamic Model for Changes in Flow and Level

Material and energy balances are commonly used to define unsteady-state models for unit operations. Consider the knockout drum we examined in detail in Section 2.3 to explain how to specify stable regulatory feedback control loops. We started that examination by defining the overall material balance of the unit operation:

$$M_I = M_{O1} + M_{O2} + M_A \qquad (2.1)$$

where
  M_I is the mass of material entering through the feed nozzle
  M_{O1} is the mass of material leaving through the top outlet nozzle
  M_{O2} is the mass of material leaving through the bottom outlet nozzle
  M_A is the accumulation (or reduction) in material in the vessel

Rearranging this equation yields:

$$M_A + M_{O1} + M_{O2} - M_I = 0 \qquad (12.1)$$

Equation 12.1 is valid for an instantaneous point in time, since every one of the terms is time dependent. If we now add that time dependence and generalize the equation, it can be written as:

$$\frac{dM(t)}{dt} + \sum_j M_j(t) - \sum_i M_i(t) = 0 \qquad (12.2)$$

where
  M is the mass in the vessel
  i represents all the inputs to the vessel
  j represents all the outputs from the vessel

Equation 12.2 is a linear ordinary differential equation (ODE) where all the variables are time dependent. The equation is completed by inserting the appropriate relationships between the inputs and outputs with time.

Unless these relationships are highly nonlinear, this class of ODE can be solved analytically using a variable transform technique. The most common transform utilized is the Laplace transform. These transforms and their application to dynamic process equations are discussed in more detail in Appendix A.

It should be noted that except for very simple models, an analytical solution is not feasible. For complex models, mathematical programs that provide numerical solutions to systems of simultaneous partial differential equations must be employed. Fortunately, these programs are readily available, so our focus is on how to set up systems of equations that can be solved to provide a model of dynamic behavior rather than on their solution.

Equation 12.2 is one of the simplest model representations we can develop. It assumes that all the material within the unit operation is homogeneous with respect to any physical property that might affect the balance. For example, when applied to the knockout drum example of Section 2.3, the equation assumes constant temperature and that the fraction of liquid in the inlet gas stream is constant. Such a model is often referred to as a *lumped parameter* model. This model can be useful for evaluating changes to the inlet flow rate or to changes in the valve opening on control valves located on either of the outlet streams.

Let's focus just on the level control loop shown in Figure 2.4. First, let's convert Equation 12.2 into a form more useful for evaluating level. The total mass in the knockout drum at any time, t, is the sum of the mass of the gas phase material and the liquid phase material. Ignoring the hemispherical ends to the knockout drum, the volume within the drum can be defined using the equation for the volume of a cylinder:

$$V_T = \frac{\pi}{4}D^2 L_T \qquad (12.3)$$

where
V_T is the total volume of the knockout drum
D is the diameter
L_T is the total length of the cylinder

In the liquid phase, the volume of liquid is simply the mass of liquid divided by its density. For the gas phase, an equation of state can be used to relate the volume of gas to the mass of the gas in the drum. For example, using a compressibility factor equation of state form yields:

$$V_G = \left(\frac{m_G}{MW}\right)\left(\frac{zRT}{P}\right) \qquad (12.4)$$

where
V_G is the volume
m_G is the mass of gas in the knockout drum
MW is the molecular weight of the gas
z is the compressibility factor for the gas
R is the universal gas constant
T is the gas temperature
P is the pressure of the gas in the knockout drum

Let $L(t)$ represent the liquid level at any time, t, in the knockout drum. Rearranging Equation 12.3 and substituting L for L_T, we calculate the volume of liquid in the vessel:

$$V_L(t) = \frac{m_L(t)}{\rho_L} = \frac{\pi}{4}D^2 L(t) \qquad (12.5)$$

where
V_L and m_L are the volume and mass of liquid in the knockout drum, respectively
ρ is the density of the liquid
D is the diameter of the drum

Rearranging Equation 12.5 yields:

$$L(t) = \left(\frac{4 m_L(t)}{\pi \rho_L D^2} \right) \quad (12.6)$$

If the temperature of the liquid is relatively constant, then we can assume the density is also constant and all of the invariant terms can be combined into a single constant, K_L, to yield:

$$m_L(t) = k_L L(t) \quad (12.7)$$

To use Equation 12.7 with Equation 12.2, we must relate the liquid mass to the total mass in the system. Within the knockout drum, the total mass is the sum of the liquid mass and the gas phase mass:

$$m_T(t) = m_L(t) + m_G(t) \quad (12.8)$$

The gas phase mass is related to the volume of the gas in the drum and the volume of gas is the difference between the total volume and the liquid volume:

$$V_G(t) = V_T - V_L(t) = \frac{\pi}{4} D^2 (L_T - L(t)) \quad (12.9)$$

Combining Equations 12.4 and 12.9 yields:

$$\frac{\pi}{4} D^2 (L_T - L(t)) = \left(\frac{m_G(t)}{MW} \right) \left(\frac{zRT}{P} \right) \quad (12.10)$$

Assuming the gas composition, gas temperature, and gas pressure are constant, all of the invariant terms can be combined into a single constant, k_G, yielding:

$$m_G(t) = k_G (L_T - L(t)) \quad (12.11)$$

Now combining Equations 12.7, 12.8, and 12.11 together yields:

$$m_T(t) = k_L L(t) + k_G (L_T - L(t)) \quad (12.12)$$

which can be rearranged to:

$$m_T(t) = L(t)(k_L - k_G) + k_G L_T \quad (12.13)$$

Now we can insert Equation 12.13 into Equation 12.2:

$$\frac{d}{dt}\left(L(t)(k_L - k_G) + k_G L_T\right) + \sum_j M_j(t) - \sum_i M_i(t) = 0 \quad (12.14)$$

Since the derivative of a constant is equal to zero, the terms:

$$\frac{d}{dt}(k_L - k_G) = 0 \quad \text{and} \quad \frac{d}{dt}(k_G L_T) = 0$$

allowing us to reduce Equation 12.14 to:

$$k\frac{dL(t)}{dt} + \sum_j M_j(t) - \sum_i M_i(t) = 0 \quad (12.15)$$

where $k = k_L - k_G$.

Equation 12.15 provides a generalized lumped parameter model relating the change in the liquid level in a knockout drum to changes in the mass of the inlet stream and/or the mass flow rate in either or both outlet streams. One way to use this expression to evaluate the dynamics of the knockout drum is to hold the outputs constant and to plot the change in liquid level over time in response to a change in the inlet mass rate.

This is an extremely simplified example. We have assumed perfect homogeneity in the drum, constant temperature, constant pressure, and constant composition. To make the model more realistic, other material and/or energy balances can be developed that can be useful in evaluating how a process reacts to various changes. For example, we can construct a component mass or mole balance for one component in the system. We'll see an example of this in Section 12.1.3.

Another level of complexity is to account for changes in key parameters within the unit operation. For example, the knockout drum can be modeled as a single-stage flash separator. A vapor-liquid equilibrium relationship can be developed that accounts for the difference in the gas phase composition compared to the liquid phase composition within the drum. We could then vary the liquid fraction in the inlet stream and evaluate the liquid level response to that variable. An example of modeling a unit operation with changes to its internal conditions is described in Section 12.1.2.

12.1.2 Dynamic Modeling of a Single Pass Shell and Tube Heat Exchanger: A Dynamic Model for Energy Balances and Temperature

Consider the heat exchanger system shown in Figure 4.3. In this unit operation, steam is condensed on the shell side of the heat exchanger to heat a

Process Dynamic Analysis

liquid as it flows through the tubes of the heat exchanger. The simplest heat exchanger would have a single tube pass through the steam condensing zone. The objective of the heat exchanger is to add a specified quantity of energy, which we characterize as duty, to the process fluid to increase its internal energy. We measure internal energy as temperature. The energy required is supplied by condensing steam, which releases energy in the form of latent heat as the fluid changes from the gas phase to the liquid phase.

Our measurement variable in Figure 4.3 is the process fluid temperature, so let's develop a series of equations to evaluate the dynamics of this system as measured by changes in the outlet process temperature. We do this by constructing a time-dependent overall energy balance around the process fluid in the heat exchanger. For any fluid, the relationship between its temperature and heat duty can be defined by:

$$Q_{p,o} = \left(C_{p,o} M_{p,o} T_{p,o} - C_{p,i} M_{p,i} T_{p,i}\right) \tag{12.16}$$

where
  M is the mass flow rate
  C_p is the heat capacity of the fluid
  T is the fluid temperature
  p,o is process outlet
  p,i is process inlet

The change in the temperature of the process stream leaving the heat exchanger with time will be based on the change of the duty of the incoming fluid plus the net rate of heat transfer from the hot fluid in the heat exchanger:

$$C_{p,o} \frac{d\left(M_{p,o} T_{p,o}\right)}{dt} = C_{p,i} \frac{d\left(M_{p,i} T_{p,i}\right)}{dt} + Q_T \tag{12.17}$$

where Q_T is the heat duty transferred from the hot stream to the process stream.

Expanding the two differential terms yields:

$$C_{p,o} M_{p,o} \frac{dT_{p,o}}{dt} + C_{p,o} T_p \frac{dM_{p,o}}{dt} = C_{p,i} M_{p,i} \frac{dT_{p,i}}{dt} + C_{p,i} T_{p,i} \frac{dM_{p,i}}{dt} + Q_T \tag{12.18}$$

Equation 12.18 is a general equation and can be used to evaluate responses due to changes in the inlet or outlet mass flow rates, the feed temperature, the heat transfer across the exchanger, and even in the desired target outlet temperature of the process fluid from the heat exchanger. However, in this form it is too general for practical use. Usually, we like to evaluate one parameter at a time.

Consider the case where we want to evaluate a change in the process outlet temperature based on changes in the inlet mass flow rate. If we also assume that the process fluid is incompressible, then a change in the inlet mass flow rate will also result in an identical change in the outlet mass flow rate. If the inlet process temperature and Q_T are held constant, Equation 12.18 reduces to:

$$C_{p,o} M_p \frac{dT_{p,o}}{dt} + \bar{C}_p (T_{p,o} - T_{p,i}) \frac{dM_p}{dt} = Q_T \qquad (12.19)$$

We can use one of the "change" models described in Section 12.3 below to provide an independent description of dM_p/dt and then calculate $M_p(t)$ at each time increment. Combined with Equation 12.19, these relationships allow evaluation of changes in mass flow rate on the outlet temperature of the process (see Section 12.3).

Another common scenario would be to hold the mass flow rate through the heat exchanger constant and evaluate the impact of a change in the outlet process temperature due to a change in the inlet process temperature. In this case, Equation 12.18 reduces to:

$$C_{p,o} M_{p,o} \frac{dT_{p,o}}{dt} = C_{p,i} M_{p,i} \frac{dT_{p,i}}{dt} + Q_T \qquad (12.20)$$

As with Equation 12.19, an independent input model is used to define the term $dT_{p,i}/dt$.

Finally, the most common use of Equation 12.18 is to evaluate how a change in the inlet steam flow rate would impact the outlet process temperature; this provides the model needed to evaluate the feedback control loop shown in Figure 4.3. In this case, both the mass flow rate and the inlet temperature are held constant but the heat flux term is now time dependent and Equation 12.18 becomes:

$$C_{p,o} M_{p,o} \frac{dT_{p,o}}{dt} = \frac{dQ}{dt} \qquad (12.21)$$

Now the overall heat transfer from the steam to the process fluid is given by:

$$Q_T = M_s \lambda \qquad (12.22)$$

where λ is the latent heat of vaporization/condensation of the steam.

If the mass flow rate of steam varies over time, then:

$$\frac{dQ_T}{dt} = \lambda \frac{dM_s}{dt} \qquad (12.23)$$

which combined with Equation 12.21 yields:

$$C_{p,o} M_{p,o} \frac{dT_{p,o}}{dt} = \lambda \frac{dM_s}{dt} \qquad (12.24)$$

These equations can also be used to model how the temperature profile of the process fluid within the heat exchanger changes with changes in each of these variables. For example, let's consider how to do this using Equation 12.24. If we measure the distance along the tube from the process (cold side) inlet then we can build an energy balance of a cross section of the heat exchanger of width Δz that starts at point z as shown in Figure 12.1:

$$Q_{p,z+\Delta z} = Q_{p,z} + Q_T \qquad (12.25)$$

where
  Q is the heat duty (J/s, Btu/h)
  p is process fluid
  T is transfer from hot side to cold side

Likewise, the heat transfer from the hot side fluid to the process fluid can be defined by:

$$Q_T = A_{lm,\Delta z} U_{\Delta z} \Delta T_{lm} \qquad (12.26)$$

where
  $A_{lm,\Delta z}$ is the log mean surface area of the tube section of length Δz (since the tube has a finite thickness, the surface area of the outside of the tube is slightly larger than the surface area of the inside of the tube)
  $U_{\Delta z}$ is the overall heat transfer coefficient for the tube section of length Δz
  ΔT_{lm} is the log mean temperature difference across the tube section of length Δz

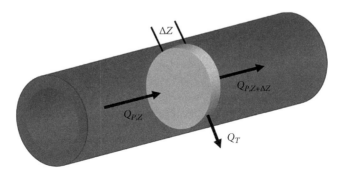

FIGURE 12.1
The energy balance around a segment of a tube in a heat exchanger.

For a sufficiently small tube length, Δz, the log mean temperature difference from the hot side to the cold side can be approximated by the temperature difference at either the inlet or outlet position of the tube section:

$$\Delta T_{lm} = T_{H,z} - T_{p,z} \tag{12.27}$$

where $T_{H,z}$ is the temperature of the hot side fluid at the point along the tube, z. For the system in Figure 4.3, this temperature is essentially the same along the entire tube unless the steam condensate is supercooled in the heat exchanger.

Combining Equations 12.15, 12.16, and 12.27 yields:

$$M_p \bar{C}_p \left(T_{p,z+\Delta z} - T_{p,z}\right) = A_{lm,\Delta z} U_{\Delta z} \left(T_{H,z} - T_{p,z}\right) \tag{12.28}$$

The surface area of the tube section can be related to the length of the tube section using the formula for the surface:

$$A_{lm} = \pi D_{lm} \Delta z \tag{12.29}$$

where D_{lm} is the log mean diameter of the tube.

Inserting Equation 12.29 into 12.28 and then dividing both sides by Δz yields:

$$\frac{M_p \bar{C}_p \left(T_{p,z+\Delta z} - T_{p,z}\right)}{\Delta z} = \pi D_m U_{\Delta z} \left(T_{H,z} - T_{p,z}\right) \tag{12.30}$$

Now if we take the limit when $\Delta z \to 0$,

$$M_p \bar{C}_p \frac{dT_p}{dz} - \pi D_m U_{\Delta z} \left(T_{H,z} - T_{p,z}\right) = 0 \tag{12.31}$$

Equation 12.31 is a linear ODE that provides us with a model of how the temperature within the process side of the heat exchanger will change with distance along the tubes. Notice that since this equation is time independent, by itself it does not provide a dynamic model of the heat exchanger.

If we use Equation 12.26 for Q_T, and now let $T_p = T_p(z,t)$, then we can combine Equations 12.24, 12.26, and 12.31 to construct a model of the dependence of the temperature of the process at any linear point within the heat exchanger tube by:

$$\frac{\partial T_p}{\partial t} + M_p \bar{C}_p \frac{\partial T_p}{\partial z} = \pi D_m U_T \left(T_H(t) - T_p\right) \tag{12.32}$$

12.1.3 Dynamic Modeling of a Fixed Bed Reactor With an External Cooling Jacket: Dynamic Modeling of Composition Change and Elemental Balance Variations

Consider a fixed bed reactor with an external cooling jacket, as shown in Figure 12.2. Process stream 1 having a mass fraction of components A and C of $x_{1,A}$ and $x_{1,c}$ is fed to the top of a reactor containing a fixed bed of catalyst. Also fed to the top of the reactor is process stream 2, which is pure B. In the reactor, A and B react to form D:

$$2A + B = D \tag{12.33}$$

which may be a reversible reaction. The rate of reaction and thus the overall conversion in the reactor is dependent upon the concentrations of A, B, and D at each point in the reactor, the activity of the catalyst, the resident time of the reactants in the reactor, and the temperature at each point in the reactor. If the reactions are in the gas phase, then pressure will also be an important parameter.

It is common to use a cooling/heating jacket or tubes to maintain the temperature throughout the reactor at the same or nearly the same point where the target reaction is optimized. This is the case in Figure 12.2. This simplifies the problem as we do not need to worry about the temperature

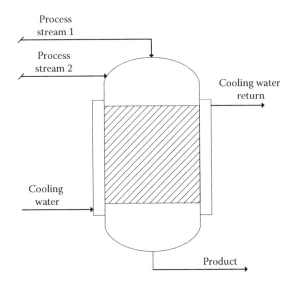

FIGURE 12.2
Fixed bed reactor with cooling jacket.

dependence of the reaction. If we did, an Arrhenius type model could be used. The Arrhenius model is:

$$k_A = k_{A,o} e^{-\Delta E_A/RT} \tag{12.34}$$

where
  k is the rate of reaction
  ΔE is the change in enthalpy between the current and reference temperatures
  R is the universal gas constant
  T is the current reaction temperature
  A is the component of interest
  o is the condition at the reference temperature

If the reactor is well designed, then we can also assume one-dimensional plug flow through the reactor. That is, at any point, z, along the length of the reactor, the composition, flow, and temperature will all be essentially the same at any point on the plane within the reactor normal to the flow direction (see Figure 12.3).

Because we are most interested in changes in the reaction rate, which is measured by changes in the outlet composition, we need to construct a mole balance rather than a mass balance of the system. Let us assume we are measuring the outlet concentration of component A (we could just as easily use

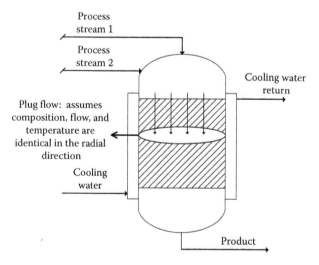

FIGURE 12.3
The plug flow condition through a fixed bed reactor.

Process Dynamic Analysis

B, D, or even the total moles). A component balance of the change in A with time over the reactor is given by:

$$\frac{d\left(M_o \dfrac{x_{A,o}}{MW_A}\right)}{dt} = \left(\frac{M_i x_{A,i}}{MW_A}\right) - \left(\frac{M_o x_{A,o}}{MW_A}\right) - \gamma_A \left(\frac{M_o x_{D,o}}{MW_D}\right) \qquad (12.35)$$

where
M is the mass flow rate
MW is the molecular weight
γ is the stoichiometric coefficient for the reaction of A into D (which for this example is 2)
i is inlet
o is outlet
A and D are defined by Equation 12.33

Expanding the left-hand side:

$$\frac{M_o}{MW_A}\frac{dx_{A,o}}{dt} + \frac{x_{a,o}}{MW_A}\frac{dM_o}{dt} = \left(\frac{M_i x_{A,i}}{MW_A}\right) - \left(\frac{M_o x_{A,o}}{MW_A}\right) - \gamma_A \left(\frac{M_o x_{D,o}}{MW_D}\right) \qquad (12.36)$$

Equation 12.36 provides a generalized description of changes in the quantity of A leaving the reactor based on changes in the mass flow rate through the reactor, the concentration of A in the inlet stream, or changes in the rate of conversion of A into D. This form is still too general for most practical applications. To simplify it further, assume that the outlet flow rate is a constant with time ($dM_o/dt = 0$). Then Equation 12.36 reduces to:

$$\frac{M_o}{MW_A}\frac{dx_{A,o}}{dt} = \left(\frac{M_i x_{A,i}}{MW_A}\right) - \left(\frac{M_o x_{A,o}}{MW_A}\right) - \gamma_A \left(\frac{M_o x_{D,o}}{MW_D}\right) \qquad (12.37)$$

which can be rearranged to:

$$\frac{dx_{A,o}}{dt} = \left(\frac{M_i x_{A,i}}{M_o}\right) - (x_{A,o}) - \gamma_A \left(\frac{MW_A x_{D,o}}{MW_D}\right) \qquad (12.38)$$

Equation 12.38 provides a model of changes in the concentration (in this case the mass concentration, but this can be easily converted into molar or volumetric concentrations). If M_i is equal to M_o, then all changes are due to changes in the rate of reaction, which we have quantified here using the outlet mass concentration of the product D.

12.2 Dynamic Models for Common Instrument and Control System Components

Dynamic models of process unit operations and groups of unit operations can be very useful in troubleshooting unstable operation. The examples in Section 12.1 demonstrate just how complicated even simple forms of these models can be. But modeling the process may not be adequate to determine why a process is unstable or provide sufficient insight for how to move a process toward more optimum operating conditions. That's because the operation of the process is also greatly influenced by the regulatory controls (actually, all of the controls) employed to manage the process.

The total model can be visualized by the simple block diagram shown in Figure 12.4. Note that just like the control loops that we are modeling, this diagram shows a closed loop process. Using the setpoints specified either manually or externally from higher-level PAS applications, the controllers generate outputs that directly change the behavior of the control variables. These variables result in changes in the process, which are then reflected by changes in the measurement variables. The measurement variables are used by the controllers to assess the condition of the process compared to the setpoints. This completes the cycle. External changes can be input anywhere in the model.

12.2.1 Control Variables

Let's start with the control variables. By far the most common control variables are throttling control valves. The response of the control valve actuator and valve position can be modeled using a first-order linear model:

$$\frac{dX}{dt} = \frac{1}{\tau_v}(X_{out} - X) \qquad (12.39)$$

where
- X is the valve position
- τ_v is the time constant corresponding to the overall response of the control valve to the input
- X_{out} is the output value of the valve position specified by the controller
- Note that X and X_{out} can be defined as the fraction of the full valve opening position specified (% open or % closed) or in units that correspond to the associated value of the dependent variable in the feedback control loop.

The value of τ_v can be obtained from the valve manufacturer, from valve tests, or approximated using literature values. Note that Equation 12.39 assumes that the actual reaction of the process to a change of the valve position is instantaneous, an assumption that is generally considered acceptable for the accuracy of these models.

The second most common control variable is the speed controller for a motor. The reactivity of motor speed to changes in the controller output is very fast, usually much faster than changes in the measurement variable or the process itself. As such, the response can be considered instantaneous. In which case, $X = X_{out}$.

The third most common control variable is the remotely operated block valve, which is used for on/off control and to change steps in sequential logic control schemes for batch and semi-batch processes. The response of the valve primarily depends on the speed of closure and on the linearity or nonlinearity of changes in the fluid flow through the valve as it is opening or closing. If a linear response occurs, then Equation 12.39 can be applied. Otherwise, a different equation must be developed that simulates the process response during the opening/closing operation.

12.2.2 Measurement Variables

As we saw in Chapter 10, there are many different devices used to determine the value of a measurement variable. In general, all of these can either be characterized as instantaneous, $Y_m = Y$, or by a first-order response using an equation comparable to Equation 12.40:

$$\frac{dY}{dt} = \frac{1}{\tau_m}(Y - Y_m) \qquad (12.40)$$

where
  Y is the measurement variable
  τ_m is the time constant that corresponds to the actual measurement device
  m is the measured value

The value of τ_m can be obtained from the instrument manufacturer, from tests, or approximated using literature values. If the most common types are employed, most modelers will model pressure, flow, and level measurements assuming an instantaneous response, while temperature and composition will be modeled using Equation 12.40. Note the time constants for temperature measuring devices will be at least one order of magnitude less than those for indirect analyzers with online analyzers having time constants approaching those of temperature devices.

Because these models are device specific, working with the manufacturer on the correct model is strongly recommended.

12.2.3 Controllers

Often during dynamic analysis, the controller action is left out of the model so that the process behavior is easier to evaluate separately from the effects of the controller. However, if we are troubleshooting unstable behavior, then

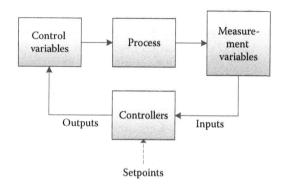

FIGURE 12.4
A simple model of the process and control system.

the controller may be the most important dynamic block in Figure 12.4. Models for controllers were introduced in Chapter 8. The most commonly applied is one of the forms of the PID controller. One use of a dynamic model is to evaluate changes in the tuning constants of the PID controller to provide improved process performance.

12.3 Incorporating Simplified Process Changes into Dynamic Models

Once our dynamic model is built, it can be used to evaluate how the process and its regulatory controls respond to process changes. This is typically accomplished by introducing one or more standard types of dynamic changes. The most common changes are the step change, the ramp change, the rectangular pulse change, the spike, and the cyclic change (Figure 12.5).

The step change models a sudden setpoint change in one or more control loop. In this case, when the model is run, the boundary condition of the controller model would include a term of the form:

$$\text{For } t < t_{chg}, \quad X_{sp} = X_{sp,o}$$

$$\text{For } t \geq t_{chg}, \quad X_{sp} = X_{sp,chg}$$

where
  t is time
  chg is the step change
  X is the control variable
  sp is the controller setpoint

Process Dynamic Analysis

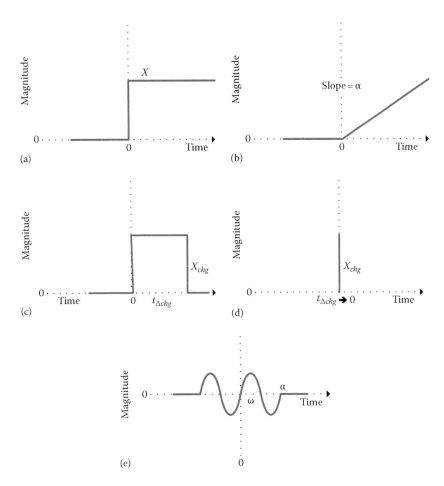

FIGURE 12.5
Simple models of disturbances or changes in a process and control system: (a) the step change, (b) the ramp change, (c) the rectangular pulse change, (d) the spike, and (e) the cyclic change.

Inputting this condition into the model allows the user to see how a measurement variable changes over time in response to the setpoint change and how quickly the overall process system settles at its new steady-state condition.

The step change can also be used to model a sudden permanent change to an input, typically a parameter that represents a disturbance variable to the system. In this case, a similar condition is introduced into the process model by using a variable specification equation. Alternately, it can be introduced via the measurement model by changing the value of an input measurement using a similar conditional statement.

Sudden changes to setpoints or to process inputs can result in unacceptable upsets in the process. As a result, the change may be introduced more gradually. In fact, modern distributed control systems give the control operator the option of ramping from the existing setpoint to a new setpoint when the change is made. The *range change* can be used to model a more gradual change from the existing setpoint or process input condition to a new setpoint or process input condition:

$$\text{For } t < t_{chg}, \quad X_{sp} = X_{sp,o}$$

$$\text{For } t \geq t_{chg}, \quad X_{sp} = X_{sp,o} + \alpha t$$

where α is the slope of the ramp.

Note that α can be either a positive or negative number depending upon the direction of the ramp change.

The *rectangular pulse* change is used to model a process upset of finite duration. For example, perhaps the inlet flow rate to a vessel suddenly increases by 10%, but then is corrected at a later time. This model is also useful in batch and semi-batch process steps where a given condition occurs for a finite period and then changes condition. A rectangular pulse of duration $t_{\Delta chg}$ is represented by:

$$\text{For } t < t_{chg}, \quad X_{sp} = X_{sp,o}$$

$$\text{For } t_{chg} \leq t \leq t_{chg+\Delta chg}, \quad X_{sp} = X_{chg}$$

$$\text{For } t > t_{chg+\Delta chg}, \quad X_{sp} = X_{sp,o}$$

When the duration of the pulse approaches zero, the model represents a *spike*. A spike is an instantaneous, very short duration change to the process. This type of behavior doesn't actually occur in process plants and even if it did, the duration is so short that the impacts would not be noticeable. However, this function is sometimes used to help set up and validate the original dynamic model.

Process or controller instabilities are often modeled using a *cyclic function*. The most common is the *sinusoidal function*. This function is represented by:

$$\text{For } t < t_{chg}, \quad X_{sp} = X_{sp,o}$$

$$\text{For } t \geq t_{chg}, \quad X_{sp} = X_{sp,o} + \alpha \sin(\omega t)$$

where
    α is the amplitude
    ω is the frequency of the wave function

Problems

12.1 To accurately model the dynamic behavior of a process, what must be included in the model?

12.2 Starting with Equation 12.2, develop a linear ODE for a generalized lumped parameter model for the overhead pressure control loop on a knockout drum, such as the one described in Section 2.3.

12.3 Starting with Equation 12.15, develop a linear ODE for a generalized lumped parameter model for the liquid flow leaving a knockout drum, such as the one described in Section 2.3.

12.4 Starting with Equation 12.15, develop a linear ODE for a generalized lumped parameter model for the vapor phase flow leaving a knockout drum, such as the one described in Section 2.3.

12.5 Extend Equation 12.24 for the case where the energy from the utility fluid is supplied by a combination of latent heat and subsequent sensible heat.

Appendix A: Transform Functions and the "s" Domain

Processes vary from steady state. These variations are the entire basis for regulatory process control, which uses the error between a process measurement variable and its desired value to determine adjustments to a control variable. The most common types of control algorithms used in process control are described in Chapter 8.

The study of process variability is known as the study of process dynamics and is introduced in Chapter 12. Process dynamics is concerned with evaluating how changes, either planned or unplanned, affect the behavior of the process. A planned change could be due to a change in production rate or a new product purity specification. Unplanned changes could be due to an equipment failure, the gradual decrease in reactivity of a catalyst, or due to an instability in a portion of the process that ripples downstream through the process.

Mathematically, we can state that all processes have a time dependency. They may also have other dependencies, but the common dependency is time. If we assume that the only significant dependency for a portion of a process is time, then we can state that:

$$\frac{dX(t)}{dt} = f(t) \tag{A.1}$$

where
  X is a dependent variable of the process such as flow, temperature, etc.
  t is the time.

Solving Equation A.1 for all times greater than $t = 0$ yields:

$$X(t) = \int_{0}^{\infty} f(t) dt \tag{A.2}$$

A.1 Laplace Transform

When $f(t)$ is:

- Bounded and has a finite number of maxima and minima,
- Has a finite number of discontinuities, and

- If there exists a real constant, a, such that the integral $\int_0^\infty e^{-iat} f(t) dt$ is convergent or
- If there exists a real constant, b, such that the function $e^{-bt}|f(t)|$ remains bounded as t goes to infinity,

then Equation A.2 can be solved using a mathematical transformation known as the Laplace transform. The Laplace transform converts a first order (in time) linear ordinary differential equation into a zero order (in "s" space) algebraic equation, thus allowing the equation to be solved. The conversion from the time domain to the "s" domain is:

$$\mathcal{F}(s) = \mathcal{L}\{f(t)\} = \int_0^\infty e^{-st} f(t) dt \qquad (A.3)$$

where
s is a complex independent variable
$\mathcal{L}$ is the Laplace operator as defined by Equation A.3

By defining the Laplace operator, we can also define the inverse Laplace operator, $\mathcal{L}^{-1}$:

$$f(t) = \frac{1}{2\Pi i} \int_{a-i\infty}^{a+i\infty} e^{st} \mathcal{F}(s) ds \qquad (A.4)$$

The derivation of Equation A.4 can be found in many references describing the solutions to ordinary differential equations such as Wylie (1975).

Equations A.3 and A.4 may look intimidating. Fortunately, tables have been compiled that provide the "s" domain solutions for many functions. Table A.1 provides those that are commonly used in the simplified descriptions of process dynamics and of process control system elements.

Another useful feature of the Laplace and inverse Laplace operators is that they are linear operators. As a linear operator, the following relationship holds:

$$\mathcal{F}\{af(t) + bg(t)\} = a\mathcal{F}\{f(t)\} + b\mathcal{F}\{g(t)\} \qquad (A.5)$$

which leads to:

$$\mathcal{L}\{af(t) + bg(t)\} = a\mathcal{F}(s) + b\mathcal{G}(s) \qquad (A.6)$$

where $\mathcal{F}$ and $\mathcal{G}$ are the individual Laplace transforms of f and g, respectively.

Appendix A: Transform Functions and the "s" Domain

TABLE A.1

Laplace Transforms for Commonly Used Process Dynamic Functions

f(t)	L{f(t)}	f(t)	L{f(t)}
Step of size "a" at $t = 0$	$\dfrac{a}{s}$	Unit impulse at $t = 0$; $\delta(t)$	1
Ramp of slope "a"; $f = (at)$	$\dfrac{a}{s}$	t^{n-1}	$\dfrac{(n-1)!}{s^n}$
at^2	$\dfrac{2a}{s^3}$	$\sin(\omega t)$	$\dfrac{\omega}{s^2 + \omega^2}$
at^n	$\dfrac{an!}{s^{n+1}}$	$\cos(\omega t)$	$\dfrac{s}{s^2 + \omega^2}$
e^{-at}	$\dfrac{1}{s+a}$	$\sin(\omega t + \phi)$	$\dfrac{\omega \cos(\phi) + s \sin(\phi)}{s^2 + \omega^2}$
$\dfrac{1}{\tau} e^{-t/\tau}$	$\dfrac{1}{\tau s + 1}$	$e^{-at} \sin(\omega t)$	$\dfrac{\omega}{(s+a)^2 + \omega^2}$
$1 - e^{-t/\tau}$	$\dfrac{1}{s(\tau s + 1)}$	$e^{-at} \cos(\omega t)$	$\dfrac{s+a}{(s+a)^2 + \omega^2}$
$t^n e^{-at}$	$\dfrac{n!}{(s+a)^{n+1}}$	$\dfrac{1}{a-b}\left(e^{-bt} - e^{-at}\right)$	$\dfrac{1}{(s+a)(s+b)}$
$\dfrac{t^{n-1} e^{-at}}{(n-1)!}$	$\dfrac{1}{(s+a)^n}$	$\dfrac{1}{\tau^n (n-1)!} t^{n-1} e^{-t/\tau}$	$\dfrac{1}{(\tau s + 1)^n}$
$\dfrac{c-a}{b-a} e^{-at} + \dfrac{c-b}{a-b} e^{-bt}$	$\dfrac{s+c}{(s+a)(s+b)}$	$\left(\dfrac{a-c}{a-b}\right)\left(\dfrac{e^{-t/a}}{a}\right) + \left(\dfrac{b-c}{b-a}\right)\left(\dfrac{e^{-t/b}}{b}\right)$	$\dfrac{cs+1}{(as+1)(bs+1)}$

The Laplace transform reduces the order of a differential equation by one level. Using Equation A.3:

$$\mathcal{F}(s) = \mathcal{L}\left\{\frac{dX}{dt}\right\} = \int_0^\infty \frac{dX}{dt} e^{-st} dt \qquad (A.7)$$

For example, if we insert Equation A.1 into Equation A.7, we get:

$$\mathcal{F}(s) = \mathcal{L}\left\{\frac{df(t)}{dt}\right\} = \int_0^\infty \frac{df(t)}{dt} e^{-st} dt \qquad (A.8)$$

Integrating Equation A.8 by parts yields:

$$\mathcal{F}(s) = \mathcal{L}\left\{\frac{df(t)}{dt}\right\} = \int_0^\infty \frac{df(t)}{dt} e^{-st} dt = \int_0^\infty f(t) e^{-st} s \, dt + f(t) e^{-st}\Big|_0^\infty \qquad (A.9)$$

The first part leads to:

$$\int_0^\infty f(t)e^{-st}s\,dt = s\mathcal{L}\{f(t)\} \qquad (A.10)$$

and the second part leads to:

$$f(t)e^{-st}\big|_0^\infty = f(0^+) \qquad (A.11)$$

which combined leads to:

$$\mathcal{L}\left\{\frac{df(t)}{dt}\right\} = s\mathcal{L}\{f(t)\} - f(0^+) \qquad (A.12)$$

where $f(0^+)$ is value of the function $f(t)$ as t approaches 0 from the positive side (recall that the Laplace transform is only valid for values of t from 0 to ∞).

Let's look at an example. In Chapter 12, we introduced the dynamic behavior of various control system elements. Let's consider the most common dynamic behavior of a control valve, which is described by Equation 12.39 as:

$$\frac{dX}{dt} = \frac{1}{\tau_v}(X_{out} - X) \qquad (12.39)$$

where
X_{out} is the output value from the controller
τ_v is the time constant of the valve

Rearranging Equation 12.39 yields:

$$\tau_v \frac{dX}{dt} + X - X_{out} = 0 \qquad (A.13)$$

If we assume that $X(0) = 1$ (or any other value that represents the value of the dependent variable at the time we choose to be "0"; but using "1" allows a simpler form of the inverse transform to be used for this example) we can transform each term. Using Equation A.12 for the first term:

$$\mathcal{L}\left\{\tau_v \frac{dX}{dt}\right\} = \tau_v s\mathcal{F}(s) - (1.0)\tau_v \qquad (A.14)$$

and the second term is:

$$\mathcal{L}\{X\} = \mathcal{F}(s) \qquad (A.15)$$

and the third term is equivalent to a step of unit size "X_{out}." Using the first entry in Table A.1, yields:

$$\mathcal{L}\{-X_{out}\} = -\left(\frac{X_{out}}{s}\right) \tag{A.16}$$

Combining Equations A.14 through A.16 back together yields:

$$\tau_v s \mathcal{F}(s) - \tau_v + \mathcal{F}(s) - \left(\frac{X_{out}}{s}\right) = \mathcal{F}(s)(\tau_v s + 1) - \tau_v - \left(\frac{X_{out}}{s}\right) = 0 \tag{A.17}$$

which can be further rearranged to solve for $\mathcal{F}(s)$:

$$\mathcal{F}(s) = \frac{\tau_v + \left(\dfrac{X_{out}}{s}\right)}{(\tau_v s + 1)} = \frac{\tau_v s + X_{out}}{s(\tau_v s + 1)} = \frac{s + c}{s(s + b)} \tag{A.18}$$

where

$$c = \left(\frac{X_{out}}{\tau_v}\right)$$

$$b = \frac{1}{\tau_v}$$

Notice that even for this very simple dynamic expression, the solution is fairly involved, which is why numerical simulation of the more complex, real-world expressions is preferred.

Equation A.12 provides an easy way to reduce the order of higher order linear ordinary differential equations. For example, let:

$$f(t) = a\frac{d^2 X}{dt^2} + b\frac{dX}{dt} + cX \tag{A.19}$$

then using Equation A.12 we get:

$$\mathcal{L}\{f(t)\} = \mathcal{L}\left\{a\frac{d^2 X}{dt^2} + b\frac{dX}{dt} + cX\right\} = a\mathcal{L}\left\{\frac{d^2 X}{dt^2}\right\} + b\mathcal{L}\left\{\frac{dX}{dt}\right\} + c\mathcal{L}\{X\} \tag{A.20}$$

and then for the first term in Equation A.20, we can apply Equation A.12:

$$\mathcal{L}\left\{\frac{d^2 X}{dt^2}\right\} = as\mathcal{L}\left\{\frac{dX}{dt}\right\} - d \tag{A.21}$$

where d is the value of $\dfrac{dX}{dt}$ as t approaches zero from the positive side.

Equation A.12 can be applied successively to higher order equations. For example:

$$\mathcal{L}\left\{\frac{d^2 f(t)}{dt^2}\right\} = s^2 \mathcal{L}\{f(t)\} - sf(0^+) - \left.\frac{df(t)}{dt}\right|_{t \to 0^+} \quad \text{(A.22)}$$

Consider a third order equation containing first, second, and third order terms:

$$\frac{d^3 X}{dt^3} + a\frac{d^2 X}{dt} + b\frac{dX}{dt} + cX = d \quad \text{(A.23)}$$

The equivalent "s" domain equation is:

$$s^3 \mathcal{F}(s) + as^2 \mathcal{F}(s) + bs\mathcal{F}(s) + c\mathcal{F}(s) = \frac{d}{s} \quad \text{(A.24)}$$

and solving for $\mathcal{F}(s)$:

$$\mathcal{F}(s) = \frac{d}{s(s^3 + as^2 + bs + c)} \quad \text{(A.25)}$$

A.2 Solving Equations in the "s" Domain

Having transformed the ordinary differential equation (in time) into an algebraic equation (in "s"), the next step is to rearrange the equation or suite of equations to solve for the transform of the dependent variable. In Equation A.1 we defined the dependent variable as "X," which is dependent upon the independent variable "t." Let us now denote the transform of "X" as "$\mathcal{X}$," which is a variable dependent on "s."

Once the algebraic equation has been rearranged, it will usually be in the form:

$$\mathcal{X}(s) = \frac{\mathcal{F}_N(s)}{\mathcal{F}_D(s)} \quad \text{(A.26)}$$

where $\mathcal{F}_N$ and $\mathcal{F}_D$ are those terms from the rearranged Laplace equation $\mathcal{F}(s)$ in the numerator and denominator, respectively.

For simple equations, we can match one of the common forms, such as those shown in Table A.1. These can be used to convert the equation back into the time domain, which provides a solution for the original dependent variable "X." For example, if we return to Equation A.18, we find that it matches the form of the last left-hand entry in Table A.1 where a = 0 and b and c have the definitions provided with Equation A.18. Applying the inverse transform then yields:

$$X(t) = \frac{\left(\dfrac{X_{out}}{\tau_v}\right)}{\dfrac{1}{\tau_v}} + \frac{\left(\dfrac{X_{out}}{\tau_v}\right) - \dfrac{1}{\tau_v}}{1 - \dfrac{1}{\tau_v}} e^{-\frac{1}{\tau_v}t} = X_{out} + \frac{X_{out} - 1}{\tau_v - 1} e^{-t/\tau_v} \quad (A.27)$$

However, for more complicated equations, additional mathematical techniques are usually needed.

A.3 Partial Fraction Expansion

For complex linear ordinary differential equations and for those equations with terms of an order higher than $\dfrac{dX(t)}{dt}$, the specific form of Equation A.26 may not match any of the common forms. For such cases, Equation A.26 can be expanded into a series of simpler expressions that do match the common forms using the method of partial fractions.

In this method, $\mathcal{F}_D$ is converted from a polynomial form into a series of first order terms. For example, the simple polynomial form:

$$\mathcal{F}_D = s^2 + s(a+b) + ab \quad (A.28)$$

can be converted into the form:

$$\mathcal{F}_D = (a+s)(b+s) \quad (A.29)$$

Inserting Equation A.29 into Equation A.26 yields:

$$X(s) = \frac{\mathcal{F}_N(s)}{\mathcal{F}_D(s)} = \frac{\mathcal{F}_N(s)}{(a+s)(b+s)} = \frac{\alpha}{(a+s)} + \frac{\beta}{(b+s)} \quad (A.30)$$

There are several common techniques that can be employed for finding the values of α and β. However, a technique commonly used by those using simple ODEs to study process dynamics is the Heaviside expansion method. In this method, both sides of the expanded equation are multiplied by one of the expanded denominator terms. For Equation A.30, this would be either (a + s) or (b + s). Next, the value of s is set such that all terms go to zero except one. If (a + s) was used, then s = −a; if (b + s) were used, then s = −b. This will yield one of the numerator terms in the expanded equation. This same process is repeated using each of the expanded denominator terms until all the numerator values have been determined. For Equation A.30, this procedure is performed twice to yield the values for α and β.

Consider the Laplace equation we derived as Equation A.25. The cubic equation in the denominator can usually be expanded if the actual values of a, b, and c are known. For example, if a = −2, b = −9, c = 18, and d = 3, then Equation A.25 becomes:

$$\mathcal{F}(s) = \frac{3}{s(s^3 - 2s^2 - 9s + 18)} \qquad (A.31)$$

which can be factored into four terms:

$$\mathcal{F}(s) = \frac{3}{s(s-2)(s-3)(s+3)} \qquad (A.32)$$

and then using the format of Equation A.30:

$$\mathcal{F}(s) = \frac{3}{s(s-2)(s-3)(s+3)} = \frac{\alpha}{(s)} + \frac{\beta}{(s-2)} + \frac{\gamma}{(s-3)} + \frac{\delta}{(s+3)} \qquad (A.33)$$

If we multiply Equation A.33 by "s" and set the value of "s" to zero, we find that α = 6. If we multiply Equation A.33 by "s − 2" and set the value of "s" to +2, we find that β = $\frac{-10}{3}$. If we multiply Equation A.33 by "s − 3" and set the value of "s" to +3, we find that γ = 6. Finally, if we multiple Equation A.33 by "s + 3" and set the value of "s" to −3, we find that δ = −30. Inserting these values into Equation A.33:

$$\mathcal{F}(s) = \frac{6}{(s)} - \frac{10}{3(s-2)} + \frac{6}{(s-3)} - \frac{30}{(s+3)} \qquad (A.34)$$

The first term matches the first form in the left-hand column of Table A.1 yielding: $f_1(t) = 6$. The remaining terms all match the fifth form in the left-hand column of Table A.1. For the second term, the constant is −10/3 and a = −2 yielding $f_2(t) = -10/3 e^{2t}$. For the third term the constant is 6 and a = −3, yielding: $f_3(t) = 6e^{3t}$. For the fourth term the constant is −30 and a = +3, yielding $f_4(t) = -30e^{-3t}$. Combining all the terms we get:

$$X(t) = 6 - \frac{10}{3}e^{2t} + 6e^{3t} - 30e^{-3t} \qquad (A.35)$$

Sometimes rearranging $\mathcal{F}_D$ into products results in repeated terms or terms that include complex numbers. There are special methods to handle these circumstances, and the reader is encouraged to consult a textbook containing more details on partial fraction expansions for how to solve these types of expressions.

A.4 General Procedure to Use Laplace Transforms to Solve Linear ODEs

Step 1: Isolate and/or simplify the process/process control system that is the focus of the study such that the system can be described by a linear ODE that satisfies the conditions given in Section A.1.

Step 2: Define an initial condition for the ODE developed in step 1.

Step 3: Transform the ODE using the transform equivalents shown in Table A.1 (or elsewhere) into an algebraic equation in the "s" domain.

Step 4: Rearrange the algebraic expression to solve for the dependent variable in the "s" domain that corresponds to the dependent variable of the original ODE developed in step 1. This equation will typically be in the form shown in Equation A.12.

Step 5: Perform partial fraction expansion, if required, for the rearranged algebraic expression developed in step 4. Solve for the numerator coefficients in the expanded form using the Heaviside expansion method or another technique.

Step 6: Transform the equation back into the time domain using the common forms found in Table A.1 (or elsewhere).

Reference

Wylie, C., *Advanced Engineering Mathematics*, 4th Edition, McGraw-Hill, New York, 1975.

Appendix B: PID Controller Tuning

The proportional-integral-derivative controller introduced in Section 8.2 can be described mathematically as:

$$CV_o = CV_{ob} + K_{PID}\left[e(t) + \frac{1}{\tau}\int_0^{t_M} e(t)dt + \tau_D \frac{de(t)}{dt}\right] \quad (8.8)$$

The controller includes three adjustable constants, known as the tuning constants:

1. The gain, K_{PID}
2. The integral time, τ
3. The derivative time, τ_D

The values selected for these constants have a substantial effect on the output calculated, which in turn will impact how the control loop behaves in response to error.

There are two aspects to the tuning constants, their magnitude and their ratio. The magnitude of the tuning constants determines how large or small the output will be in response to a given value of the error. For example, if the gain is much larger than one (in dimensionless terms), then the output to the controller will be much larger than the error.

The second factor is the relative ratio of the three constants. A well-tuned controller will balance the responsive, dampening, and reactive contributions of the three terms in Equation 8.4. The gain should be adequate to respond to changes in setpoint or process upsets. The integral time should allow the control loop to reach new steady-state conditions, either due to a setpoint change or a long-term change in the process conditions, via a strongly dampened oscillating output as shown in Figure B.1. When the rate of change in the error is significant, the derivative time should provide adequate contributions to react to the rate of change in order to minimize the long-term error in the control loop. This balance is sometimes referred to as the responsiveness-robustness balance.

No selection of tuning constants can completely overcome poor process control scheme design, although the impacts of such schemes can be minimized. However, loop tuning should only be seen as a temporary solution when poor control scheme design is encountered and steps should be taken to improve the scheme permanently.

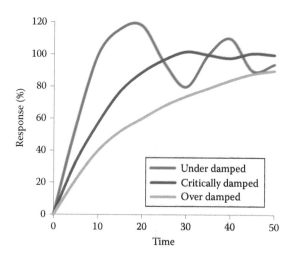

FIGURE B.1
The measurement response for an ideally tuned PID controller.

B.1 Auto-Tuning Systems

Virtually every modern stand-alone controller includes loop tuning algorithms in their firmware while distributed control systems (DCS) often offer multiple options for automatic loop tuning. Starting from a set of default constants embedded in the software, these auto-tuning functions adjust the tuning constants to achieve a reasonable responsiveness-robustness balance based on the process conditions that the controller experiences during the time that the auto-tune function is enabled. For most simple feedback control loops, this level of loop tuning is all that is required for stable operation of the control scheme.

Many of these auto-tuning systems are based on the relay auto-tuning method developed by Åström and Hägglund (1984). In this method, the PID controller is temporarily replaced by an on-off controller. The controller will generate a sustained oscillating output that will lead to an oscillating measurement value, as shown in Figure B.2. The ultimate gain is calculated by:

$$K_u = \frac{4d}{\pi a} \qquad (B.1)$$

where
  d is the relay amplitude of the auto-tune controller
  a is the amplitude of the measurement variable versus time function resulting from the on-off controller action

Appendix B: PID Controller Tuning

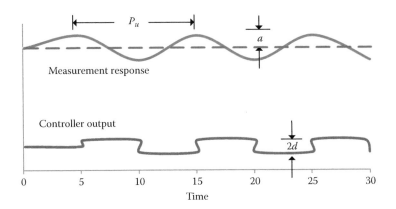

FIGURE B.2
The controller output and measurement behavior for the relay auto-tuning method. The variables are defined by Equation B.1.

The ultimate period, P_u, is equal to the period of oscillation of the measurement variable versus time function. The auto-tuning algorithm calculates values for the three tuning constants based on the ultimate gain and period values following one of a number of different methods. A common method, the Ziegler-Nichols method, is described in Section B.2 as this method is also used to manually determine initial estimates of the tuning constants.

B.2 Manual Methods to Determine Initial Tuning Constants

In addition to their auto-tuning features, stand alone controllers and DCS include the capability to manually specify the tuning constants. This is important because auto-tune systems may not be able to provide good initial estimates of the tuning constants for more complicated process control schemes such as those that include cascade or combination FF/FB loops. Auto-tune programs typically also lack the ability to determine the constants for each region of a scheduled tuning parameter system such as those described in Section 8.4.2.

For large projects, system configurators will use case tools to help select good starting values for the constants. These values are usually adequate for system commissioning. However, these configuration tools may not be available for smaller projects, so it may be useful to know how to manually specify initial tuning constants. Manual estimating methods can be divided into two categories: experimental methods and model-based methods.

B.2.1 Experimental Tuning Constant Estimation Methods

Experimental methods to determine initial values for PID controller tuning constants use one or more simple open- or closed-loop tests to evaluate measurement variable response to changes in the output. The results are typically evaluated by one or more response parameters, such as the ultimate gain and ultimate period described in Section B.1 for the auto-tuning method. Values for the tuning constants are then determined by correlation from the response parameters.

An early version of this approach, and the most widely described method, is the *Ziegler-Nichols method* (Ziegler and Nichols 1942), also known as the *continuous cycling method*. It is very similar to the auto-tuning method described in Section B.1. First, the PID algorithm is temporarily replaced with a simple proportional controller with a small gain value, typically around 0.5. Next, a small setpoint change (on the order of 3%–5%) is made and the gain value is increased in small increments until a sustained oscillation (i.e., continuous cycling) of the measurement variable is achieved. The gain value at the point of sustained oscillation is the ultimate gain, K_u, while the period of oscillation is the ultimate period, P_u.

The values of K_u and P_u are correlated to initial tuning constants for the full PID controller. In addition to Ziegler and Nichols original correlations, other correlations have been proposed. For example, Tyreus and Luyben (1992) felt that the original correlations were too responsive and proposed slightly different values that increased the dampening action of the controller. Both Ziegler-Nichols and Tyreus-Luyben correlations are summarized in Table B.1.

There are many additional manual experimental methods available that are well described in the literature. Good summaries and comparisons of the various methods are available from Seborg et al. (2011) and by Riggs and Karim (2006).

B.2.2 Model-Based Tuning Constant Estimation Methods

These methods may be useful if there are only one or two control loops being addressed. The first step is to develop a model of the process and control system dynamics as described in Chapter 12. For the purposes of using these models for tuning constant estimation, simplified models are often used.

TABLE B.1

PID Controller Settings Based on the Ziegler-Nichols Ultimate Gain and Period

Correlation	K_{PID}	τ	τ_D
Ziegler-Nichols values	$0.6 K_u$	$P_u/2$	$P_u/8$
Tyreus-Luyben values	$0.45 K_u$	$2.2 P_u$	$P_u/6.3$

Once the model is developed, simple simulated upsets are performed to generate measurement response functions. These functions are then correlated to the tuning constants. The use of these methods in commercial processes is rare and limited to extremely complex systems where none of the easier methods can provide even a reasonable starting estimate of the tuning constants. Since their use is rare, the reader is directed to the previously referenced works by Seborg and coworkers and by Riggs and Karim for further details about these methods.

B.3 Final Loop Tuning During Operation

Due to the complexities of most chemical processes, controller performance can often be improved by making minor adjustments to one or more of the three PID controller tuning constants when the control loops are placed into service. This activity will be a major part of any new or retrofit project commissioning phase. It also should be performed whenever a new control loop is put into service, even if it is a direct replacement of a previous control loop. Further, there may be times when it is appropriate to retune an existing controller due to changes in process conditions.

Here are some general tips that may be useful during final loop tuning:

1. Evaluate the responsiveness of the process and control loop.
    1.1 When pressure, flow, temperature, or direct analysis are used as measurement variables in the control loop, the dynamics are usually responsive or very responsive. The only time these types of control loops won't be responsive is when there is a time lag in the control loop either because the measurement and control points are physically located far apart or because they are separated by a significant liquid volume, such as a drum or tank.
    For responsive loops, the magnitude of the gain relative to the inverse integral time should be fairly high. Changes will occur quickly and the PID controller output also needs to react quickly to keep up with the changes.
    1.2 When level or sampled-type analysis are used as measurement variables or when there is a significant time lag in the control loop, the magnitude of the gain relative to the inverse integral time should be fairly low since the process and control system are nonresponsive. If responsiveness is needed, a cascade scheme should be employed rather than trying to make a nonresponsive loop more responsive by overadjusting the gain.

1.3 If frequent significant changes in the setpoint are expected, increase the magnitude of the derivative time relative to the gain. This will insure that the control loop reacts quickly to these changes but then returns to the more stable conditions that result from moderate gain and integral time constants.

2. Identify the process objective of the control loop.

   2.1 If the control loop impacts a product specification, increase the magnitude of the derivative time constant to minimize the possibility of going off-specification due to a process upset.

   2.2 If the control loop is part of a process recycle, increase the magnitude of the inverse integral time so that the control loop helps to smooth out instabilities in the recycle loop and avoids introducing new instabilities by control variable actions.

   2.3 If the control loop is part of an auxiliary or utility system, increase the magnitude of the inverse integral time to minimize dynamics from this control loop on the process. It is better to have a slightly less efficient auxiliary/utility if this will avoid inefficiencies in the process streams. Conversely, many of the control loops on process streams can have higher magnitude gain constants to maximize process efficiency.

3. Commission and tune control loops backwards.

   3.1 If your process control scheme is similar to those shown in Chapters 3 through 6, you can avoid propagating tune test dynamics through the process by commissioning the system starting at the end of the process and working backwards to the beginning of the process. This will allow you to perform better tuning tests and get to the best tuning constants quickly.

   3.2 Use multiple simple experiments to determine the best values of the tuning constants. Simple dynamic tests described in Chapter 12 include the step change, the ramp change, the rectangular pulse, the spike, and the cyclic change. Different tests simulate different scenarios. For example, the step and ramp changes simulate a setpoint change. The other three simulate process upset conditions that could occur.

References

Åström, K.J. and Högglund, T., Automatic tuning of simple regulators with specification on the gain and phase margins, *Automatica*, 20:645, 1984.

Riggs, J.B. and Karim, M.N., *Chemical and Bio-Process Control*, 3rd Edition, Ferret Publishing, Lubbock, TX, 2006.

Seborg, D.E., Edgar, T.F., Mellichamp, D.A., and Doyle, F.J., *Process Dynamics and Control*, 3rd Edition, Wiley & Sons, Hoboken, NJ, 2011.

Tyreus, B.D. and Luyben, W.L., Tuning PI controllers for integrator/deadtime processes, *Ind. Eng. Chem. Res.*, 31:2625, 1992.

Ziegler, J.G. and Nichols, N.B., Optimum settings for automatic controllers, *Trans. ASME*, 64:759, 1942.

Appendix C: Controller Script

Advanced automation functions are usually performed in control computers, although some of the less critical functions are beginning to be available remotely using cloud-based apps. These include:

- Supervisory controls (Sections 7.1 and 9.2)
- Area and plant-wide optimizers (Section 9.3.2)
- Fuzzy logic controllers (Section 9.2.2)

Most vendors have one or more programming languages that they offer for designing or customizing these types of functions. Common languages include BASIC (one or more of the many versions), FORTRAN, C, and C++.

The example controller script included in this textbook is based on a simplified version of the commands, functions, and syntax from BASIC. Each line is numbered and processed sequentially. Only one function is performed per line unless a special type of connector is used to allow multiple functions to be performed.

Below are definitions and examples of the common BASIC keywords used in the textbook:

    LET: assigns an external value to a variable

        Example: *LET th=TI-100(t)*   *Let the value of "th" be the measurement variable contained in Indicator TI-100 at time "t"

    IF [condition 1] THEN: [condition 2, assignment 2, or operation 2]: When the criterion defined as condition 1 is satisfied or is true, then make the condition 2 or assignment 2 or perform operation 2.

        Example: *IF th.GT.thmax THEN talarm=1*   *If the value of "th" is greater than the value of "thmax," assign a value of 1 to "talarm." This might be used to activate an alarm when a temperature measurement exceeds a maximum value. Note: if "th" is not greater than "thmax," then the value of talarm is unchanged.

    IF [condition 1] THEN: [condition 2, assignment 2, or operation 2] ELSE: [condition 3, assignment 3, or operation 3]: If the criterion defined as condition 1 is satisfied or is true, then make the condition 2 or assignment 2 or perform operation 2. However, if the criterion defined as condition 1 is not satisfied or is false, then make the condition 3 or assignment 3 or perform operation 3.

        Example: *IF th.GT.thmax THEN talarm=1 ELSE talarm=0*   *If the value of "th" is greater than the value of "thmax," assign a value of 1 to "talarm." If the value of "th" is not greater than (less than or equal to) the value of "thmax," assign a value of 0 to "talarm."

GOTO [line X]: Instead of advancing to the next line number in sequence, jump to line number X.

Example: *3 GOTO 6   *instead of advancing and performing the function specified on line 4, go to line 6.*

IF [condition 1] GOTO [line]: If the criterion defined as condition 1 is satisfied or true, jump to the line number "line". Note: if condition 1 is not satisfied, the program sequences to the next line of the script.

Example: *2 IF th.GT.thmax GOTO 6   *If the value of "th" is greater than the value of "thmax," go to line number 6*

.
.
.

6 talarm=1   *Assign a value to "talarm" of 1

.
.

IF [condition 1] GOTO [line] ELSE: [condition 3, assignment 3, or operation 3]: If the criterion defined as condition 1 is satisfied or is true, then go to "line". However, if the criterion defined as condition 1 is not satisfied or is false, then make the condition 3 or assignment 3 or perform operation 3.

Example: *2 IF th.GT.thmax GOTO 6 ELSE talarm=0   **If the value of "th" is greater than the value of "thmax," go to line 6. If the value of "th" is not greater than (less than or equal to) the value of "thmax," assign a value of 0 to "talarm."*

.
.
.

6 talarm=1   *Assign a value to "talarm" of 1

.
.
.

Mathematical Operators

= or .eq.	Make the variable on the left equal to the variable or expression on the right
+	Add the values of two variables together

Appendix C: Controller Script

−	Subtract the value of the second variable from the value of the first variable
* or ×	Multiply the values of two variables together
/	Divide the left-hand variable by the right-hand variable
.GT.	Determine if the value of the variable on the left is greater than the value of the variable on the right
.GE.	Determine if the value of the variable on the left is greater than or equal to the value of the variable on the right
.LT.	Determine if the value of the variable on the left is less than the value of the variable on the right
.LE.	Determine if the value of the variable on the left is less than or equal to the value of the variable on the right

Index

A

Absolute pressure measurement, 290
Absorption system, 164–166
Access security, 326
Adaptive controllers, 259–261
Adiabatic flash separator, 145–148
Advanced process control (APC)
   calculation blocks, 269
   control scripts, 269–270
   fuzzy logic controller, 272–275
   MIMO controls, 275–276
   MISO controls, 270–273
   MVC, 276–279
AFC valve, *see* Air-fail-closed valve
AFO valve, *see* Air-fail-open valve
Air-fail-closed (AFC) valve, 72, 247, 309–310
Air-fail-open (AFO) valve, 72, 246–247, 309–310
Air operated valve (AV), 73
Allocation control, 235–236
Analytical measurement method, 11
Anti-reset windup, 249
Antisurge control, 96–98
APC, *see* Advanced process control
Applications integration infrastructure, 319
Applications model, 319–320
Application wrappers, *see* Applications integration infrastructure
Area classification concept, 330
Area data reconciliation, 268
Area optimizer, 268
Arrhenius equation model, 259, 354
Associated gas, 145
Auto-ignition point, 328–329
Automatic level measurement sensors, 302
Automation and control system projects; *see also* Plant automation system projects
   guidelines, 334–336
   specification and design, 317–334

B

Band gap, 241
Batch unit operations, 4, 101, 105
   vs. continuous process, 326–327
   higher-level automation techniques, 282–284
   independent variables, 42
   override control, 233
   PAS projects, 326–327
   process drawings, 19–22
   reactors control
      bioreactors, 207–209
      continuous processes, 209–215
      rectangular pulse change models, 360
   SLC, 75, 77–78
   split range control, 233
   tuning parameter scheduling, PID controller, 258
Bernoulli equation, 83–84, 92, 219, 303
Bernoulli-type flowmeters, 258–259, 291, 293–295
BFD, *see* Block flow diagram
Biological reaction systems, 208–209
Biological transformations, 4
Blending unit operation, CALC blocks in, 221–226
Block flow diagram (BFD), 13–15
Block valves, 311
Blowers, 92, 110
Bourdon tube pressure gauge, 287–288
Bubble-type level measuring device, 303
Business/administrative functions, 320
Butterfly valves, 311–312
By-products, definition of, 1

C

Calculation blocks (CALC block)
   vs. APC, 269–270
   blending applications, 221–226
   flowmeter, 219–221
   heat exchangers, 199, 223–228
   high selector, 231–232

385

low selector, 231–232
mathematical operation, 220
safety controls, of continuous flow
reactors, 201
Capacitance sensor, 302
Cascade control
loops, 61–67
symbols, 34
Casual user, 323
Chemical reactions, 12, 40, 195
Commercial off-the-shelf (COTS)
software package, 323
Composition measurement
fixed bed reactor, 353–355
measuring devices, 305–307
Compressible fluid flow dynamics, 9
Compressible fluid motive force devices
compressors, 92–97
expanders and turbines, 97–100
Compressors
inferential control, 228–229
motive force, 92–97, 110
Conductive tapes, 302
Confidential security, 326
Constraint control, 235–238
Contingency, 330
Continuous cycling method, see Ziegler-Nichols method
Continuous emission monitors (CEMs), 306
Continuous flow reactors control, 195
CSTR system, 203–207
heterogeneous binary reaction, 197–201
isothermal reaction conditions, 196–197
safety controls
reactor overpressure, 203
temperature excursions, 201
typical safety system, 202–203
Continuous-stirred-type reactor (CSTR), 203
distillation system, 204–207
mass balance, 204
Continuous unit operations, 3–4
vs. batch process, 326–327
batch reactors control, 209–215
fluidized bed solid sorbent systems
complete control scheme, 182
control strategy design, 177, 179
mass balance, 180

Controller blocks
CALC block, 220
symbols, 33–34
Controller gain constant, 246
Controller parameter scheduling, see Tuning parameter scheduling
Controller saturation, 247
Control scripts
APC, 269–270
BASIC keywords, 381–382
Control valves (CVs), 40
adiabatic flash separator, 146–148
continuous flow reactors, 198–199
split range control, 233, 235
two-phase separator, 149–150
Control variable (CV)
derivative controller, 250
integral control, 248
PI control, 249
PID controllers, 245, 251
proportional response control, 246–247
Control variables, see Independent process variables
Coriolis-type mass flowmeter, 293–294
Coupled/cross coupled control system, 48–49
Critical security, 326
Cross heat exchanger control, 122–125
with phase change, 125–130
CSTR, see Continuous-stirred-type reactor
CV, see Control valves, see Control variable
Cyclic function models, 359–360

D

Data/information model, 320
Data reconciliation system, 279–280
DCS, see Distributed control systems
Decoupled control system, 48
Dedicated human user interface (DHUI), 323–324
Dependent process variables, 42–43, 357
Derivative time, 250
Detonation, 329
DHUI, see Dedicated human user interface

Diaphragms, 287
Direct fired heater control, 134–138
Direct measurement type online
   analyzer, 305
Direct mechanical flowmeters, 296–297
Direct mixing heat transfer control, 130–133
Displacer-type level measurement
   device, 303–304
Distillation system controls
   bottoms system controls, 160–163
   cascade control loop, 64, 66
   constraint control, 236–238
   CSTR, 204–207
   goal, 150–151
   liquid level measurement, 10
   mass balance, 155–156
   overhead configuration, 62–65
   overhead system controls
      partial condenser, 154, 157–158
      total condenser, 154, 158–160
   rectifying column, 152–153
   reflux accumulator, 63–66
   stripping column, 151–152
   VLE changes, 151
Distributed control systems (DCS),
   219–220
   check-pointing, 22
   elements, 26–28
   generic hierarchy, 26–27
   PLCs, 28–29
Disturbance variables, 49–53
Double contingency, 330
Dual redundancy, 333–334
Dynamic matrix control system, 276
Dynamic modeling
   fixed bed reactor, 353–355
   liquid knockout drum, 344–348
   process and control system
      controllers, 357–358
      control variables, 356–357
      measurement variables, 357
      simple model, 356, 358
   process changes
      cyclic change, 359–360
      rectangular pulse change, 359–360
      spike, 359–360
      step change, 358–360
   single pass shell and tube heat
      exchanger, 348–352

E

Electrical resistance heat transfer control
   control scheme, 133–135
   on/off controllers, 241–242
Electromagnetic flowmeter, 296
Electromechanical pressure sensors,
   288–289
Electronic control systems, 35–36
Electronic speed controller (ESC), 310
Emergency isolation system (EIS), 332
Emergency management systems,
   symbols used in, 35
Emergency shutdown system (ESD),
   332–333
Energy balances
   single pass shell and tube heat
      exchanger, 349, 351
Enterprise management, in plant-wide
   and process area automation,
   282
Enterprise resource planning system
   (ERP), 323
Equipment protection system (EPS),
   107–108, 332–333
Error, 8, 47, 241, 243, 245
Error-squared controller, 254–255
ESC, *see* Electronic speed controller
Expanders, 97–100
External setpoint (ESP), 34, 64
Extraction system, 166–168

F

Fans, 92, 110
Feedback control loop, 43–49
Feed forward control, 53–56
Feed forward/feedback cascade control
   loops, 68–69
Fired heater control, 134–138
Fire triangle, 328
Fixed bed regenerated solid sorbent
   systems
   complete control scheme, 188–189
   control strategy design, 181, 183, 185
   energy inputs, 190
   logic flow diagram, 184
   mass balance, 185
   sequential events table, 185–188

Fixed catalyst bed reactors control, 195–196, 353–355; *see also* Continuous flow reactors control
Fixed speed compressors, 92–93
Fixed speed pump control
 flow/pressure, 85–86
 minimum flow recycle, 85–87
 on/off control scheme, 89–90
 pump curve, 84–85
Flash drum, 145–148
Flash separator, dynamic modeling of, 348
Float systems, 302
Flow measurement, 9
 calculation block, 219–221
 measuring devices, 291–299
Fluid-fluid heat transfer control
 cross exchanger control, 122–125
  with phase change, 125–130
 gaseous process stream condensation, 119–122
 incompressible utility liquid, 114–117
 vaporizing liquid process stream, 117–119
Fluidized bed solid sorbent systems
 complete control scheme, 182
 control strategy design, 177, 179
 mass balance, 180
Fluidized catalyst bed reactors control, 195–196
Force-summing devices, 287
Fouling, 138–139
Fuzzy logic controller, 272–275

G

Gain, *see* Controller gain constant
Gain scheduling, 255
Gas absorption with solvent recovery
 control scheme, 173–174
 mass balance, 170–171
 primary objective, 172
 process scheme, 168–170
 secondary objective, 172
Gate valves, 312
Gauge pressure measurement, 291
Gravity separator, *see* Phase separator

H

Hazards analysis, 330
HAZOP analysis, 330
Heat exchangers
 allocation control, 235–236
 calculation block, 199, 223–228
 feedback control, 53–54
 feed forward control loop, 55
 feed forward/feedback control scheme, 55–57
 fouling/scaling monitoring system, 138–139
Heat transfer control, 39
 direct mixing, 130–133
 electrical resistance, 133–135
 fired heater, 134–138
 fluid-fluid, 113–130
 fouling/scaling monitoring, 138–139
 on/off controllers, 241–242
 parallel heat exchangers, 139–141
Heuristically developed model, 343
Heuristic controllers, *see* Adaptive controllers
Higher-level automation techniques
 APC
  calculation blocks, 269
  control scripts, 269–270
  fuzzy logic controller, 272–275
  MIMO controls, 275–276
  MISO controls, 270–273
  MVC, 276–279
 batch processes, 282–284
 PAS
  area data reconciliation, 268
  area optimizer, 268
  automation elements, 263–264
  data/information model, 264–265
  field instrumentation, 265–267
  online models, 268
  plant-wide operational control applications, 268
  process facility objectives, 263
  regulatory controls, 267
  supervisory controls, 267
 plant-wide and process area automation
  data reconciliation system, 279–280
  enterprise and supply chain management, 282

Index 389

planning and scheduling, 281–282
RTO, 280–281
semi-batch processes, 282–284
symbols, 35
High select CALC block controller, 231–232
HUI, *see* Human user interface
Human user interface (HUI), 2, 323–324
Hydrostatic head measurements, 302–304

I

Incompressible fluid flow dynamics, 9
Incompressible fluid motive force devices, *see* Pumps
Independent process variables, 40–42, 356–357
Inferential control, 227–231
Infrared sensors, 301
Input/output (I/O) diagram, 12–13
Instrumentation
  composition, 305–307
  flow/mass, 291–299
  level, 301–305
  pressure, 287–291
  pressure relief valves, 313–315
  remotely operated block valves, 311–313
  speed control systems, 310
  temperature, 299–301
  throttling control valves, 307–310
  vibration, 307
Integral windup, *see* Reset windup
Intensive user, 323
International Society of Automation (ISA), 29
I/O diagram, *see* Input/output diagram
ISA, *see* International Society of Automation

K

kiloPascals (kPa), 290–291

L

Lactic acid production, 209–215
Laplace transform, 363–368
  linear ODEs, 371
  liquid knockout drum, 345
  "s" domain, 368–369
Level alarm, low-low (LALL) safety system, 73
Level measurement
  liquid knockout drum, 345
  measuring devices, 301–305
  on/off controllers, 242–243
Linear ODEs, 371
Liquid knockout drum
  disturbance variables, 49, 55
  dynamic modeling, 344–348
  feedback control, 44–45, 47
  feed forward/feedback control scheme, 57
  process and safety systems, 71, 73–74
Lock hoppers, *see* Pressurized lock hoppers
Logarithmic controller, 258–259
Logic flow diagram, 19, 21
Low select CALC block controller, 231–232
Lumped parameter model, 345, 348

M

Magnetic flowmeters, 296
Maintenance functions, 320
Manometers
  for flow measurement, 293
  for pressure measurement, 290
Mass balance/material balance
  continuous flow reactors, 198
  distillation systems, 155–156
  fixed bed regenerated solid sorbent systems, 185
  fluidized bed solid sorbent systems, 180
  liquid knockout drum, 344–345
  once-through fixed bed solid sorbent systems, 176
  phase separator, 149
  pressure vessel, 43
  single-stage gas absorber system, 164
  single-stage once-through liquid-liquid extraction system, 167–168
Mass measuring devices, 291–299
Mass transfer control, 40

Mathematical operators, 382–383
Measurement error, see Error
Measurement variables, 258–259; see also Dependent process variables
Model-based control, see Inferential control
Model predictive control system, 276
Modified PID control algorithms, 254–255
Momentum transfer, 39
Motive force units, 39
   compressible fluid devices, 92–100
   equipment protection systems, 107–108
   incompressible fluid devices, 83–92
   solids transport systems
      categories, 100
      gases and liquids mixture, 100–102
      physical devices, 102–103
      physical pressure/gravity, 103–106
   startup and shutdown events, 107
   switching controls, 108–110
Motor operated valve (MV), 73, 76
Multifunction control blocks, 34–35
Multiple input/multiple output (MIMO) APC, 275–276
Multiple input/single output (MISO) APC, 270–273
Multistage absorption with solvent recovery, see Gas absorption with solvent recovery
Multistage distillation, 150–163
Multistage reactor system, 195–196
Multivariable control system (MVC), 276–279, 343
MVC, see Multivariable control system

N

NFPA, see U.S. National Fire Protection Association
No redundancy, 333–334
Normal security, 326

O

Objectives-based process control paradigm, 321–322
Offset error, 247
Once-through fixed bed solid sorbent systems
   control scheme, 178
   mass balance, 176
   process scheme, 175–177
Online analyzers, 305–307
Online models, 268
On/off controllers
   fixed speed pump control, 89–90
   heat transfer, 241–242
   level measurement, 242–243
      single level switch, 242, 244
      two switch system, 242–244
Open/close controllers, see On/off controllers
Open/hold/close (O/H/C) controller, 243
Open loop control, 55
Ordinary differential equation (ODE)
   liquid knockout drum, dynamic modeling of, 345
   shell and tube heat exchanger, dynamic modeling of, 352
Orifice meters
   flow rate measurement, 291–292
   on/off controllers, 219–221
Output bias, 245
Override control, 233–234

P

Parallel heat exchangers control
   allocation control, 235–236
   switching controls, 139–141
Parameter-based process control paradigm, 321
Partial fraction expansion, 369–370
PAS projects, see Plant automation system projects
PFD, see Process flow diagram
Phase separator, 149–150
Physical automation systems, 322–323
Physical motive force devices, 102–103
Physical security, 325
PID controller, see Proportional-integral-derivative controller
Piezomagnetic sensor, 305
Piping and instrument diagrams (P&ID), 15, 18–19, 22, 25

Index 391

Pitot tube flowmeters, 294–295
Planning app, 281–282
Plant automation system (PAS) projects, 2–3
   applications, 319–322
   area data reconciliation, 268
   area optimizer, 268
   automation elements, 263–264
   batch process automation, 326–327
   data/information model
      enterprise management layer, 264
      process and field instruments layer, 265
      process management layer, 265
      production management layer, 264–265
   design guidelines, 335–336
   facility objectives, 317
   field instrumentation, 265–267
   functional view, 318
   HUI, 323–324
   online models, 268
   physical and systems security, 324–326
   physical view, 318
   plant-wide operational control applications, 268
   process facility objectives, 263
   process safety, 327–334
   project life cycles, 335
   regulatory controls, 267
   supervisory controls, 267
   systems integration infrastructure, 319, 322–323
Plant-wide and process area automation
   data reconciliation system, 279–280
   enterprise and supply chain management, 282
   planning and scheduling, 281–282
   RTO, 280–281
PLCs, *see* Programmable logic control systems
Plug-flow-type reactors, 195–196
Plug valves, 312–313
Pneumatically actuated throttling control valve, 308
Pneumatically driven actuators, 309–310
Polylactic acid (PLA) production, 209
Positive-displacement flowmeter, 298

Pounds per square inch (PSIA), 291
Preliminary hazards analysis, 330
Pressure measuring devices, 9, 287–291
Pressure relief valves, 313–315
Pressure vessel
   coupled regulatory control system design, 47–49
   decoupled regulatory control system design, 48
   disturbance variables, 50, 52
   feed forward controller, 56
   process, 43–44
   safety codes, 73
   use, 43
Pressurized lock hoppers, 103–105
Process
   and control system
      controllers, 357–358
      control variables, 356–357
      measurement variables, 357
      simple model, 356, 358
   definition, 1
   drawings, 19–22
   dynamics, 8–11, 363
   goal, 1
   and safety systems, 69–74
Process control, basis of, 343
Process flow diagram (PFD), 14, 16–17, 21–23
Process-process heat transfer control, 122–130
Products, definition of, 1
Programmable logic control systems (PLCs), 28–29
Proportional band (PB), 247
Proportional-integral-derivative (PID) controller
   advantages, 251–252
   auto-tuning systems, 374–375
   control equations, 251
   derivative control, 250–251
   derivative kick, 252–253
   error, 243, 245
   final loop tuning, 377–378
   integral control, 248–250
   limitations, 252
   manual methods
      experimental methods, 376
      model-based methods, 376–377

measurement response, 373–374
nonlinear control strategies
  error-squared controller, 254–255
  logarithmic controller, 258–259
  truncating effect, 253–254
  tuning parameter scheduling, 255–258
  proportional response control, 246–247
  tuning constants, 373
Pumps
  Bernoulli equation, 83–84
  flow/pressure, 85–86
  minimum flow recycle, 85–87
  on/off control scheme, 89–90
  piping configuration, 87–89, 91
  pump curve, 84–85
  variable speed pump, 91–92

Q

Qualitative BFDs, 13
Quantitative BFDs, 13–14

R

Ramp change models, 359–360
Range change models, 360
Rate of deviation alarm, 70–71
Ratio control loops, 58–61
Ratio with feedback trim control, 67, 165
Reactions control
  batch reactions, 207–215
  continuous flow reactors
    CSTR system, 203–207
    heterogeneous binary reaction, 197–201
    isothermal reaction conditions, 196–197
    safety controls, 201–203
Real-time heuristic models, 343
Real-time optimizers (RTOs), 280–281
Reciprocating globe valves, 308–309
Rectangular pulse change models, 359–360
Recyclable and/or consumable materials, 1
Recycle loops, 11–12

Redundancy
  vs. assurance, 334
  vs. availability, 333–334
Reflux accumulator, 63–66
Relief valve, 73
Remotely operated block valves, 311–313, 357
Remote setpoint (ESP), 34
Reset windup, 249
Residence times, 10
Resistance temperature device (RTD), 300–301
Rotameters, 295–296
Rotary globe valves, 309
RTD, see Resistance temperature device
RTOs, see Real-time optimizers
Run/stop controllers, see On/off controllers
Rupture disks, 314–315

S

Safety automation systems
  continuous flow reactors, 201–203
  emergency systems, 331–333
  fire triangle, 328
  hazards analysis, 330
  NFPA specification, 329
  process safety layers, 330–331
  redundancy, 333–334
  two-phase separator, 150
Sampled type online analyzers, 306–307
Scaling, 138–139
Scheduling app, 281–282
Scheduling variable, 255
Self-adaptive controller, 260
Semi-batch unit operations, 4, 7
  fixed bed regenerated solid sorbent systems
    complete control scheme, 188–189
    control strategy design, 181, 183, 185
    energy inputs, 190
    logic flow diagram, 184
    mass balance, 185
    sequential events table, 185–188
  higher-level automation techniques, 282–284
  hoppers, 103

Index 393

process drawings, 19–22
rectangular pulse change models, 360
SLC, 75, 78
split range control, 233
Separation controls
　absorption system, 164–166
　distillation systems, 150–163
　extraction system, 166–168
　multistage absorption, 168–175
　single-stage separation
　　flash drum, 145–148
　　phase separator, 149–150
　solid sorbent systems
　　fixed bed regenerated systems, 181–190
　　fluidized bed systems, 177–181
　　once-through fixed bed systems, 175–177
Sequential events table
　pressurized lock hoppers, 106
　semi-batch seed reactor, 213
Sequential logic controller (SLC), 27, 34, 74–78
Sequential logic table, 20–21, 24
Setpoints, 8, 47
Shell and tube heat exchanger
　dynamic modeling, 348–352
　liquid level measurement, 10
Shutdown events, 107
Single-stage gas absorber system
　control scheme, 166
　mass balance, 164
　primary objective, 165–166
　secondary objectives, 166
Single-stage once-through liquid-liquid extraction system
　control scheme, 169
　mass balance, 167–168
　primary objective, 168
　process scheme, 166–167
　secondary objectives, 168
Single-stage separation
　flash drum, 145–148
　phase separator, 149–150
Sinusoidal function model, 360
SLC, see Sequential logic controller
Soft sensor technique, 230
Solids handling devices
　categories, 100

gases and liquids mixture, 100–102
physical devices, 102–103
physical pressure/gravity, 103–106
Solid sorbent systems
　fixed bed regenerated systems, 181–190
　fluidized bed systems, 177–181
　once-through fixed bed systems, 175–177
Speed control systems, 310
Spike models, 359–360
Split range control, 233–235
Spring-loaded pressure relief valve, 313–314
Start/stop controllers, see On/off controllers
Startup events, 107
Step change models, 358–360
Strain gauge transducers, 289–290
Straps, see Conductive tapes
Supply chain management, in plant-wide and process area automation, 282
Surge phenomenon, 94–96
Switching controls, in motive force units, 108–110
Symbology for control systems
　device designations, 29, 32
　higher-level control system, 34–35
　mathematical symbols, 33
　measurement parameter designations, 29, 31
　process drawings, 29–31
Systems integration infrastructure, 319, 322–323

T

Technical support functions, 320
Temperature excursions, 201
Temperature measurement, 9
　measuring devices, 299–301
　single pass shell and tube heat exchanger, 349–352
Thermistors, 300–301
Thermocouple, 299–300
Throttling control valves, 41, 307–310
Time lag, 8
Triple redundancy, 334

Tuning parameter scheduling, 255–258
Turbines, 97–100
Two-phase separator, 149–150
Two-stage CSTR reactor system, 204–207
Two-step batch process, 4–6
Tyreus-Luyben correlations, 376

U

UHUI, see Universal human user interface
Ultrasonic flowmeter, 296–297
Unit operations
  chemical reactors, see Reactions control
  compressible fluid motive force devices, 92–100
  definition, 7
  heat transfer, see Heat transfer control
  incompressible fluid motive force devices, see Pumps
  separation, see Separation controls
Universal human user interface (UHUI), 324
U.S. National Fire Protection Association (NFPA), 329

V

Vacuum pumps, 92
Vapor-liquid equilibrium (VLE) separation system, 145–146, 148, 150–152, 155, 157
Variable speed compressors, 93–95
Variable speed pump control
  control scheme, 91–92
  pump curve, 84–85
Venturi meters
  calculation block, 219–221
  flow rate measurement, 292–293
Vibration measuring devices, 307
VLE system, see Vapor-liquid equilibrium separation system
Vortex flowmeters, 298

W

Wastes, definition of, 1
Wave-based level sensor, 303–305
Weigh-in-motion mass flowmeter, 298–299

Z

Ziegler-Nichols method, 376

PGSTL 09/01/2017